ADVANCED APPLIED STRESS ANALYSIS

ELLIS HORWOOD SERIES IN MECHANICAL ENGINEERING

Series Editor: J.M. ALEXANDER, Professor of Mechanical Engineering, University of Surrey, and Stocker Visiting Professor of Engineering and Technology, Ohio University

The series has two objectives: of satisfying the requirements of post-graduate and mid-career engineers, and of providing clear and modern texts for more basic undergraduate topics. It is also the intention to include English translations of outstanding texts from other languages, introducing works of international merit. Ideas for enlarging the series are always welcomed.

Alexander, J.M.	**Strength of Materials: Vol 1: Fundamentals; Vol. 2: Applications**
Alexander, J.M., Brewer, R.C. & Rowe, G.	**Manufacturing Technology Volume 1: Engineering Materials**
Alexander, J.M., Brewer, R.C. & Rowe, G.	**Manufacturing Technology Volume 2: Engineering Processes**
Atkins A.G. & Mai, Y.W.	**Elastic and Plastic Fracture**
Beards, C.	**Vibration Analysis and Control System Dynamics**
Beards, C.	**Structural Vibration Analysis**
Beards, C.	**Noise Control**
Besant, C.B. & C.W.K. Lui	**Computer-aided Design and Manufacture, 3rd Edition**
Borkowski, J. and Szymanski, A.	**Technology of Abrasives and Abrasive Tools**
Borkowski, J. and Szymanski, A.	**Uses of Abrasives and Abrasive Tools**
Brook, R. and Howard, I.C.	**Introductory Fracture Mechanics**
Cameron, A.	**Basic Lubrication Theory, 3rd Edition**
Collar, A.R. & Simpson, A.	**Matrices and Engineering Dynamics**
Cookson, R.A. & El-Zafrany, A.	**Finite Element Techniques for Engineering Analysis**
Cookson, R.A. & El-Zafrany, A.	**Techniques of the Boundary Element Method**
Ding, Q.L.	**Surface Modelling for CAD/CAM**
Dowling, A.P. & Ffowcs-Williams, J. E.	**Sound and Sources of Sound**
Edmunds, H.G.	**Mechanical Foundations of Engineering Science**
Fenner, D.N.	**Engineering Stress Analysis**
Fenner, R.T.	**Engineering Elasticity**
Ford, Sir Hugh, FRS, & Alexander, J.M.	**Advanced Mechanics of Materials, 2nd Edition**
Gallagher, C.C. & Knight, W.A.	**Group Technology Production Methods in Manufacture**
Gohar, R.	**Elastohydrodynamics**
Gosman, B.E., Launder, A.D. & Reece, G.	**Computer-aided Engineering: Heat Transfer and Fluid Flow**
Haddad, S.D. & Watson, N.	**Principles and Performance in Diesel Engineering**
Haddad, S.D. & Watson, N.	**Design and Applications in Diesel Engineering**
Haddad, S.D.	**Advanced Operational Diesel Engineering**
Hunt, S.E.	**Nuclear Physics for Engineers and Scientists**
Irons, B.M. & Ahmad, S.	**Techniques of Finite Elements**
Irons, B.M. & Shrive, N.	**Finite Element Primer**
Johnson, W. & Mellor, P.B.	**Engineering Plasticity**
Juhasz, A.Z. and Opoczky, L.	**Mechanical Activation of Silicates by Grinding**
Kleiber, M.	**Incremental Finite Element Models in Non-linear Solid Mechanics**
Kleiber, M.	**Finite Element Methods in Engineering**
Leech, D.J. & Turner, B.T.	**Engineering Design for Profit**
Lewins, J.D.	**Engineering Thermodynamics**
Malkin, S.	**Materials Grinding: Theory and Applications**
McArthur, H.	**Motor Vehicle Corrosion**
McCloy, D. & Martin, H.R.	**Control of Fluid Power: Analysis and Design 2nd (Revised) Edition**
Osyczka, A.	**Multicriterion Optimisation in Engineering**
Oxley, P.	**Mechanics of Machining**
Piszcek, K. and Niziol, J.	**Random Vibration of Mechanical Systems**
Polanski, S.	**Bulk Containers: Design and Engineering of Surfaces and Shapes**
Prentis, J.M.	**Dynamics of Mechanical Systems, 2nd Edition**
Renton, J.D.	**Applied Elasticity**
Richards, T.H.	**Energy Methods in Vibration Analysis**
Richards, T.H.	**Energy Methods in Stress Analysis: with Introduction to Finite Element Techniques**
Ross, C.T.F.	**Computational Methods in Structural and Continuum Mechanics**
Ross, C.T.F.	**Finite Element Programs for Axisymmetric Problems in Engineering**
Ross, C.T.F.	**Finite Element Methods in Structural Mechanics**
Ross, C.T.F.	**Applied Stress Analysis**
Ross, C.T.F.	**Advanced Applied Stress Analysis**
Roznowski, T.	**Moving Heat Sources in Thermoelasticity**
Sawczuk, A.	**Mechanics of Plastic Structures**
Sherwin, K.	**Engineering Design for Performance**
Szczepinski, W. & Szlagowski, J.	**Plastic Design of Complex Shape Structured Elements**
Thring, M.W.	**Robots and Telechirs**
Tseung, A.C.C.	**Lotus 123 in Science and Technology**
Walshaw, A.C.	**Mechanical Vibrations with Applications**
Williams, J.G.	**Fracture Mechanics of Polymers**
Williams, J.G.	**Stress Analysis of Polymers 2nd (Revised) Edition**

ADVANCED APPLIED STRESS ANALYSIS

C. T. F. ROSS, B.Sc.(Hons.), Ph.D.
Reader in Applied Mechanics
Portsmouth Polytechnic

ELLIS HORWOOD LIMITED
Publishers · Chichester

Halsted Press: a division of
JOHN WILEY & SONS
New York · Chichester · Brisbane · Toronto

First published in 1987 by
ELLIS HORWOOD LIMITED
Market Cross House, Cooper Street,
Chichester, West Sussex, PO19 1EB, England
The publisher's colophon is reproduced from James Gillison's drawing of the ancient Market Cross, Chichester.

Distributors:
Australia and New Zealand:
JACARANDA WILEY LIMITED
GPO Box 859, Brisbane, Queensland 4001, Australia
Canada:
JOHN WILEY & SONS CANADA LIMITED
22 Worcester Road, Rexdale, Ontario, Canada
Europe and Africa:
JOHN WILEY & SONS LIMITED
Baffins Lane, Chichester, West Sussex, England
North and South America and the rest of the world:
Halsted Press: a division of
JOHN WILEY & SONS
605 Third Avenue, New York, NY 10158, USA

British Library Cataloguing in Publication Data
Ross, C. T. F.
Advanced applied stress analysis. —
(Ellis Horwood series in mechanical engineering).
1. Strains and stresses
I. Title
620.1'123 TA407

Library of Congress Card No. 87–9280

ISBN 0–7458–0076–9 (Ellis Horwood Limited — Library Edn.)
ISBN 0–7458–0260–5 (Ellis Horwood Limited — Student Edn.)
ISBN 0–470–20874–0 (Halsted Press)

Printed in Great Britain by R. J. Acford, Chichester

Table of Contents

Author's preface

Advanced Applied Stress Analysis is a sequel to *Applied Stress Analysis*, an earlier book by the author, and it is intended for use by senior undergraduates and postgraduates in engineering and architecture. Like its predecessor, *Applied Stress Analysis*, *Advanced Applied Stress Analysis* has been written in a style which caters for the mathematical difficulties experienced by many modern undergraduates.

Although, of course, in the industrial world, much structural design is carried out by computer methods, it is the present author's belief that structural designers should have a thorough knowledge of the principles and concepts of stress analysis, so that they can satisfactorily interpret the results presented in their computer output. The book, therefore, takes this into account, by removing those sections of traditional stress analysis that are better replaced by computer methods, but retaining those sections that introduce or enhance fundamental concepts and principles. In any case, for many structural designs, all that is required are the use of closed-loop trivial solutions, and if such solutions exist for the structure in question, then the structural designer might as well use these solutions.

The book contains a large selection of worked examples in every chapter, and all chapters contain a section entitled "Examples for Practice", which the reader might find useful for consolidating his/her newly acquired skills.

Chapter 1 is on "Unsymmetrical Bending of Beams", and covers both symmetrical section beams, loaded unsymmetrically, and also the behaviour of laterally loaded beams of unsymmetrical section. Mohr's circle of intertia is introduced, and through the use of worked examples, the method of calculating the principal second moments of area of unsymmetrical sections is demonstrated. Other worked examples include the calculation of bending stresses and bending deflections of unsymmetrical beams, loaded with concentrated and distributed loads.

Chapter 2 is on "Shear Stresses in Bending and Shear Deflections". The chapter commences by introducing vertical and horizontal shearing stress distributions, and this theory is extended to determine the shear stress distributions in curved thin-walled open and closed tubes.

The concept of the shear centre position is introduced, and worked examples are used to demonstrate how the shear centre position can be calculated for thin-walled open and closed tubes.

Shear deflections of laterally loaded beams are also introduced.

Chapter 3 is on "Theories of Elastic Failure". The five major theories of elastic failure are introduced, and through the use of two worked examples, some of the differences between the five major theories are demonstrated.

Chapter 4 is on "Plasticity" and commences by describing the plastic hinge. The statical and kinematical methods are introduced, and the latter method is used extensively to demonstrate the structural design of beams and rigid-jointed plane frames through plastic considerations. The method is extended to plastic torsion, and worked examples are used to calculate residual stresses.

Chapter 5 is on "Torsion of Non-circular Sections". After deriving the torsion equation for non-circular sections, applications are made to some simple non-circular sections. Prandtl's membrane analogy is introduced and the torsional constant is calculated for a number of thin-walled open sections. The Batho–Bredt theory is introduced, and applications are made to single and multi-cell tubes under torsion. The plastic theory of the torsion of non-circular sections is also introduced through the sand-hill analogy.

Chapter 6 is on "The Buckling of Struts" and it commences with discussing the Euler theory for axially loaded struts. This theory is extended to investigate the inelastic instability of axially loaded struts, through the use of the Rankine–Gordon formula. The chapter also considers eccentrically loaded and laterally loaded struts, and the effects of initial curvature on axially loaded struts.

In this chapter, an introduction is made of the use of Laplace Transforms for analysing axially and laterally loaded struts, and, in particular, of struts under complex loading.

Chapter 7 is on "Thick Curved Beams", and it commences by introducing the basic theory of thick curved beams. This theory is applied to two thick curved beams, one of which resembles a crane hook.

Chapter 8 is on "Circular Plates", and it commences by determining the differential equation for the small deflection theory of flat circular plates. This differential equation is solved for a number of circular plates with different boundary conditions and loads, some of which are quite complex. The large deflections of thin plates, together with the shear deflections of very thick plates, are also discussed.

Chapter 9 is on "Thick Cylinders and Spheres", and it commences by determining the equations for the hoop and radial stresses of thick cylinders under pressure. The Lamé line is introduced, and it is applied to a number of cases, including thick compound tubes with interference fits.

Thermal stresses in thick cylinders are introduced, and the plastic theory of thick tubes is discussed. Applications are made to a few cases involving either thermal stresses or plasticity.

The theories for thick spherical shells and for rotating rings and discs, together with thermal effects are introduced, and worked examples are included on these topics.

Chapter 10 is on "Finite Difference Methods", and it commences by introducing central, forward and backward differences. Application is then made of the theory of central differences to a few beams. A solution for the longitudinal strength of ships through the finite difference method is given, together with central difference equations for laterally loaded plates and for the torsion of non-circular sections.

Chapter 11 is on "The Matrix Displacement Method", and commences by introducing the finite element method. A stiffness matrix is obtained for a rod element, and application is made to a statically indeterminate plane pin-jointed truss, with the aid of a worked example.

Acknowledgements

The author would like to thank the following of his colleagues for the helpful comments and contributions they have made to him on the subject of stress analysis over many years:

Harry Brown, Jim Byrne, Mick Devane, John Gibbs, Brian Lord, Harry Newman and Phil Thompson.

In particular, the author is grateful to Professor Terry Duggan and to Graham White for the continued interest they have shown in the author's work over a period of twenty years.

A special thanks to Dave Hewitt and Terry Johns for their contributions.

Finally, he would like to thank Mrs Lesley Jenkinson for the considerable care and devotion she showed in typing the manuscript.

Notation

Unless otherwise, stated, the following symbols are used:

E	Young's modulus of elasticity
F	shearing force (SF)
G	shear or rigidity modulus
g	acceleration due to gravity
I	second moment of area
J	torsional constant
K	bulk modulus
l	length
M	bending moment (BM)
R, r	radius
T	torque or temperature change
t	thickness
W	concentrated load
w	load/unit length
$\hat{x}$	maximum value of x
α	coefficient of linear expansion
γ	shear strain
ε	direct or normal strain
θ	angle of twist
ν	Poisson's ratio
ρ	density
σ	direct or normal stress
τ	shear stress
P	load or pressure
P_e	Euler buckling load

P_R	Rankine buckling load
λ	load factor
σ_{yp}	yield stress
W_c	plastic collapse load
τ_{yp}	yield stress in shear
$[k]$	= elemental stiffness matrix in local co-ordinates
$[k^0]$	= elemental stiffness matrix in global co-ordinates
$\{p_i\}$	= a vector of internal nodal forces
$\{q^0\}$	= a vector of external nodal forces in global co-ordinates
$\{u_i\}$	= a vector of nodal displacements in local co-ordinates
$\{u_i^0\}$	= a vector of nodal displacements in global co-ordinates
$[K_{11}]$	= that part of the system stiffness matrix that corresponds to the "free" displacements
$[\Xi]$	= a matrix of directional cosines
$[I]$	= identity matrix
[ ]	= a square of rectangular matrix
{ }	= a column vector
⌊ ⌋	= a row vector
[0]	= a null matrix
NA	neutral axis
KE	kinetic energy
PE	potential energy
UDL	uniformly distributed load
WD	work done
2E11	2×10^{11}
3.2E-3	3.2×10^{-3}
*	multiplier
⇒	vector defining the direction of rotation, according to the *right-hand screw rule*. The direction of rotation, according to the right-hand screw rule, can be obtained by pointing the right hand in the direction of the double-tailed arrow, and rotating it *clockwise*.

SOME SI UNITS IN STRESS ANALYSIS

s	second (time)
m	metre
kg	kilogram (mass)
N	Newton (force)
Pa	Pascal (pressure) = 1 N/m^2
MPa	megapascal (10^6 pascals)
bar	(pressure), where 1 bar = 10^5 N/m^2 = 14.5 lbf/in^2
kg/m^3	kilograms/cubic metre (density)
W	watt(power), where 1 watt = 1 ampere * volt = N m/s = 1 joule/s
h.p.	horse-power (power), where 1 h.p. = 745.7 W

PARTS OF THE GREEK ALPHABET COMMONLY USED IN MATHEMATICS

α	alpha
β	beta
γ	gamma
δ	delta
Δ	delta (capital)
ε	epsilon
ζ	zeta
η	eta
θ	theta
κ	kappa
λ	lambda
μ	mu
ν	nu
ξ	xi
Ξ	xi (capital)
π	pi
σ	sigma
Σ	sigma (capital)
τ	tau
φ	phi
χ	chi
ψ	psi
ω	omega
Ω	omega (capital)

To the memory of my late father,
Thomas Vincent Ross
(19th July 1891–2nd March 1977)
Formerly, Chief Draughtsman
of the
Bengal Nagpur Railway

1

Unsymmetrical Bending of Beams

1.1.1 ASYMMETRICAL BENDING

The two most common forms of unsymmetrical (or asymmetrical) bending of straight beams are as follows:

(a) when symmetrical section beams are subjected to unsymmetrical loads, as shown by Fig. 1.1;

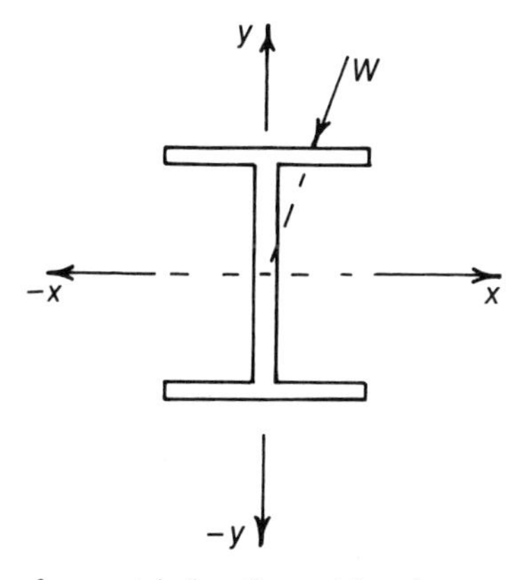

Fig. 1.1. A beam of symmetrical section, subjected to an asymmetrical load.

(b) when unsymmetrical section beams are subjected to either symmetrical or unsymmetrical loading, as shown in Fig. 1.2.

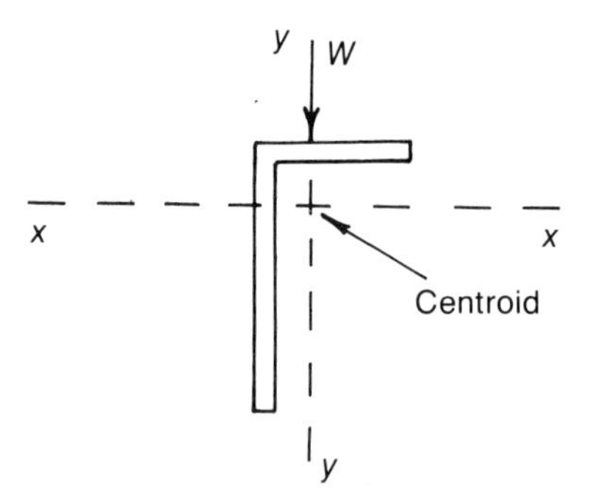

Fig. 1.2. A beam of unsymmetrical section, subjected to a vertical load.

1.2.1 SYMMETRICAL SECTION BEAMS, LOADED ASYMMETRICALLY

In the case of the symmetrical section beam which is loaded asymmetrically, the skew load of Fig. 1.3(a) can be resolved into two components, mutually perpendicular to each other, and acting along the axes of symmetry, as shown in Figs. 1.3(b) and 1.3(c). Assuming that the beam behaves in a linear elastic manner, the effects of bending can be considered separately, about each of the two axes of symmetry, namely x–x and y–y, and later, the effects of each component of W can be superimposed, to give the resultant stresses and deflections.

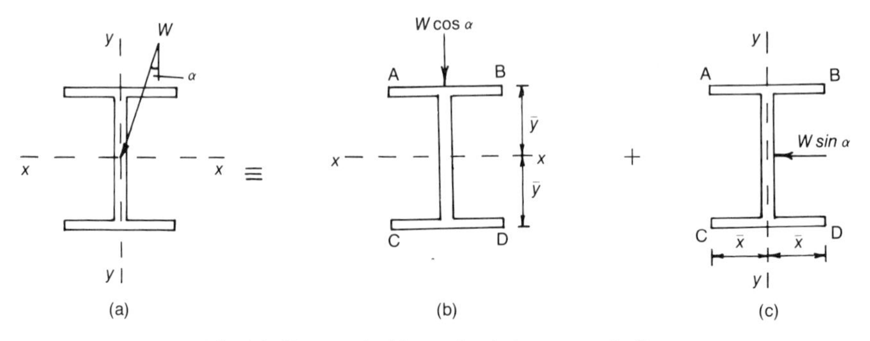

Fig. 1.3. Symmetrical beam, loaded asymmetrically.

1.2.2

To demonstrate the process, let us assume that the beam of Fig. 1.3(a) is of length l, and is simply-supported at its ends, with W at mid-span.

It can readily be seen that the components of W are $W \cos \alpha$, acting along the y axis, and $W \sin \alpha$, acting along the x axis, where the former causes the beam to bend about its x–x axis, and the latter causes bending about the y–y axis.

The effect of $W \cos \alpha$ will be to cause the stress in the flange AB, namely $\sigma_{y(AB)}$, to be compressive, whilst the stress in the flange CD, namely $\sigma_{y(CD)}$, will be tensile, so that:

$$\sigma_{y(AB)} = -\frac{W\cos(\alpha)l\bar{y}}{4I_{xx}} \tag{1.1}$$

and

$$\sigma_{y(CD)} = \frac{W\cos(\alpha)l\bar{y}}{4I_{xx}} \tag{1.2}$$

Similarly, owing to $W \sin\alpha$, the stress on the flange edges B and D, namely $\sigma_{x(BD)}$, will be compressive, whilst the stress on the flange edges A and C, namely $\sigma_{x(AC)}$, is tensile, so that:

$$\sigma_{x(BD)} = -\frac{W\sin(\alpha)l\bar{x}}{4I_{yy}} \tag{1.3}$$

and

$$\sigma_{x(AC)} = \frac{W\sin(\alpha)l\bar{x}}{4I_{xx}} \tag{1.4}$$

The combined effects of $W\cos\alpha$ and $W\sin\alpha$ will be such that the magnitude of the maximum stresses will be largest at the points B and C, where at the point B, the stress is compressive, whilst at the point C, the stress is tensile, and of the same magnitude as the stress at the point B.

At the points A and D, the effects of $W\cos\alpha$ will be to cause stresses of opposite sign to the stresses caused by $W\sin\alpha$, so that the magnitude of these stresses will be less than those at the points B and C.

Thus, in general, the stress at any point in the section of the beam is given by:

$$\sigma = \frac{M\cos(\alpha)\cdot y}{I_{xx}} + \frac{M\sin(\alpha)\cdot x}{I_{yy}} \tag{1.5}$$

where, $\bar{x}$ and $\bar{y}$ are perpendicular distances of the outermost fibres from y–y and x–x, respectively.

I_{xx} = 2nd moment of area of section about x–x

I_{yy} = 2nd moment of area of section about y–y

M is a bending moment due to W and l, and x and y are as defined in Fig. 1.1.

1.3.1 EXAMPLE 1.1 UNSYMMETRICALLY LOADED BEAM

A cantilever of length 2 m and of rectangular section 0.1 m × 0.05 m is subjected to a skew load of 5 kN, at its free end, as shown in Fig. 1.4. Determine the stresses at A, B, C and D at its fixed end.

1.3.2

$$M_y = \text{bending moment about } x\text{–}x$$
$$= W\sin 30 \times 2 = 5\,\text{kN m}$$

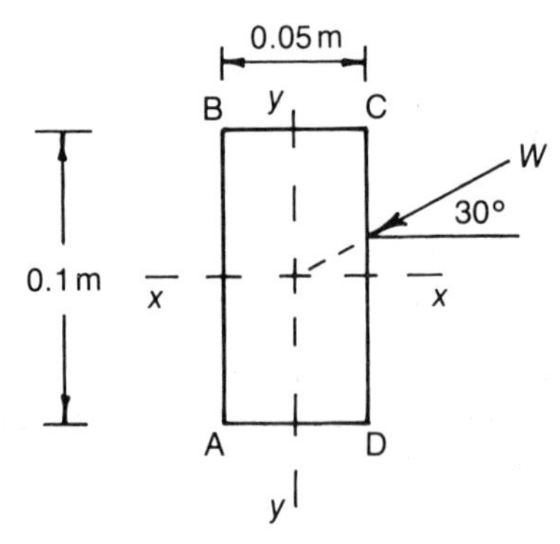

Fig. 1.4. Rectangular section beam.

$$M_x = \text{bending moment about } y\text{–}y$$
$$= W\cos 30 \times 2 = 8.66\,\text{kN m}$$

I_{xx} = second moment of area of the cross-section about x–x

$$= \frac{0.05 \times 0.1^3}{12} = 4.167\text{E-6}\,\text{m}^4$$

I_{yy} = second moment of area of the cross-section about y–y

$$= \frac{0.1 \times 0.05^3}{12} = 1.042\text{E-6}\,\text{m}^4$$

σ_A = maximum stress at the point "A"

$$= -\frac{5\text{E}3 \times 0.05}{4.167\,\text{E-6}} - \frac{8.66\text{E}3 \times 0.025}{1.042\text{E-6}}$$

$$\underline{\sigma_A = -60 - 207.9 = -267.9\,\text{MN/m}^2}$$

$$\underline{\sigma_B = +60 - 207.9 = -147.9\,\text{MN/m}^2}$$

$$\underline{\sigma_C = \quad 60 + 207.9 = 267.9\,\text{MN/m}^2}$$

$$\underline{\sigma_D = -60 + 207.9 = 147.9\,\text{MN/m}^2}$$

1.4.1 UNSYMMETRICAL SECTIONS

To demonstrate the more complex problem of the bending of an unsymmetrical section, consider the cantilever of Fig. 1.5, which is of "Z" section.

If the symmetrical theory of bending is applied to the cantilever of Fig. 1.5, then at the built-in end, the stress in the top flange would be uniform and tensile, whilst the stress in the bottom flange would be uniform and compressive.

Thus, according to the theory of bending of symmetrical sections, the resisting couple, due to the stresses, would balance the bending moment about

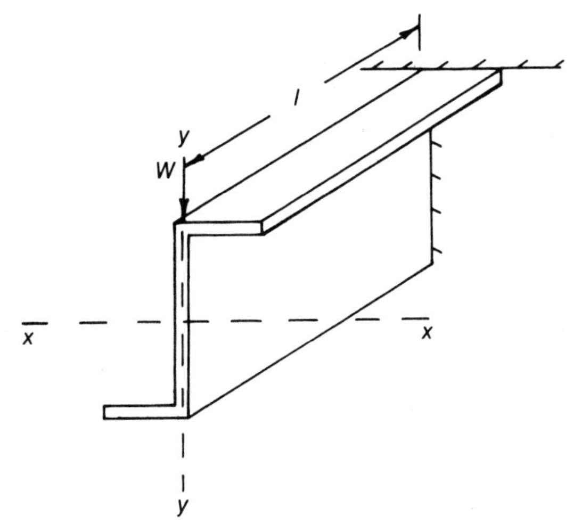

Fig. 1.5. Cross-section that is unsymmetrical about y–y.

x–x due to W, but if such a stress system existed, they would also cause a resisting couple about y–y, which is impossible, as there is no applied bending moment about y–y.

It is evident, therefore, that the theory of bending for symmetrical sections cannot be applied to unsymmetrical sections, and the mathematical explanation for this is as follows:

If,

$$M_y = \text{bending moment about } x\text{–}x$$

and

$$M_x = \text{bending moment about } y\text{–}y$$

then according to simple bending theory

$$M_x = \sum \sigma * y * \delta a \tag{1.6}$$

and

$$M_y = \sum \sigma * x * \delta a \tag{1.7}$$

where,

$$\sigma = \text{stress due to bending } = Ey/R$$

$$= \text{constant} * y \tag{1.8}$$

$$\delta a = \text{elemental area}$$

Substituting equation (1.8) into equations (1.6) and (1.7):

$$M_x = \sum \text{constant} * y^2 * \delta a = \text{constant} * \sum y^2 * \delta a$$

$$= \text{constant} * I_{xx} \tag{1.9}$$

and

$$M_y = \sum \text{constant} * x * y * \delta a$$

$$= \text{constant} * I_{xy} \tag{1.10}$$

where,

$$I_{xy} = \text{the product of inertia}$$

However, from the heuristic arguments of Section 1.4.1,

$$M_y = 0$$

therefore

$$I_{xy} = 0$$

Thus the only way that simple bending theory can be satisfied for unsymmetrical sections is for the beam to bend about those two mutually perpendicular axes where the product of inertia is zero.

It will now be shown that these two axes are in fact the *principal axes of bending* of the section, rather similar to the axes of principal stresses and principal strains, as discussed in reference [1].

1.5.1 CALCULATION OF I_{xy}

Two elements will be considered in this section, namely a rectangle and the quadrant of a circle.

Consider a rectangle of area A in the positive quadrant of the Cartesian co-ordinate system of Fig. 1.6.

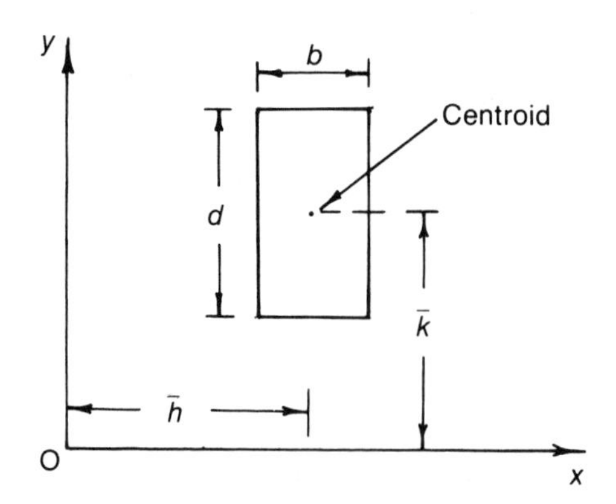

Fig. 1.6. Rectangular element.

Let,

$\bar{h}$ = distance of the centroid of the rectangular element from Oy

$\bar{k}$ = distance of the centroid of the rectangular element from Ox

Now, from equation (1.10):

$$I_{xy} = \int xy \, \mathrm{d}a$$

which for the rectangular element for Fig. 1.6 becomes:

$$I_{xy} = \iint (\mathrm{d}x * \mathrm{d}y)x * y$$

$$= \int_{\bar{k}-d/2}^{\bar{k}+d/2} \int_{\bar{h}-b/2}^{\bar{h}+b/2} xy \cdot \mathrm{d}x \cdot \mathrm{d}y$$

$$= bd\overline{hk}$$

therefore

$$\underline{I_{xy} = A\overline{hk}} \tag{1.11}$$

where, A = area of rectangle. Hence, for a built-up section, consisting of n rectangles, equation (1.11) becomes:

$$I_{xy} = \sum_{i=1}^{n} A_i \bar{h}_i \bar{k}_i \tag{1.12}$$

1.5.2

For the *quadrant of a circle* of Fig. 1.7,

$$I_{xy} = \int xy * \mathrm{d}a$$

$$= \int_0^{\pi/2} \int_0^R r\cos\theta * r\sin\theta * \mathrm{d}r * r \cdot \mathrm{d}\theta$$

$$= \int_0^{\pi/2} \int_0^R r^3 \cos\theta \sin\theta \,\mathrm{d}r\,\mathrm{d}\theta$$

$$= \frac{R^4}{4} \int_0^{\pi/2} \cos\theta \,\mathrm{d}(-\cos\theta)$$

$$= \frac{R^4}{4} * \left[-\frac{\cos^2\theta}{2} \right]_0^{\pi/2}$$

$$\underline{I_{xy} = R^4/8}$$

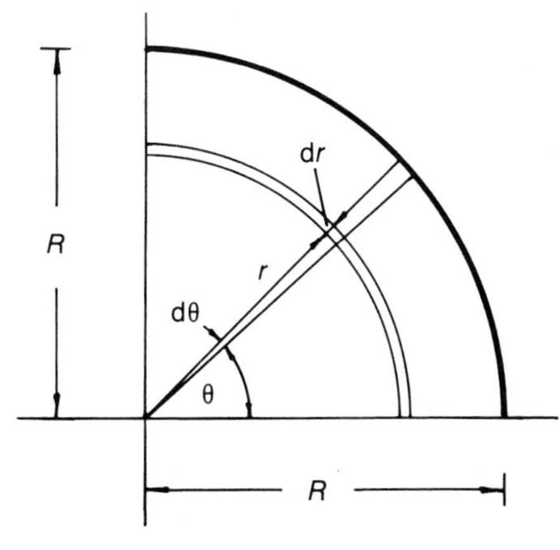

Fig. 1.7. Quadrant of a circle.

1.6.1 PRINCIPAL AXES OF BENDING

Let OU and OV be the principal axes, and Ox and Oy be the reference (or global) axes, as shown in Fig. 1.8.

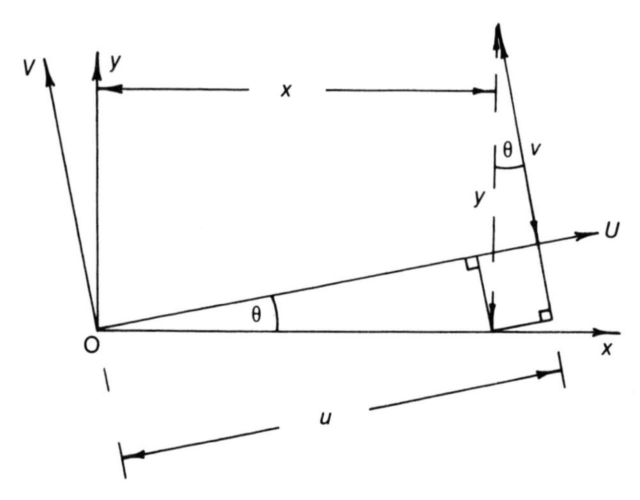

Fig. 1.8. Principal and reference axes.

From Fig. 1.8, it can be seen that:

$$u = x\cos\theta + y\sin\theta \tag{1.13}$$

$$v = y\cos\theta - x\sin\theta \tag{1.14}$$

Equations (1.13) and (1.14) are in fact *co-ordinate transformations*, as these equations transform one set of orthogonal co-ordinates to another set of orthogonal co-ordinates.

1.6.2

Now to satisfy simple bending theory and equilibrium, it is necessary for the product of inertia to be zero, with reference to the principal axes of bending, namely the axes OU and OV, i.e.

$$I_{UV} = \int uv\cdot \mathrm{d}a = 0$$

or,

$$\int (x\cos\theta + y\sin\theta) * (y\cos\theta - x\sin\theta) * \mathrm{d}a = 0$$

or

$$\int xy\cdot\cos^2\theta\cdot \mathrm{d}a - \int x\cdot{}^2\cos\theta\cdot\sin\theta\cdot \mathrm{d}a$$

$$+ \int y\cdot{}^2\sin\theta\cdot\cos\theta\cdot \mathrm{d}a - \int xy\cdot\sin\cdot{}^2\theta\cdot \mathrm{d}a$$

$$= (\cos^2\theta - \sin^2\theta)\int xy\cdot da + \cos\theta\sin\theta\left[\int y\cdot^2 da - \int x\cdot^2 da\right]$$

$$= \cos 2\theta I_{xy} + \tfrac{1}{2}\sin 2\theta (I_x - I_y) = 0$$

therefore

$$\underline{\tan 2\theta = \frac{2I_{xy}}{(I_y - I_x)}} \tag{1.15}$$

1.6.3 To determine I_U and I_V, the principal second moments of area.

Now,

$$I_U = \int v.^2 da \tag{1.16}$$

Substituting equation (1.14) into (1.16):

$$I_U = \int (y\cos\theta - x\sin\theta)^2\cdot da$$

$$= \cos^2\theta\int y\cdot^2 da + \sin^2\theta\int x\cdot^2 da - 2\cos\theta\cdot\sin\theta\int xy.da$$

$$= \cos^2\theta\cdot I_x + \sin^2\theta\cdot I_y - \sin 2\theta\cdot I_{xy}$$

$$= \frac{(1+\cos 2\theta)}{2}I_x + \frac{(1-\cos 2\theta)}{2}I_y - \sin 2\theta\cdot\frac{(I_y - I_x)}{2}\tan 2\theta$$

$$\underline{I_U = \tfrac{1}{2}(I_x + I_y) + \tfrac{1}{2}(I_x - I_y)\sec 2\theta} \tag{1.17}$$

Similarly,

$$I_V = \int u\cdot^2 da$$

$$= \cos^2\theta\cdot I_y + \sin^2\theta\cdot I_x + \sin 2\theta\cdot I_{xy}$$

$$\underline{I_V = \tfrac{1}{2}(I_x + I_y) - \tfrac{1}{2}(I_x - I_y)\sec 2\theta} \tag{1.18}$$

1.6.4

If equations (1.17) and (1.18) are added together, it can be seen that:

$$I_U + I_V = I_x + I_y \tag{1.19}$$

Equation (1.19) is known as the *INVARIANT OF INERTIA*.

1.7.1 MOHR'S CIRCLE OF INERTIA

Equations (1.15), (1.17) and (1.18) can be represented by a circle of inertia, as shown by Fig. 1.9, rather similar to Mohr's circles for stress and strain, as discussed in reference [1].

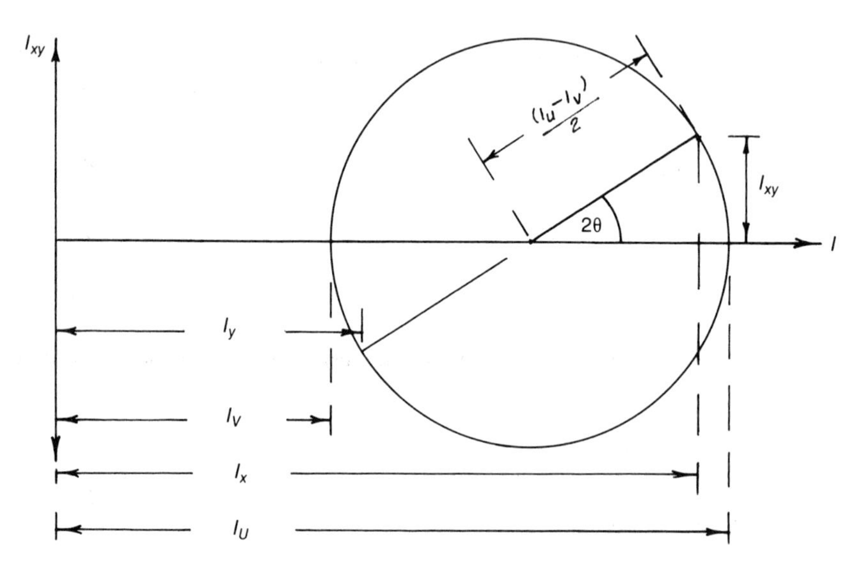

Fig. 1.9. Mohrs circle of inertia.

From Fig. 1.9 it can readily be seen that I_U and I_V are maximum and minimum values of second moments of area, and this is why I_U and I_V are called principal second moments of area.

1.7.2

Plots of the variation of the radius of gyration in any direction are called *momental ellipses*, and typical momental ellipses are shown in Fig. 1.10.

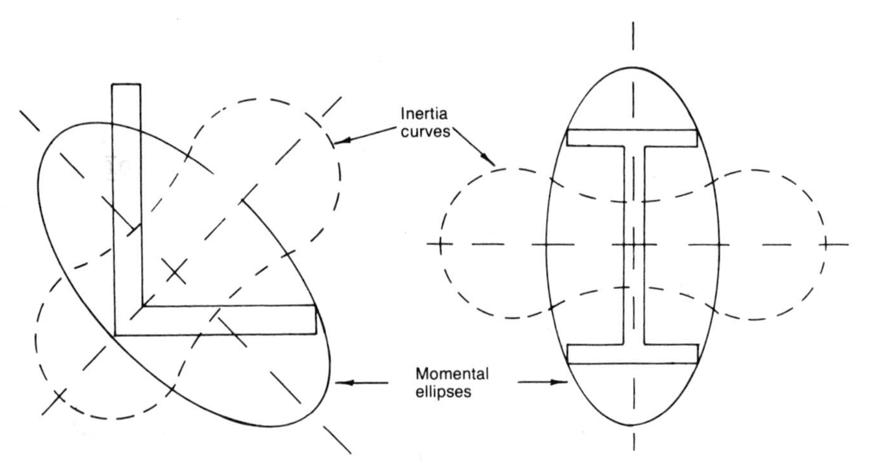

Fig. 1.10. Momental ellipses and inertia curves.

1.7.3

Structural engineers sometimes find another set of curves useful, which are known as *inertia curves*. The inertia curve of a section is obtained by plotting a radius vector equal to the second moment of area of the section at any angle θ, as shown by the dashed lines of Fig. 1.10. From Fig. 1.10, it can be seen that the inertia curves are effectively plotted in a direction perpendicular to the momental ellipses.

1.8.1 EXAMPLE 1.2 PRINCIPAL SECOND MOMENTS OF AREA OF UNSYMMETRICAL SECTIONS

Determine the directions and values of the principal second moments of area of the asymmetrical sections of Figs. 1.11 and 1.12.

(a)

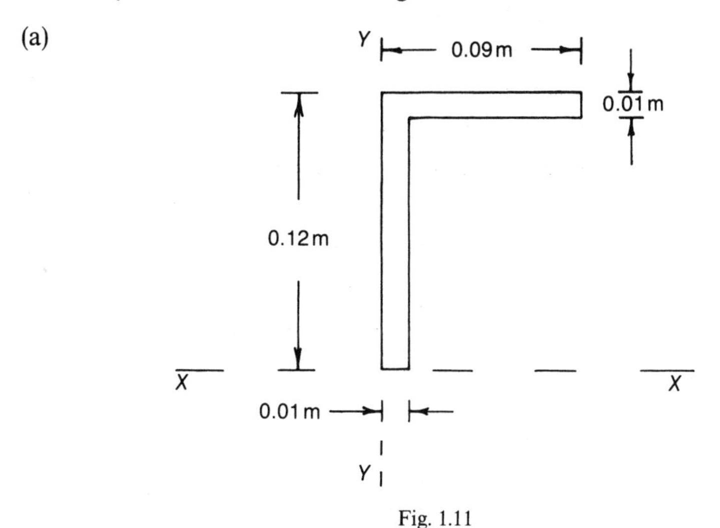

Fig. 1.11

(b)

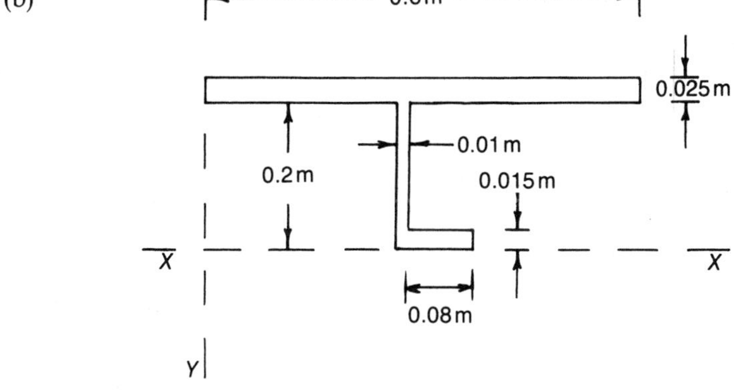

Fig. 1.12

1.8.2

Consider the angle bar of Fig. 1.11 and assume that its centroid is at O, as shown in Fig. 1.13. For convenience, the angle bar can be assumed to be composed of two rectangles, and the geometrical properties of the angle bar can be calculated with the aid of Tables 1.1 and 1.2, where,

a = area of an element

y = distance of the local centroid of the element from X–X

x = distance of the local centroid of the element from Y–Y

i_H = second moment of area of an element about an axis passing through its local centroid and parallel to X–X

i_V = second moment of area of an element about an axis passing through its local centroid and parallel to Y–Y

ay = the product $a * y$

ay^2 = the product $a * y * y$

ax = the product $a * x$

ax^2 = the product $a * x * x$

$\sum$ = summation of column, where appropriate

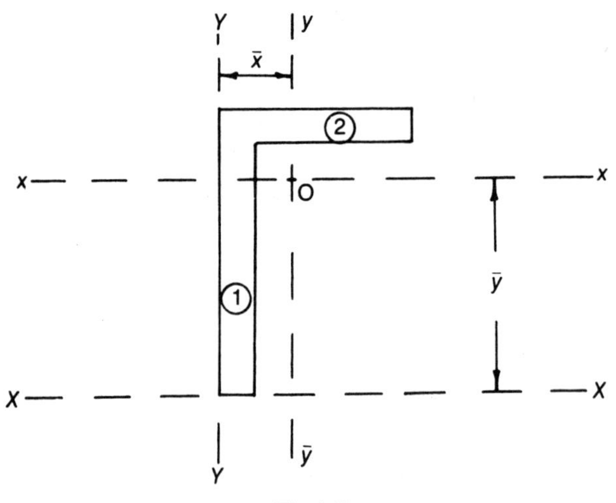

Fig. 1.13

Table 1.1 Calculation of I_{xx}

Section	a	y	ay	ay^2	i_H
①	1.2E-3	0.06	7.2E-5	4.32E-6	$\frac{0.01 \times 0.12^3}{12} = 1.44\text{E-6}$
②	8E-4	0.115	9.2E-5	1.058E-5	$\frac{0.08 \times 0.01^3}{12} = 6.7\text{E-9}$
$\sum$	2E-3	—	1.64E-4	1.49E-5	1.447E-6

From Table 1.1,

$$\bar{y} = \frac{\sum ay}{\sum a} = \frac{1.64\text{E-}4}{2\text{E-}3} = \underline{0.082\,\text{m}}$$

$$I_{XX} = \sum i_H + \sum ay^2 = 1.49\text{E-}5 + 1.447\text{E-}6$$

$$\underline{I_{XX} = 1.635\text{E-}5\,\text{m}^4}$$

$$I_{xx} = I_{XX} - \bar{y}^2 \sum a = 1.635\text{E-}5 - 0.082^2 \times 2\text{E-}3$$

$$\underline{I_{xx} = 2.902\text{E-}6\,\text{m}^4} \tag{1.20}$$

Table 1.2. Calculation of I_{yy}

Section	a	x	ax	ax^2	i_V
①	1.2E-3	0.005	6E-6	3E-8	$\frac{0.12 \times 0.01^3}{12} = 1\text{E-}8$
②	8E-4	0.05	4E-5	2E-6	$\frac{0.01 \times 0.08^3}{12} = 4.27\text{E-}7$
$\sum$	2E-3	—	4.6E-5	2.03E-6	4.37E-7

From Table 1.2,

$$\bar{x} = \frac{\sum ax}{\sum a} = \frac{4.6\text{E-}5}{2\text{E-}3} = \underline{0.023\,\text{m}}$$

$$I_{YY} = \sum i_V + \sum ax^2 = 4.37\text{E-}7 + 2.03\text{E-}6$$

$$\underline{I_{YY} = 2.467\text{E-}6\,\text{m}^4}$$

$$I_{yy} = I_{YY} - \bar{x}^2 \sum a = 2.467\text{E-}6 - 0.023^2 \times 2\text{E-}3$$

$$\underline{I_{yy} = 1.409\text{E-}6\,\text{m}^4} \tag{1.21}$$

1.8.3 To Calculate I_{xy}

From equation (1.12),

$$I_{xy} = \sum A\overline{hk}$$

$$= 1.2\text{E-}3 * (-(\bar{x} - 0.005)) * (-(\bar{y} - 0.06))$$

$$+ 8\text{E-}4 * (0.05 - \bar{x}) * (0.115 - \bar{y})$$

$$= 1.2\text{E-}3 * 0.018 * 0.022 + 8\text{E-}4 * 0.027 * 0.033$$

$$\underline{I_{xy} = 1.188\text{E-}6\,\text{m}^4} \tag{1.22}$$

1.8.4 To calculate θ

Substituting the appropriate values from equations (1.20) to (1.22) into equation (1.15):

$$\tan(2\theta) = \frac{2 * 1.188\text{E-6}}{(1.409\text{E-6} - 2.902\text{E-6})}$$

$$2\theta = -57.86^\circ$$

$$\underline{\theta = -28.93^\circ} \tag{1.23}$$

where θ is shown in Fig. 1.14.

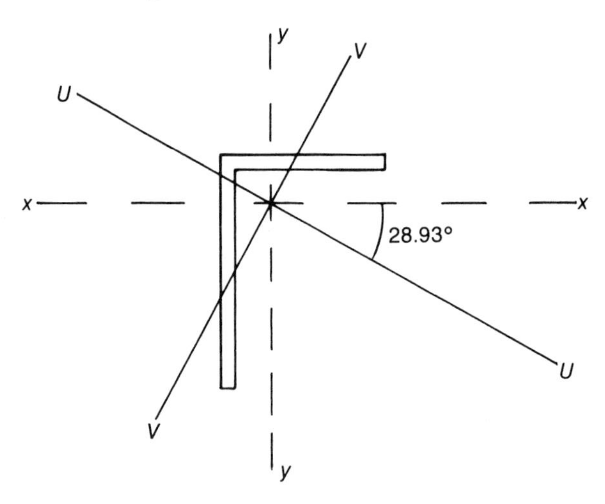

Fig. 1.14. Directions of the principal axes.

Now,

$$I_U = \tfrac{1}{2}(I_x + I_y) + \tfrac{1}{2}(I_x - I_y)\sec 2\theta$$

and

$$I_V = \tfrac{1}{2}(I_x + I_y) - \tfrac{1}{2}(I_x - I_y)\sec 2\theta$$

Hence, by substituting the values for I_x, I_y, I_{xy} and θ into the above,

$$I_U = \tfrac{1}{2}(2.902 + 1.409)\text{E-6} + \tfrac{1}{2}(2.902 - 1.409)\text{E-6} * \frac{1}{\cos(-57.86)}$$

$$= 2.156\text{E-6} + 7.465\text{E-7} * 1.88$$

$$\underline{I_U = 3.56\text{E-6}\,\text{m}^4} \tag{1.24}$$

and

$$I_V = 2.156\text{E-6} - 1.403\text{E-6}$$

$$\underline{I_V = 7.53\text{E-7}\,\text{m}^4} \tag{1.25}$$

1.8.5

Let the centroid of the section of Fig. 1.12 be at O, as shown in Fig. 1.15.

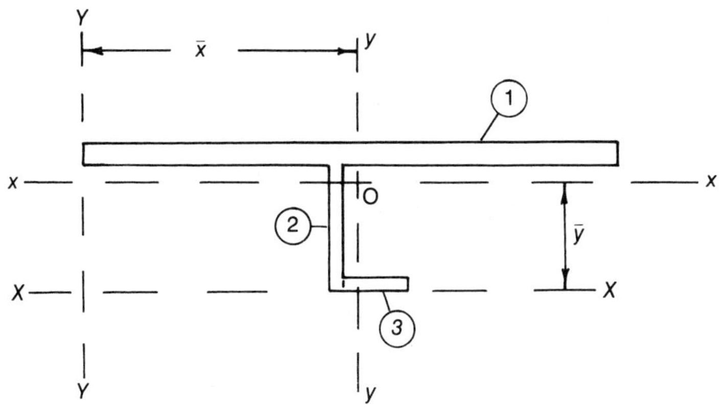

Fig. 1.15. Cross-section of beam.

Tables 1.3 and 1.4 show the calculations for determining I_x and I_y of this section, where the symbols have the same meanings as described in Section 1.8.2.

Table 1.3. Calculation for I_x.

Section	a	y	ay	ay^2	i_H
①	0.015	0.2125	3.188E-3	6.773E-4	7.81E-7
②	2E-3	0.1	2E-4	2E-5	6.667E-6
③	1.2E-3	7.5E-3	9E-6	6.75E-8	2.25E-8
$\sum$	0.0182	—	3.397E-3	6.974E-4	7.471E-6

From Table 1.3,

$$\bar{y} = \frac{\sum ax}{\sum a} = \underline{0.1866\,\text{m}}$$

$$I_{XX} = \sum i_H + \sum ay^2 = \underline{7.049\text{E-4}\,\text{m}^4}$$

$$I_{xx} = I_{XX} - \bar{y}^2 \sum \text{a} = \underline{7.118\text{E-5}\,\text{m}^4} \tag{1.26}$$

Table 1.4. Calculation for I_y.

Section	a	x	ax	ax^2	i_V
①	0.015	0.3	4.5E-3	1.35E-3	4.5E-4
②	2E-3	0.3	6E-4	1.8E-4	1.67E-8
③	1.2E-3	0.345	4.14E-4	1.428E-4	6.4E-7
$\sum$	0.0182	—	5.514E-3	1.673E-3	4.51E-4

From Table 1.4,

$$\bar{x} = \frac{\sum ax}{\sum a} = \underline{0.303\,\text{m}}$$

$$I_{YY} = \sum i_V + \sum ax^2 = \underline{2.124\text{E-3}\,\text{m}^4}$$

$$I_{yy} = \mathbf{I}_{YY} - \bar{x}^2 \sum \text{a} = \underline{4.531\text{E-4}\,\text{m}^4} \tag{1.27}$$

$$\begin{aligned} I_{xy} &= \sum A\overline{hk} \\ &= 0.015 * (0.3 - \bar{x}) * (0.2125 - \bar{y}) \\ &\quad + 2\text{E-3} * (0.3 - \bar{x}) * (0.1 - \bar{y}) \\ &\quad + 1.2\text{E-3} * (0.345 - \bar{x}) * (7.5\text{E-3} - \bar{y}) \\ &= -1.166\text{E-6} + 5.196\text{E-7} - 9.02\text{E-6} \end{aligned}$$

$$\underline{I_{xy} = -9.673\text{E-6}\,\text{m}^4} \tag{1.28}$$

1.8.6

Now,

$$2\theta = \tan^{-1}\left(\frac{2I_{xy}}{I_y - I_x}\right) = \tan^{-1}\left(\frac{1.935\text{E-5}}{3.819\text{E-4}}\right)$$

$$\underline{\theta = -1.45^\circ}$$

Now,

$$I_U = \tfrac{1}{2}(I_x + I_y) + \tfrac{1}{2}(I_x - I_y)\sec 2\theta$$

$$= \tfrac{1}{2}(7.118\text{E-5} + 4.531\text{E-4}) + \tfrac{1}{2}(7.118\text{E-5} - 4.531\text{E-4}) * \frac{1}{\cos(-2.9)}$$

$$\underline{I_U = 7.095\text{E-5}\,\text{m}^4}$$

Similarly,

$$\underline{I_V = 4.533\text{E-4}\,\text{m}^4}$$

1.9.1 STRESSES IN BEAMS OF ASYMMETRICAL SECTION

Equation (1.5) can be extended to the bending of unsymmetrical sections, as shown by equation (1.29), which gives the value of bending stress at any point "P" in the positive quadrant for the orthogonal axes OU and OV, where θ, u and v are defined in Fig. 1.16.

$$\sigma = \frac{M.\cos\theta \cdot v}{I_{UU}} + \frac{M \cdot \sin\theta \cdot u}{I_{VV}} \tag{1.29}$$

From equation (1.29), it can be seen that M is due to a load that acts perpendicular to the Ox axis, where in Fig. 1.16, M is shown according to

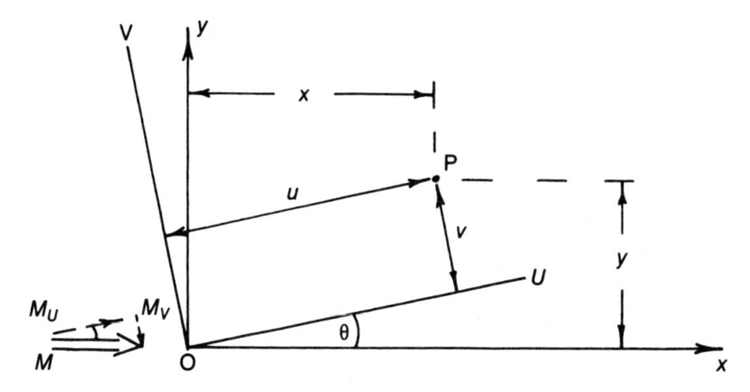

Fig. 1.16. Principal axes of bending, *Ou* and *Ov*.

the right-hand screw rule. The components of M, namely M_U and M_V, cause bending about the principal axes Ou and Ov, respectively, where

$$M_U = \mathrm{M}\cdot\cos\theta$$

and

$$M_V = M\cdot\sin\theta$$

1.10.1 EXAMPLE 1.3 BENDING STRESS IN A CANTILEVER

A cantilever of length 2 m is subjected to a concentrated load of 5 kN at its free end, as shown in Fig. 1.17. Assuming that the cantilever's cross-section is as shown in Example 1.2(a), determine the direction of its neutral axis and the position and magnitude of the maximum stress.

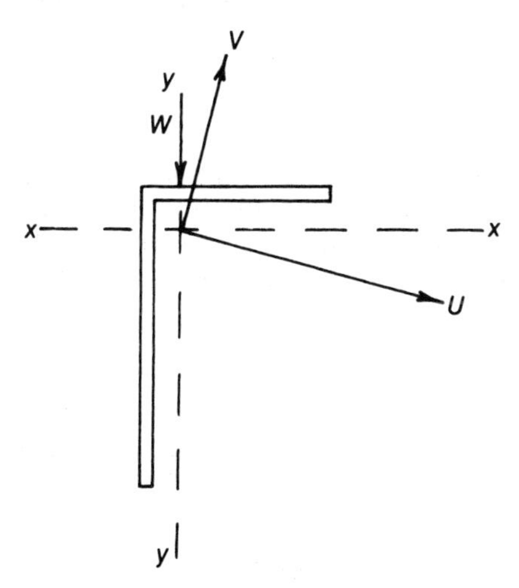

Fig. 1.17. Cross-section of cantilever.

The maximum bending moment ($\hat{M}$) occurs at the built-in end, where,

$$\hat{M} = Wl = 5\,\mathrm{kN} \times 2\,\mathrm{m} = 10\,\mathrm{kN\,m} \tag{1.30}$$

At the neutral axis, the bending stress is zero, i.e.

$$\frac{M\cos\theta\cdot v}{I_{UU}}+\frac{M\cdot\sin\theta\cdot u}{I_{VV}}=0 \tag{1.31}$$

or

$$\frac{v}{u}=-\frac{I_{UU}}{I_{VV}}\tan\theta$$

$$=\tan\beta$$

where β is defined in Fig. 1.18. Therefore

$$\underline{\beta=\tan^{-1}(-I_{UU}*\tan\theta/I_{VV})} \tag{1.32}$$

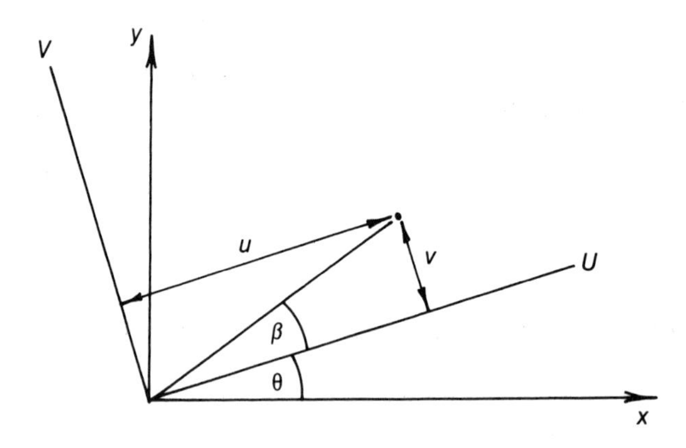

Fig. 1.18. Definition of β.

Substituting the appropriate values from Section 1.8.4 into equation (1.32):

$$\beta=\tan^{-1}(-3.56\text{E-}6*(-0.5527)/7.53\text{E-}7)$$

$$\underline{\beta=69.06^\circ}$$

The direction of the neutral axis is shown in Fig. 1.19, and the largest stress due to bending will occur at a point on the section which is at the furthest perpendicular distance from the neutral axis (NA).

From Fig. 1.19, it can be seen that the furthest perpendicular distance from NA is at the point "B".

From equations (1.13) and (1.14):

$$u_B=x_B\cos\theta+y_B\sin\theta$$

$$=-0.013\times0.875-0.083\times(-0.483)$$

$$\underline{u_B=0.029\text{ m}}$$

and

$$v_B=y_B\cos\theta-x_B\sin\theta$$

$$=-0.083\times0.875-0.013\times0.483$$

$$\underline{v_B=-0.0789\text{ m}}$$

where, u_B, v_B, etc. are defined in Fig. 1.19.

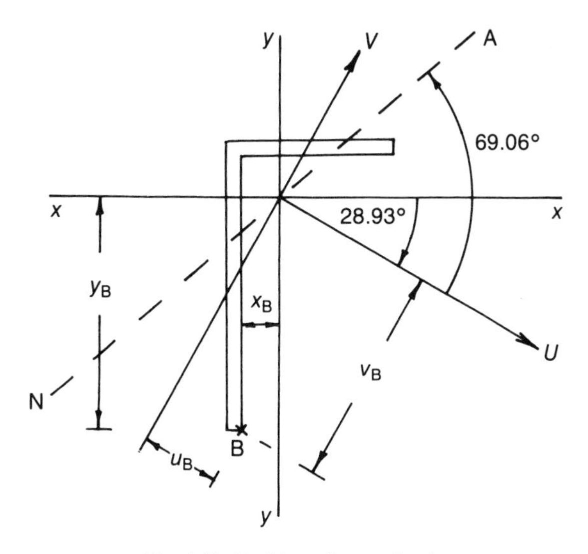

Fig. 1.19. Position of neutral axis.

From equation (1.27),

$$\sigma_B = \text{stress at the point B}$$

$$= \frac{10\,\text{kN m} \times 0.875 \times (-0.0789)\,\text{m}}{3.56\text{E-6}\,\text{m}^4}$$

$$+ \frac{10\,\text{kN m} \times (-0.483) \times (0.029)\,\text{m}}{7.53\text{E-7}\,\text{m}^4}$$

$$= -(193.9 + 186)\,\text{MN/m}^2$$

$$\underline{\sigma_B = -\,380\,\text{MPa (compressive)}}$$

1.11.1 EXAMPLE 1.4 BENDING OF A BEAM WITH A UNIFORMLY DISTRIBUTED LOAD

If an encastré beam of length 2 m and with a cross-section as in Example 1.2(b) is subjected to the uniformly distributed load shown in Fig. 1.20, determine the position and value of the maximum stress.

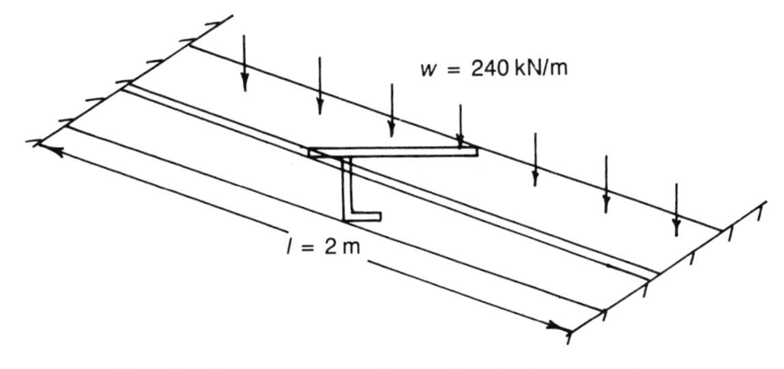

Fig. 1.20. Encastré beam with a uniformly distributed load.

The maximum value for bending moment ($\hat{M}$) occurs at the ends [1], and is given by:

$$\hat{M} = \frac{wl^2}{12} = \frac{240 \times 2^2}{12} = \underline{80\,\text{kN m}}$$

From equation (1.32):

$$\beta = \tan^{-1}(-I_{UU} * \tan\theta / I_{VV})$$
$$= \tan^{-1}(-7.095\text{E-}6 * (-0.0253)/4.533\text{E-}4)$$
$$\underline{\beta = 0.23^\circ} \tag{1.33}$$

The position of the neutral axis (NA) is shown in Fig. 1.21.

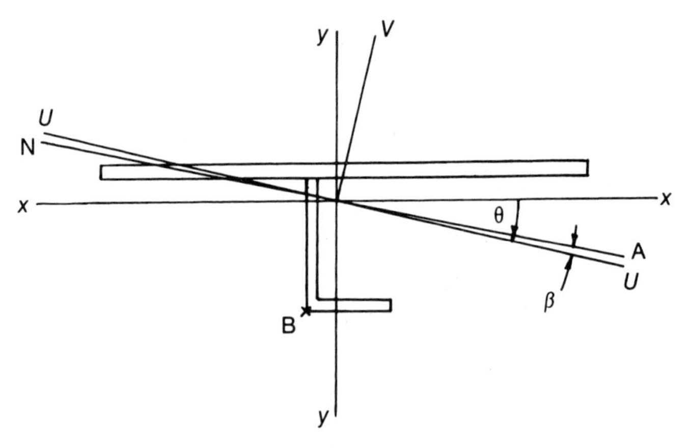

Fig. 1.21. Position of neutral axis for encastré beam.

By inspection, it can be seen that the point of maximum stress occurs at "B", which is the furthest perpendicular distance from the neutral axis.

Prior to calculating σ_B, the stress at the point "B", the distance u_B and v_B have to be determined, as follows:

$$u_B = x_B \cos\theta + y_B \sin\theta$$
$$= (0.295 - 0.303) * 0.9997 + 0.1866 * 0.0253$$
$$= -7.998\text{E-}3 + 4.721\text{E-}3$$
$$\underline{u_B = -3.277\text{E-}3\,\text{m}} \tag{1.34}$$

$$v_B = x_B \sin\theta - y_B \cos\theta$$
$$= -8\text{E-}3 * (-0.0253) - 0.1866 * (0.9997)$$
$$\underline{v_B = -0.1867\,\text{m}} \tag{1.35}$$

where, u_B, v_B, x_B and y_B are co-ordinates of the point "B", as defined in Fig. 1.19.

From equation (1.29):

$$\sigma_B = 80\,\text{kN m}\left[\left(\frac{-0.1867 * 0.9997}{7.095\text{E-}5}\right) - \frac{3.277\text{E-}3 * (-0.0253)}{4.533\text{E-}4}\right]\frac{\text{m}}{\text{m}^4}$$

$$= \frac{80\,\text{kN}}{\text{m}^2} * (-2631.3 + 0.18)$$

$$\underline{\sigma_B = -210.5\,\text{MN/m}^2\ \text{(compressive)}}$$

1.12.1 EXAMPLE 1.5 DEFLECTION OF AN ASYMMETRICAL SECTION CANTILEVER

Determine the end deflection of the cantilever of Example 1.3, given that,

$$E = 2\text{E}11\ \text{N/m}^2$$

In this case, the components of load can be resolved along the two principal axes of bending, and each component of deflection can then be calculated, and hence, the resultant deflection can be obtained.

Now, from reference [1], the maximum deflection (δ) of an end-loaded cantilever is given by:

$$\delta = \frac{Wl^3}{3EI}$$

where,

W = load

l = length of cantilever

E = elastic modulus

I = second moment of area of cantilever section about its axis of bending

For the present problem,

$$\delta_u = \text{deflection under load in the } u \text{ direction}$$

$$= -\frac{W \sin\theta \cdot l^3}{3EI_V}$$

and

$$\delta_v = \text{deflection under load in the } v \text{ direction}$$

$$= -\frac{W \cos\theta \cdot l^3}{3EI_U}$$

therefore

$$\delta_u = \frac{10 \times 10^3 \times 0.8752 \times 8}{3 \times 2 \times 10^{11} \times 3.56\text{E-}6} = \underline{0.0328\ \text{m}}$$

$$\delta_v = -\frac{10 \times 10^3 \times 0.4837 \times 8}{3 \times 2 \times 10^{11} \times 7.53\text{E-}7} = \underline{-0.0816\ \text{m}}$$

These two components of deflection can be drawn, as shown in Fig. 1.22. Hence, from Pythagoras' theorem and from elementary trigonometry,

$$\underline{\delta = \sqrt{\delta_u^2 + \delta_v^2} = 0.0917}$$

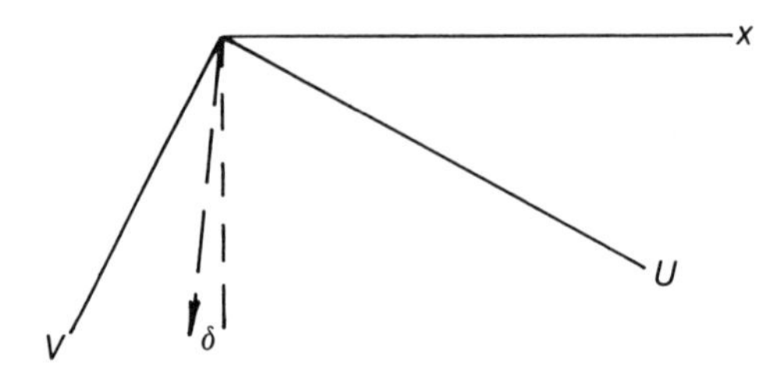

Fig. 1.22

$$@ \text{ an angle of } 90° - 28.93° - \tan^{-1}\left(\frac{\delta_v}{\delta_u}\right)$$

$$= 90° - 28.93° - 69.03°$$

$$= \underline{-7.96°} \text{ [to the vertical (clockwise)]}$$

1.13.1 EXAMPLE 1.6 DEFLECTION OF AN ASYMMETRICAL SECTION ENCASTRÉ BEAM

Determine the central deflection of the encastré beam of Example 1.4, given that $E = 2 \times 10^{11}$ N/m^2.

From reference [1], the maximum deflection of an encastré beam with a uniformly distributed load is given by:

$$\delta = \frac{wl^4}{384EI}$$

where,

w = load/unit length

l = length

E = Young's modulus

I = second moment of area

For the present problem,

$$\delta_u = \text{central deflection in the } u \text{ direction} = \frac{w \sin\theta \cdot l^4}{384EI_V}$$

$$\delta_v = \text{central deflection in the } v \text{ direction} = -\frac{w \cos\theta \cdot l^4}{384EI_U}$$

$$\delta_u = \frac{240 \times 10^3 \times (-0.0253) \times 16}{384 \times 2 \times 10^{11} \times 4.533\text{E-4}} = \underline{2.791\text{E-3 mm}}$$

$$\delta_v = -\frac{240 \times 10^3 \times 0.9997 \times 16}{384 \times 2 \times 10^{11} \times 7.095\text{E-}5} = \underline{-0.7045\,\text{mm}}$$

Resultant deflection $= \delta = 0.7045\,\text{m}$ @ $1.22°$ (clockwise) from the vertical

N.B. The theory in this chapter does not include the additional effects of shear stresses due to torsion and bending that occur when unsymmetrical beams are loaded through their centroids, but these theories are dealt with in some detail in Chapters 2 and 5.

EXAMPLES FOR PRACTICE 1

1. Determine the direction and magnitudes of the principal second moments of area of the angle bar shown in Fig. Q.1.1.

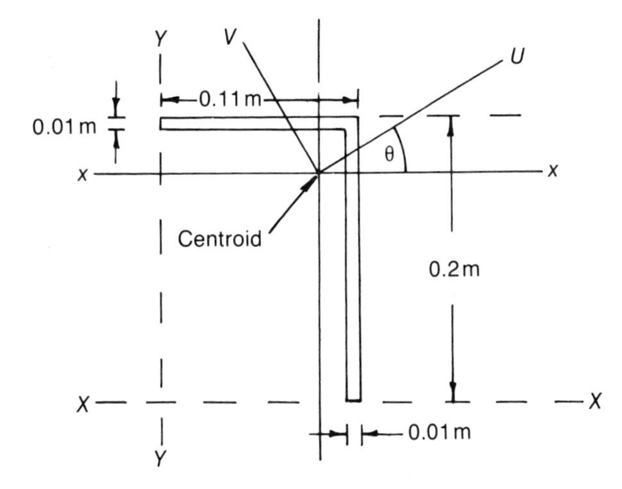

Fig. Q.1.1

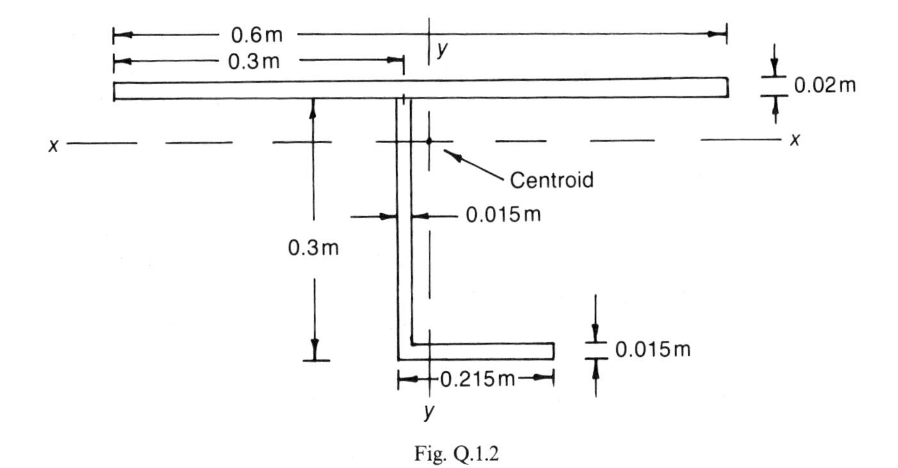

Fig. Q.1.2

$\{I_{xx} = 1.27\text{E-}5\,\text{m}^4;\ \ I_{yy} = 2.849\text{E-}6\,\text{m}^4;\ \ I_{xy} = -3.843\text{E-}6\,\text{m}^4;\ \ \theta = 17.64°;$ $I_{UU} = 1.381\text{E-}5\,\text{m}^4;\ I_{VV} = 1.741\text{E-}6\,\text{m}^4\}$

2. Determine the direction and magnitudes of the principal second moments of area of the section of Fig. Q.1.2.

   $\{I_{xx} = 2.844\text{E-}4\,\text{m}^4;\ I_{yy} = 3.994\text{E-}4\,\text{m}^4;\ I_{xy} = -7.097\text{E-}5\,\text{m}^4;\ I_U = 1.592\text{E-}4\,\text{m}^4;\ I_V = 5.246\text{E-}4\,\text{m}^4;\ \theta = -25.49°\}$

3. Determine the stresses at the corners and the maximum deflection of a cantilever of length 3 m, loaded at its free end with a concentrated load of 10 kN, as shown in Fig. Q.1.3.

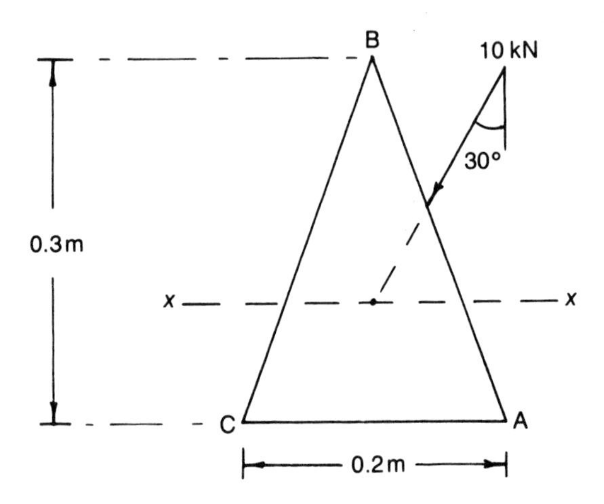

Fig. Q.1.3. Cross-section of cantilever.

   $\{\sigma_A = 12.68\,\text{MPa};\ \sigma_B = 34.64\,\text{MPa};\ \sigma_C = -47.32\,\text{MPa}$; neutral axis is 60° clockwise from $x-x\}$

4. If a cantilever of length 3 m, and with a cross-section as shown in Fig. Q.1.1, is subjected to a vertically applied downward load at its free end, of magnitude 4 kN, determine the position and value of the maximum bending stress.

   $\{$NA is $-68.38°$ from U–U; $\sigma_B = -163.63\,\text{MPa}\}$

5. A simply-supported beam of length 4 m, and with a cross-section as shown in Fig. Q.1.2, is subjected to a centrally placed concentrated load of 20 kN, acting perpendicularly to the x–x axis, and through the centroid of the beam. Determine the stress at mid-span at the point "B" in the cross-section of the beam.

   $\{$NA is 8.23° from U–U; $\sigma_B = -19.82\,\text{MPa}\}$

6. Determine the components of deflection under the load, in the directions

of the principal axes of bending, for the beam of Example 4, given that:

$$E = 2\text{E}11\ \text{N/m}^2$$

$\{\delta_u = -0.031\ \text{m};\ \delta_v = -0.0124\ \text{m}\}$

7. Determine the components of the central deflection in the directions of the principal axes of bending for the beam of Example 5, given that:

$$E = 1\text{E}11\ \text{N/m}^2$$

$\{\delta_u = 2.19\text{E-}4\ \text{m};\ \delta_v = -1.51\text{E-}3\ \text{m}\}$

2

Shear Stresses in Bending and Shear Deflections

2.1.1 SHEAR STRESSES DUE TO BENDING

If a horizontal beam is subjected to transversely applied vertical loads, so that the bending moment changes, then there will be vertical shearing forces at every point along the length of the beam, where the bending moment changes [1].

Furthermore, if the cross-section of the beam is of a built-up section, such as a rolled steel joist (RSJ), or a tee beam or a channel section, then there will be horizontal shearing stresses in addition to the vertical shearing stresses, where both are caused by the same vertically applied shearing forces. Vertical shearing stresses, due to bending, are those that act in a vertical plane, and horizontal shearing stresses are those that act in a horizontal plane, and later in this chapter, it will be shown that for curved sections, such as split tubes etc., the shearing stresses due to bending act in the planes of the curved sections. Similar arguments apply to horizontal beams subjected to laterally applied horizontal shearing forces.

To demonstrate the concept of shearing stresses due to bending, the variation of vertical shearing stresses across the section of a beam will be first considered, followed by a consideration of the variation of horizontal shearing stresses due to the same vertical shearing forces.

2.1.2 Vertical Shearing Stresses

Consider a beam, subjected to a system of vertical loads, so that the bending moment changes along the length of the beam, as shown in Fig. 2.1.

Consider an elemental length "dx" of the beam, as shown in Fig. 2.2(a).

Consider the sub-element of Fig. 2.2(b), and also the sub-sub-element in

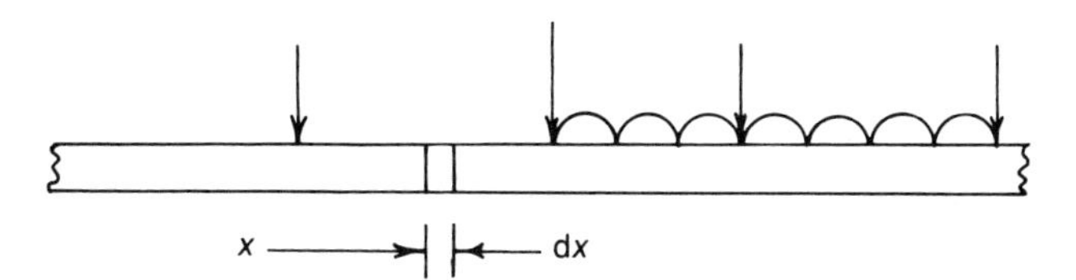

Fig. 2.1. Beam subjected to vertical loads.

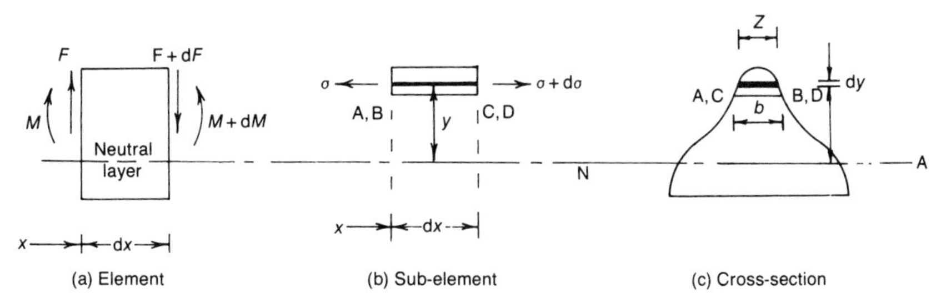

Fig. 2.2. Beam element under shearing force.

the same figure, which is shown shaded. Let the stress on the left of the sub-sub-element of Fig. 2.2(b) be σ, and the stress on the right of the sub-sub-element be $\sigma + \mathrm{d}\sigma$, due to M and $M + \mathrm{d}M$, respectively.

From Fig. 2.2(b), it can be seen that there is an apparent unbalanced force in the x direction, of magnitude $\mathrm{d}\sigma * Z * \mathrm{d}y$, but as the sub-sub-element is in equilibrium, this is not possible, i.e. the only way equilibrium can be achieved is for a longitudinal horizontal stress to act tangentially along the rectangular face ABCD. It is evident that as this stress acts tangentially along the face ABCD, it must be a shearing stress.

Let,

$$\tau = \text{shearing stress on the face ABCD}$$

Resolving horizontally,

$$\tau * b * \mathrm{d}x = \int \mathrm{d}\sigma * \mathrm{d}A$$

$$= \int \frac{\mathrm{d}M * y * \mathrm{d}A}{I}$$

where,

$$\mathrm{d}A = Z * \mathrm{d}y$$

or

$$\tau = \frac{\mathrm{d}M}{\mathrm{d}x} \cdot \frac{1}{bI} \int y \cdot \mathrm{d}A$$

but,

$$\frac{\mathrm{d}M}{\mathrm{d}x} = F = \text{shearing force at } x$$

therefore

$$\tau = \frac{F}{bI}\int y \cdot \mathrm{d}A \tag{2.1}$$

where, in this case,

$\int y \cdot \mathrm{d}A$ = first moment of area of the section *above the plane ABCD*, and about the neutral axis, NA.

Owing to complementary shearing stresses [1], the shearing stress on the face ABCD will be accompanied by three other shearing stresses of the same magnitude, as shown in Fig. 2.3.

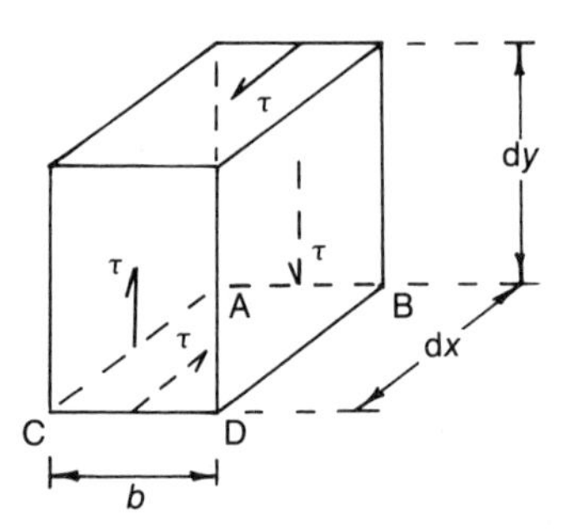

Fig. 2.3. Complementary shearing stresses in a vertical plane.

From Fig. 2.3, it can be seen that these four complementary shearing stresses act together in a vertical plane, and this is why these shearing stresses are called vertical shearing stresses.

2.1.3 Horizontal Shearing Stresses

Consider a horizontal top flange, of thickness t, on a horizontal beam under the action of vertical shearing forces, as shown in Fig. 2.4.

Let,

M = bending moment at AB

$M + \mathrm{d}M$ = bending moment at CD

σ = bending stress in flange at AB

$\sigma + \mathrm{d}\sigma$ = bending stress in flange at CD

Consider the equilibrium of an element of the flange, ABCD, in the x direction. From the plan view of Fig. 2.4, it can be seen that the apparent unbalanced force in the longitudinal direction on this element is $\mathrm{d}\sigma * t * (B - Z)$. However, as the beam is in equilibrium, no such unbalanced force can exist, and the only way that equilibrium can be achieved in this flange element is for a horizontal shearing stress to act tangentially along BD.

Let,

τ = shearing stress on the face BD

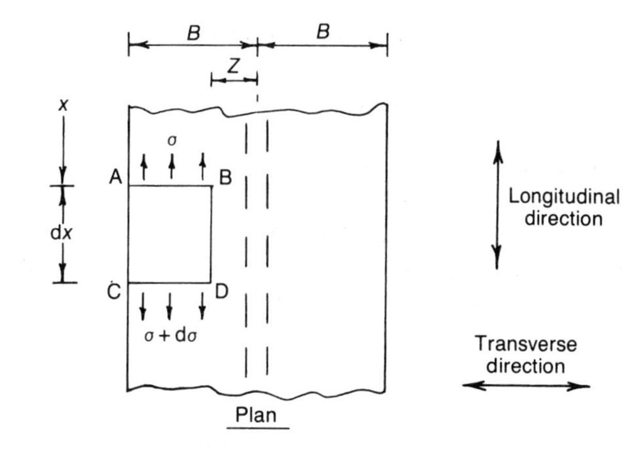

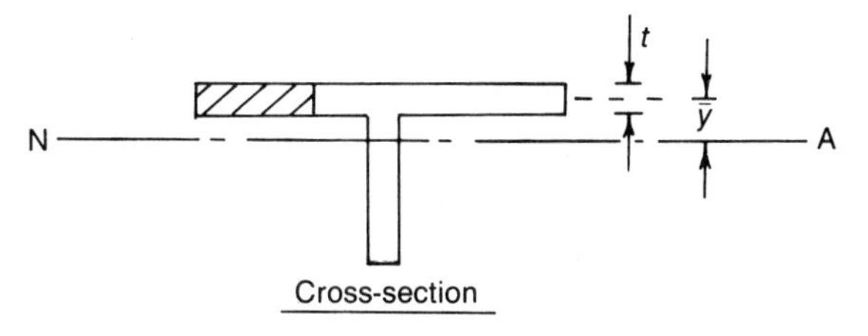

Fig. 2.4. Horizontal flange on a beam.

Considering horizontal equilibrium of the element ABCD in the x direction:

$$\mathrm{d}\sigma * t * (B - Z) = \tau * \mathrm{d}x * t$$

or,

$$\frac{\mathrm{d}M * \bar{y} * t * (B - Z)}{I} = \tau * \mathrm{d}x * t$$

or,

$$\tau = \frac{\mathrm{d}M}{\mathrm{d}x} \frac{(B - Z)\bar{y}}{I}$$

but,

$$\frac{\mathrm{d}M}{\mathrm{d}x} = F = \text{the vertical shearing force}$$

therefore

$$\tau = \frac{F(B - Z)\bar{y}}{I} \tag{2.2}$$

Equation (2.2) can be obtained directly from equation (2.1), as follows

$$\tau = \frac{F \int y \cdot \mathrm{d}A}{bI}$$

which, for the horizontal flange element (ABCD) of Fig. 2.4, becomes,

$$\tau = \frac{F}{tI} * (B - Z) * t * \bar{y} = \frac{F(B - Z)\bar{y}}{I}$$

which is identical to equation (2.2).

It should be noted that, in this case, $\int y \cdot dA$ was the first moment of area of the flange element ABCD about NA, and b was, in fact, the flange thickness t. The reason for applying equation (2.1) in this way was because the shearing stresses in the flange act in a horizontal plane, as shown in Fig. 2.5.

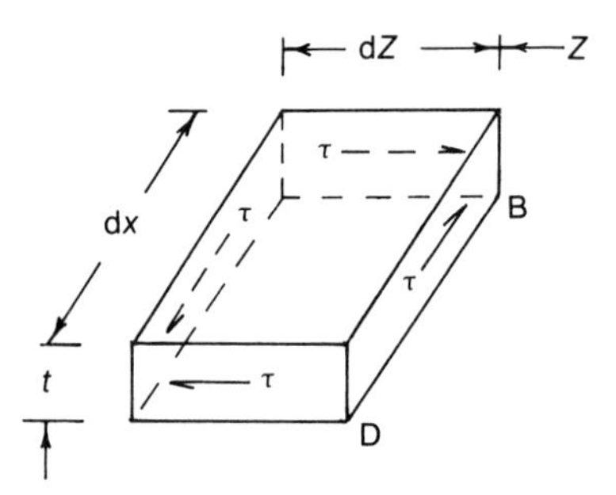

Fig. 2.5. Complementary shearing stresses in a horizontal plane.

The horizontal shearing stress distribution of equation (2.2) can be seen to vary linearly along the width of the flange, having a maximum value at the centre of the flange, and zero values at the flange edges.

In fact, owing to the effects of complementary shearing stresses, this shearing stress will be accompanied by three other shearing stresses, where all four shearing stresses act together in a horizontal plane, as shown in Fig. 2.5.

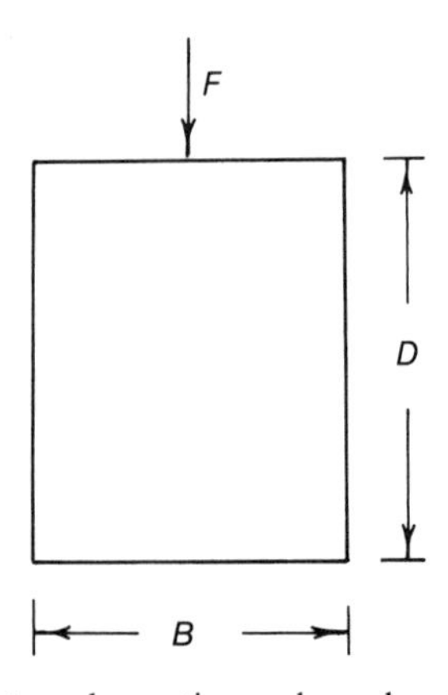

Fig. 2.6. Rectangular section under a shearing force F.

2.2.1 EXAMPLE 2.1 VERTICAL SHEARING STRESS DISTRIBUTION IN A RECTANGULAR SECTION

Calculate and sketch the vertical shearing stress distribution in a horizontal beam of rectangular section, subjected to a transverse vertical shearing force F, as shown in Fig. 2.6.

2.2.2 From equation (2.1),

$$\tau = \text{vertical shearing stress at any point } y \text{ from NA (Fig. 2.7)}$$

$$= \frac{F \int y \, \mathrm{d}A}{bI}$$

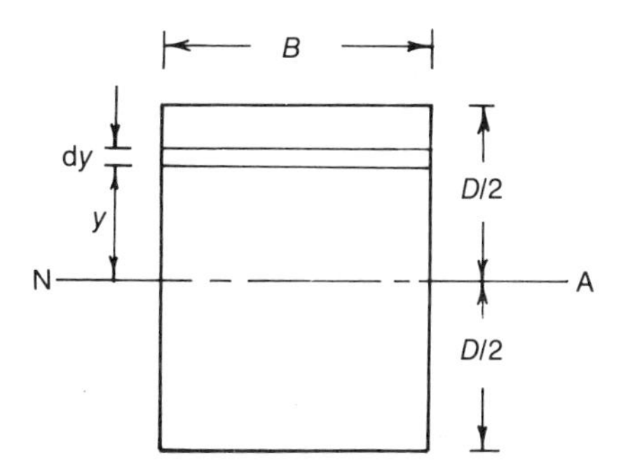

Fig. 2.7. Rectangular section.

Now,

$$I = BD^3/12$$

and for this case,

$$b = B$$

Therefore,

$$\tau = \frac{12 * F}{B * BD^3} \int_y^{D/2} B * \mathrm{d}y * y$$

$$= \frac{12F}{BD^3} \left[\frac{y^2}{2} \right]_y^{D/2}$$

$$\underline{\tau = \frac{6F}{BD^3} \left(\frac{D^2}{4} - y^2 \right)} \tag{2.3}$$

Equation (2.3) can be seen to vary parabolically, having a maximum value at NA, and being zero at the top and bottom.

Let,

$$\hat{\tau} = \text{maximum shearing stress}$$

$$= \frac{6F}{BD^3} * \frac{D^2}{4}$$

$$\hat{\tau} = \frac{1.5F}{BD} \tag{2.4}$$

Equation (2.4) shows that the maximum value of shearing stress ($\hat{\tau}$) is 50% greater than the average value (τ_{av}) where,

$$\tau_{av} = \frac{F}{BD}$$

A plot of the variation of vertical shearing stress for this section is shown in Fig. 2.8.

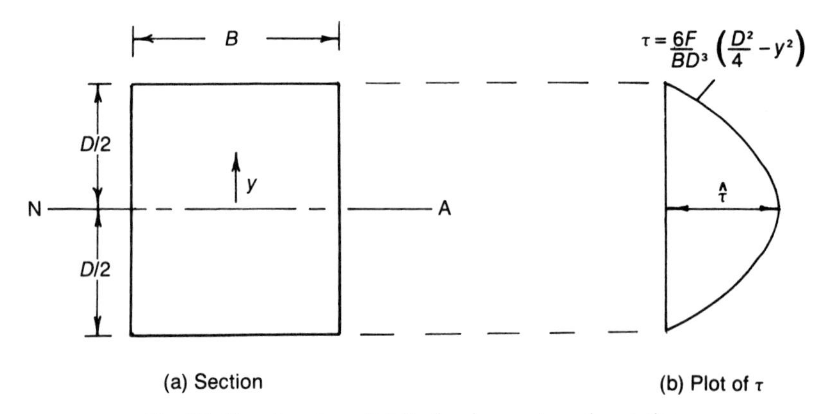

Fig. 2.8. Shearing stress distribution in a rectangular section.

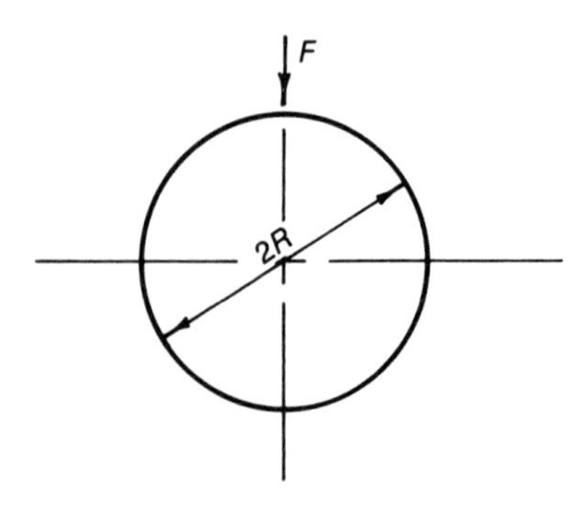

Fig. 2.9. Circular section.

2.3.1 EXAMPLE 2.2 VERTICAL SHEARING STRESS DISTRIBUTION IN A CIRCULAR SECTION

Calculate and sketch the vertical shearing stress distribution in a horizontal beam of circular section, subjected to the shearing force F, shown in Fig. 2.9.

2.3.2

Let,

$$\tau = \text{vertical shearing stress at any distance } y \text{ from NA (Fig. 2.10)}$$

$$= \frac{F}{bI} \int y \cdot \mathrm{d}A$$

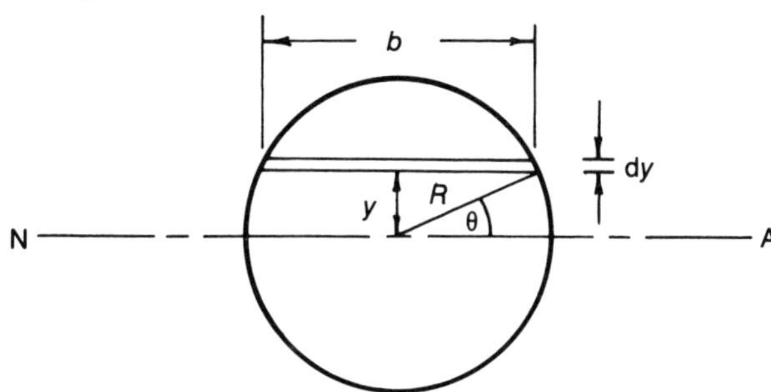

Fig. 2.10. Circular section

$$\tau = \frac{F}{bI} \int_y^R b \cdot \mathrm{d}y \cdot y$$

but,

$$I = \pi R^4/4$$

$$b = 2R\cos\theta$$

and,

$$y = R\sin\theta$$

By differentiation,

$$\mathrm{d}y = R\cos\theta \cdot \mathrm{d}\theta$$

therefore

$$\tau = \frac{4F}{2\pi R^5 \cos\theta} \int_\theta^{\pi/2} 2R\cos\theta \cdot R\sin\theta \cdot R\cos\theta \cdot \mathrm{d}\theta$$

$$= \frac{-4F}{\pi R^2 \cos\theta} \int_\theta^{\pi/2} \cos^2\theta(\mathrm{d}\cos\theta)$$

$$= \frac{-4F}{\pi R^2 \cos\theta} \left[\frac{\cos^3\theta}{3}\right]_\theta^{\pi/2}$$

$$\tau = \frac{4F\cos^2\theta}{3\pi R^2} = \frac{4F}{3\pi R^2}(1 + \sin^2\theta)$$

but, $\sin\theta = y/R$;

therefore

$$\tau = \frac{4F}{3\pi R^2}\left[1 + \left(\frac{y}{R}\right)^2\right] \tag{2.5}$$

which, again, is a parabolic distribution, as shown in Fig. 2.11.

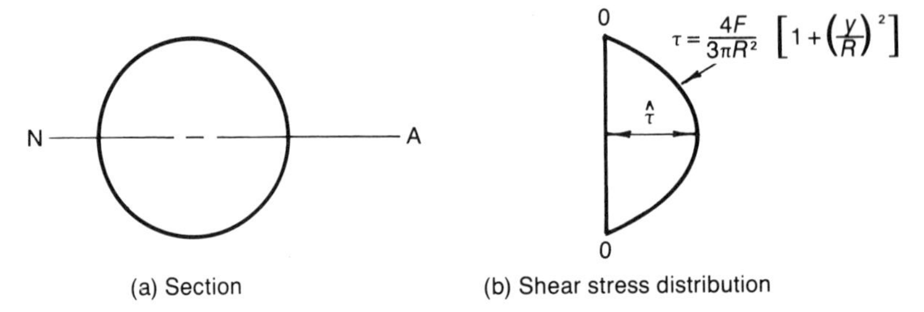

Fig. 2.11. Shear stress distribution in a circular section.

$\hat{\tau}$ = the maximum value of shearing stress

$$= \frac{4F}{3\pi R^2}$$

which can be seen to be about 33.3% larger than the average shearing stress τ_{av}, where,

$$\tau_{av} = \frac{F}{\pi R^2}$$

2.4.1 EXAMPLE 2.3 VERTICAL SHEARING STRESS DISTRIBUTION IN A TRIANGULAR SECTION

Calculate and sketch the distribution of vertical shearing stress in the triangular section of Fig. 2.12.

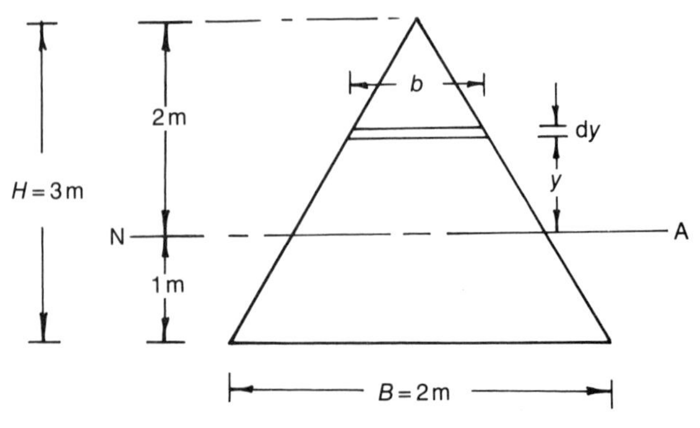

Fig. 2.12. Triangular section.

2.4.2

At any distance y above NA in the triangular section of Fig. 2.12.

$$\tau = \frac{F\int y \cdot \mathrm{d}A}{bI}$$

but,

$$I = BH^3/36 = 2 \times 3^3/36 = 1.5$$

and,

$$b = B\left(\tfrac{2}{3} - \frac{y}{H}\right) = 2(0.667 - 0.333y)$$

therefore

$$\tau = \frac{F}{2(0.667 - 0.333y) \times 1.5}\int b \cdot \mathrm{d}y \cdot y$$

$$= \frac{F}{(2-y)}\int 2(0.667y - 0.333y^2)\mathrm{d}y$$

$$= \frac{2F}{(2-y)}\left[\frac{0.667\,y^2}{2} - 0.111y^3\right]_y^2$$

$$= \frac{2F}{(2-y)}\{[1.333 - 0.88889] - [0.333y^2 - 0.111y^3]\}$$

$$\tau = \frac{2F}{(2-y)}[0.444 - (0.333y^2 - 0.111y^3)] \tag{2.6}$$

For maximum τ,

$$\frac{\mathrm{d}\tau}{\mathrm{d}y} = 0$$

therefore

$$(2-y)[-(0.667y - 0.333y^2)] - [0.444 - (0.333y^2 - 0.111y^3)](-1)$$

or

$$-(2-y)(0.667y - 0.333y^2) + (0.444 - 0.333y^2 + 0.111y^3) = 0$$

$$-1.333y + 0.667y^2 + 0.667y^2 - 0.333y^3 + 0.444 - 0.333y^2 + 0.111y^3 = 0$$

$$-0.222y^3 + y^2 - 1.333y + 0.444 = 0$$

$$0.222y^3 - y^2 + 1.333y - 0.444 = 0 \tag{2.7}$$

Equation (2.7) has three roots, but for this case, the root of interest is the lowest positive root, which can be obtained by the Newton–Raphson [2] iterative process, as follows:

$$y_{(i)} = y_{(i-1)} - \frac{f(y_{i-1})}{f'(y_{i-1})}$$

where,

$$y_i = i\text{th approximation—to be determined}$$
$$y_{i-1} = \text{known approximation at the } i-1 \text{ stage}$$

In this case,

$$f(y) = 0.222y^3 - y^2 + 1.333y - 0.444$$

and

$$f'(y) = \frac{\mathrm{d}[f(y)]}{\mathrm{d}y} = 0.667y^2 - 2y + 1.333$$

For *1st approximation*, let

$$y = 0$$

or

$$y(0) = 0$$

therefore

$$y(1) = 0 + \frac{0.444}{1.333} = 0.333$$

2nd approximation

$$y(2) = 0.333 - \frac{[8.198\text{E-3} - 0.111 + 0.444 - 0.444]}{[0.0741 - 0.667 + 1.333]}$$

$$= 0.333 + \frac{0.103}{0.741} = 0.472$$

3rd approximation

$$y(3) = 0.472 - \frac{[0.023 - 0.222 + 0.629 - 0.444]}{[0.149 - 0.944 + 1.333]}$$

$$= 0.472 + \frac{0.014}{0.538} = 0.498$$

4th approximation

$$y(4) = 0.498 - \frac{[0.027 - 0.248 + 0.664 - 0.444]}{[0.165 - 0.996 + 1.333]}$$

$$= 0.498 + \frac{1\text{E-3}}{0.502} = \underline{0.5\,\text{m}}$$

As the relative difference between $y(4)$ and $y(3)$ is small, it can be assumed that,

$$\underline{y = 0.5\,\text{m}}$$

N.B. An alternative method of determining y is by use of the computer program presented in Appendix I. Substituting this value of y into equation (2.6),

$$\hat{\tau} = \frac{2F}{1.5}[0.444 - (0.0833 - 0.01388)] = \underline{0.5F}$$

@ NA (i.e. $y = 0$),

$$\underline{\tau = 0.444F}$$

In this case, the average shear stress,

$$\tau_{av} = F/(2 \times 3/2)$$
$$= \underline{0.333F}$$

A sketch of the distribution of vertical shearing stress across the section is shown in Fig. 2.13.

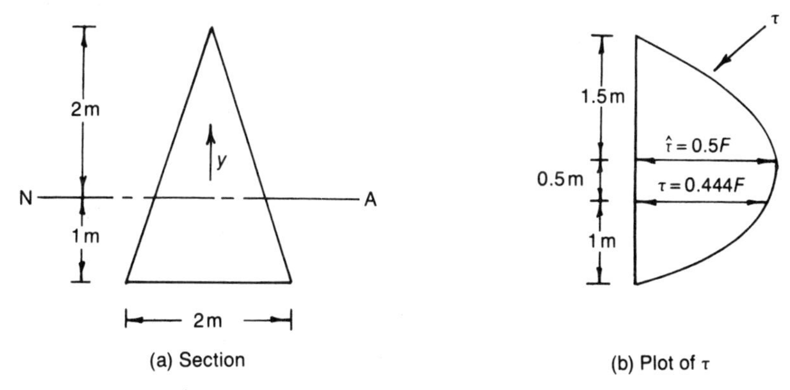

Fig. 2.13.. Stress distribution in a triangular section.

2.4.3

N.B. In practice, shearing stresses in horizontal beams, due to transverse vertical shearing forces, will not have the distributions assumed in Examples 2.2 and 2.3 and as shown in Fig. 2.14, but the more complex forms of Fig. 2.15. The determination of the correct forms of Fig. 2.15 is beyond the scope of this book, and the reader is referred to more advanced works on this topic [3–7]. In both figures, the size of the arrows is related to the magnitude of the shearing stresses.

Fig. 2.14. Incorrect shear stress distributions due to bending.

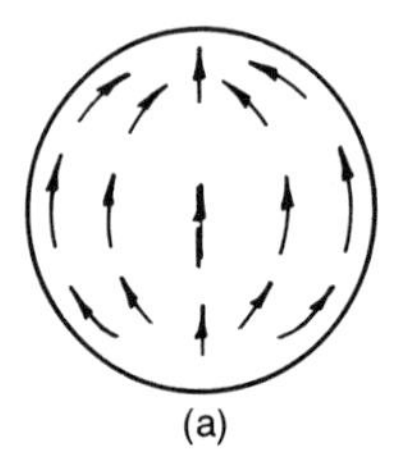

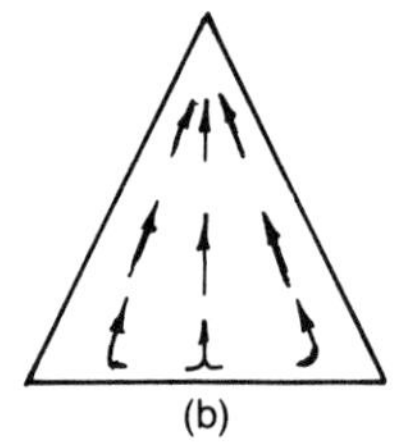

Fig. 2.15. Correct shear stress distributions due to bending.

2.5.1 EXAMPLE 2.4 VERTICAL SHEARING STRESS IN A SECTION OF COMPLEX SHAPE

A beam of the section in Fig. 2.16 is subjected to a bending moment of 0.5 MN.m, so that the section bends about a horizontal plane NA. If the maximum principal stress due to this bending moment lies on the axis A–A, and it is not to exceed 80% of the greatest bending stress, determine the value of the shearing force that acts on this section.

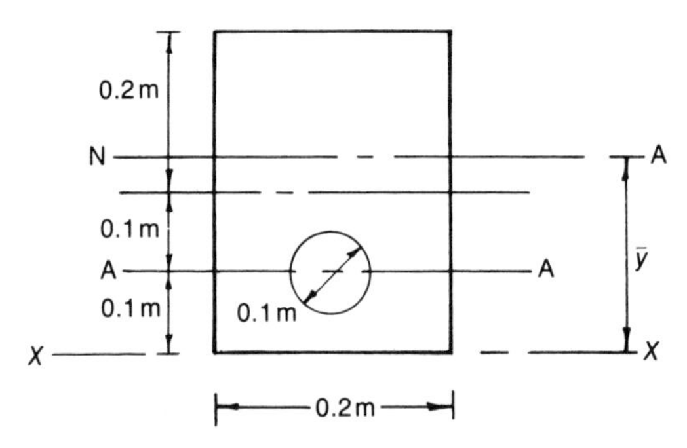

Fig. 2.16. Complex section.

2.5.2

To determine $\bar{y}$, *I*, *etc*., let us use Table 2.1, where the symbols are as defined in Chapter 1.

Table 2.1

Section	a	y	ay	ay^2	i_H
①	0.08	0.2	0.016	3.2E-3	1.0667E-3
②	−7.854E-3	0.1	−7.854E-4	−7.854E-5	−4.909E-6
$\sum$	0.0721	—	0.0152	3.121E-3	1.062E-3

$$\bar{y} = \frac{\sum ay}{\sum a} = 0.211\ \text{m}$$

$$I_{XX} = \sum ay^2 + \sum i_H = 4.183\text{E-3 m}^4$$

$$\underline{I = I_{XX} - \bar{y}^2 \sum a = 9.73\text{E-4 m}^4}$$

$$\hat{\sigma} = \frac{0.5 \times 0.211}{9.73\text{E-4}} = \underline{108.4\ \text{MN/m}^2}$$

@ A–A, $\sigma_1 = 0.8 \times 108.4 =$ maximum principal stress

$$\underline{\sigma_1 = 86.75\ \text{MN/m}^2}$$

@ A–A, $\sigma_x = \dfrac{0.5 \times (0.211 - 0.1)}{9.73\text{E-4}} = 57\ \text{MN/m}^2$

and

$$\underline{\sigma_y = 0}$$

Now from reference [1],

$$\sigma_1 = \tfrac{1}{2}(\sigma_x + \sigma_y) + \tfrac{1}{2}\sqrt{[(\sigma_x - \sigma_y)^2 + 4\tau_{xy}^2]}$$

i.e.

$$86.75 = \frac{57}{2} + \tfrac{1}{2}\sqrt{[57^2 + 4\tau_{xy}^2]}$$

or,

$$7525.6 = 812.3 + \tfrac{1}{4}[57^2 + 4\tau_{xy}^2]$$

or,

$$4\tau_{xy}^2 = 23\,604$$

$$\underline{\tau_{xy} = 76.8\ \text{MN/m}^2} \text{ @ A–A}$$

Now,

$$\tau_{xy} = \frac{F \int y\, dA}{bI}$$

which, when applied to A–A, becomes,

$$76.8 = \frac{F \int y\, dA}{0.1 \times 9.73\text{E-4}}$$

where,

$$\int y\, dA = 0.211 \times 0.2 \times \frac{0.211}{2} - \frac{\pi \times 0.1^2}{4} \times (0.211 - 0.1)$$

$$= 4.452\text{E-3} - 8.718\text{E-4} = 3.58\text{E-3}$$

therefore

$$76.8 = \frac{F \times 3.58\text{E-3}}{9.73\text{E-5}}$$

or,

$$\underline{F = 2.1\ \text{MN}}$$

i.e. the *vertical shearing force* at the section

$$= F = 2.1 \text{ MN}$$

2.6.1 EXAMPLE 2.5 VERTICAL AND HORIZONTAL SHEARING STRESSES

Calculate and sketch the distribution of vertical and horizontal shearing stresses, due to bending, which occur on a beam with the cross-section shown in Fig. 2.17, when it is subjected to a vertical shearing force of 30 kN through its centroid.

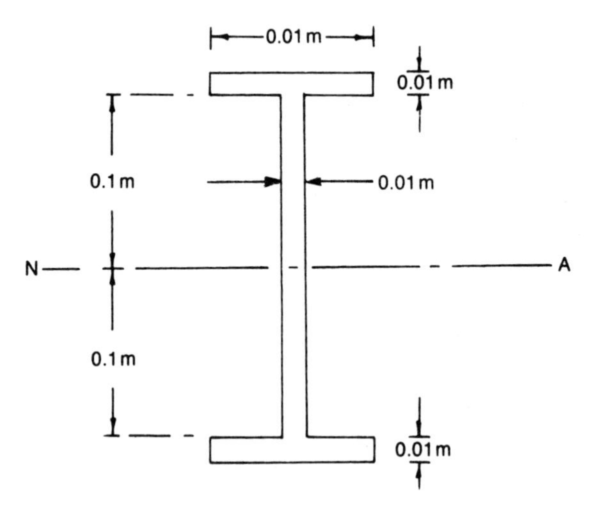

Fig. 2.17. Beam cross-section.

2.6.2

$$I = \text{second moment of area about NA}$$

$$= \frac{0.1 \times 0.22^3}{12} - \frac{0.09 \times 0.2^3}{12}$$

$$I = 2.873\text{E-5 m}^4$$

2.6.3 Vertical Shearing Stress

At the *top of the flange*,

$$\int y \cdot \mathrm{d}A = 0$$

therefore

$$\tau_1 = 0$$

At the *bottom of the flange*,

$$\int y \cdot dA = 0.1 \times 0.01 \times 0.105 = 1.05\text{E-4}\ \text{m}^3$$

therefore

$$\tau_2 = \frac{30\text{E3} \times 1.05\text{E-4}}{0.1 \times 2.873\text{E-5}}$$

$$\underline{\tau_2 = 1.096\ \text{MN/m}^2}$$

At the *top of the web*,

$$\int y \cdot dA = 1.05\text{E-4}\ \text{m}^3$$

therefore

$$\tau_3 = \frac{30\text{E3} \times 1.05\text{E-4}}{0.01 \times 2.873\text{E-5}}$$

$$\underline{\tau_3 = 10.96\ \text{MN/m}^2}$$

From equation (2.1), it can be seen that the maximum shear stress, namely $\hat{\tau}$, occurs where $(\int y \cdot dA)/b$ is a maximum, which in this case is at NA.

@ NA,

$$\int y \cdot dA = 1.05\text{E-4} + 0.1 \times 0.01 \times 0.05$$

$$= 1.55\text{E-4}$$

therefore

$$\hat{\tau} = \frac{30\text{E3} \times 1.55\text{E-4}}{0.01 \times 2.873\text{E-5}}$$

$$\underline{\hat{\tau} = 16.19\ \text{MN/m}^2}$$

2.6.4 Horizontal Shear Stress

From equation (2.2), it can be seen that the horizontal shearing stress varies linearly along the flanges of the RSJ, from zero at the free edges to a maximum value $\hat{\tau}_F$ at the centre.

$$\tau_F = \frac{F(B - Z)\bar{y}}{I}$$

and,

$$\hat{\tau}_F = \frac{FB\bar{y}}{I} = \frac{30\text{E3} \times 0.05 \times 0.105}{2.873\text{E-5}}$$

$$\underline{\hat{\tau}_F = 5.48\ \text{MN/m}^2}$$

2.6.5

A plot of the vertical and horizontal shearing stresses is shown in Fig. 2.18.

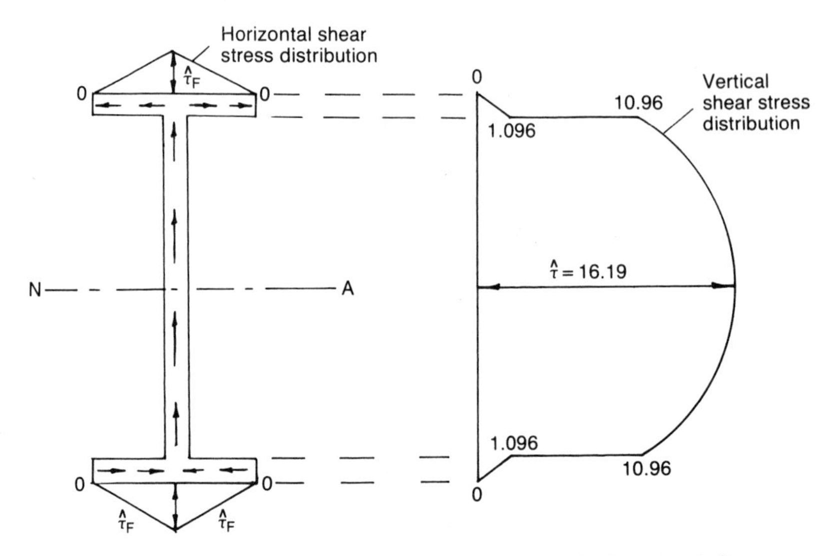

Fig. 2.18. Vertical and horizontal shearing stress distributions (MN/m^2).

2.7.1 EXAMPLE 2.6 SHEARING STRESS DISTRIBUTION IN A CHANNEL BAR

Calculate and sketch the distribution of vertical and horizontal shear stress, in the channel bar of Fig. 2.19, when it is subjected to a vertical shearing force of 30 kN.

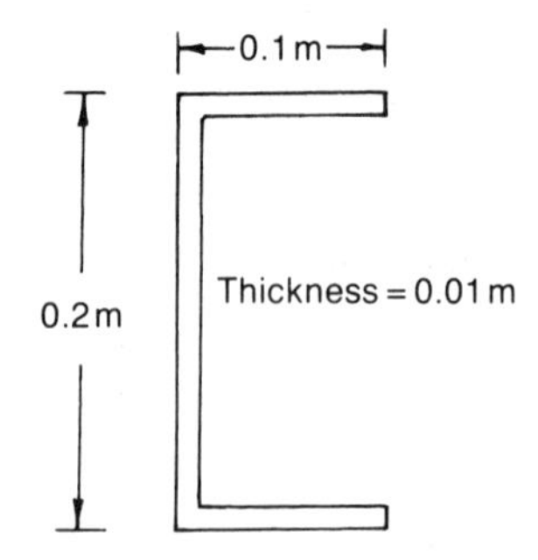

Fig. 2.19. Channel section.

2.7.2

$$I = \frac{0.1 \times 0.2^3}{12} - \frac{0.09 \times 0.18^3}{12} = 2.293\text{E-5 m}^4$$

2.7.3 Vertical Shearing Stress

At the *bottom of the flange*,

$$\int y \cdot \mathrm{d}A = 0.1 \times 0.01 \times 0.095 = \underline{9.5\text{E-}5\,\mathrm{m}^3}$$

and,

$$\tau_1 = \frac{30\text{E}3 \times 9.5\text{E-}5}{0.1 \times 2.293\text{E-}5} = \underline{1.24\,\mathrm{MN/m}^2}$$

At the *top of the web*,

$$\tau_2 = \frac{30\text{E}3 \times 9.5\text{E-}5}{0.01 \times 2.293\text{E-}5} = \underline{12.43\,\mathrm{MN/m}^2}$$

The maximum shear stress, $\hat{\tau}$, occurs at NA, because this is the point at which $(\int y \cdot \mathrm{d}A)/b$ is a maximum.

@ NA,

$$\int y \cdot \mathrm{d}A = 9.5\text{E-}5 + 0.095 \times 0.01 \times 0.095/2$$

$$= \underline{1.4\text{E-}4\,\mathrm{m}^3}$$

and,

$$\hat{\tau} = \frac{30\,\text{E}3 \times 1.4\text{E-}4}{0.01 \times 2.293\text{E-}5} = \underline{18.33\,\mathrm{MN/m}^2}$$

2.7.4 Horizontal Shearing Stress

From equation (2.2), it can be seen that the horizontal shearing stress varies linearly along the flange width, from zero at the right edge to a maximum value of $\hat{\tau}_F$ at the intersection between the flange and the web.

From equation (2.2),

$$\tau = \frac{F(B - Z)\bar{y}}{I}$$

and,

$$\hat{\tau}_F = FB\bar{y}/I = \frac{30\text{E}3 \times 0.09 \times 0.095}{2.293\text{E-}5} = \underline{11.19\,\mathrm{MN/m}^2}$$

2.7.5

A plot of the vertical and horizontal shearing stress distributions is shown in Fig. 2.20, where the arrows are used to indicate the direction and magnitude of the vertical and horizontal shearing stresses acting on the beam's cross-section.

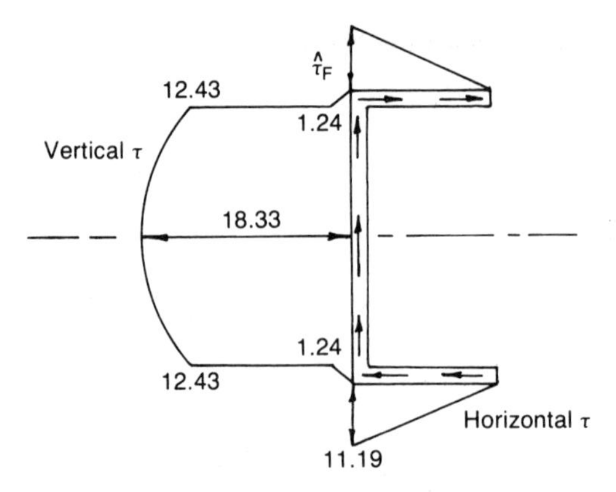

Fig. 2.20. Plot of vertical and horizontal shear stress distributions.

2.8.1 SHEAR CENTRE

When slender symmetrical section cantilevers are subjected to transverse loads in a laboratory, good agreement is usually found between the predictions of simple bending theory and experimental observations. If, however, when similar tests are carried out on cantilevers with unsymmetrical sections, such as channel bars, angle irons, etc., comparison between theoretical predictions and experimental observations is usually poor.

The explanation for this can be obtained by considering the channel section of Fig. 2.21, and assuming that the shearing force F is applied through its centroid.

From Fig. 2.20, the shear stress distribution due to bending will be as shown in Fig. 2.21(a), where the magnitude and direction of the arrows are used to indicate the magnitude and direction of the shearing stresses due to bending.

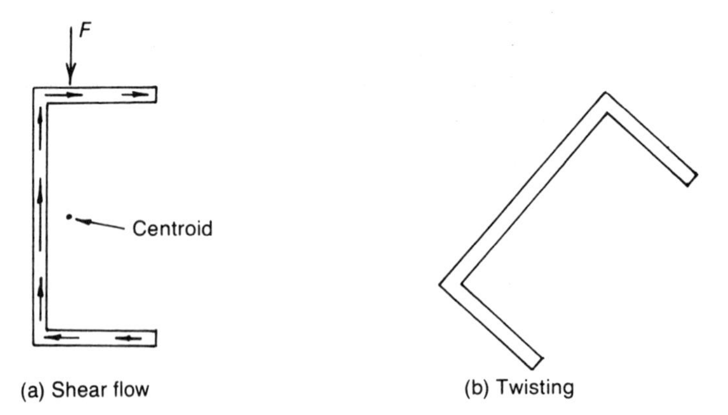

Fig. 2.21. Shear flow in the cross-section of a channel bar.

From Fig. 2.21(a), it can be seen that horizontal equilibrium is achieved by the horizontal shearing force in the top flange being equal, but opposite, to the horizontal shearing force in the bottom flange.

Similarly, vertical equilibrium is achieved by the internal resisting vertical shearing force in the web being equal and opposite to the applied external vertical shearing force F. However, from Fig. 2.21(a), it can be seen that rotational equilibrium is not achieved, so that if F is applied through the centroid of the section, the beam will twist, as shown in Fig. 2.21(b), and as a result, shearing stresses due to torsion will occur in addition to shearing stresses due to bending.

In general, it is advisable to eliminate shearing stresses due to torsion, as these can be relatively large, and to achieve this, it is necessary to ensure that F acts at a point where rotational equilibrium is achieved. This point is called the shear centre, and some typical positions of shear centre are shown in Fig. 2.22, together with distributions of shear stress due to bending, caused by F.

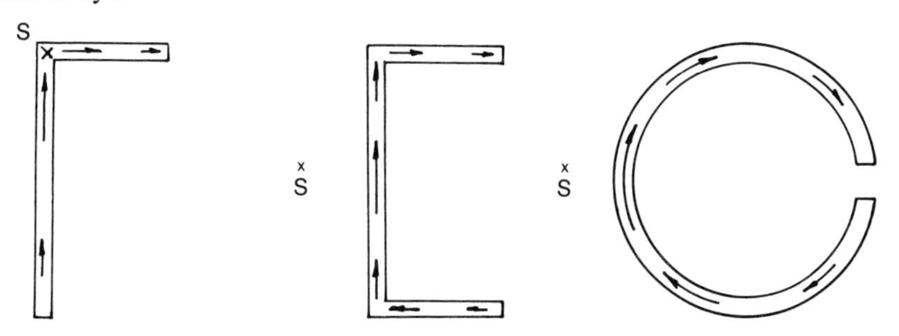

Fig. 2.22. Some shear centre positions "S".

N.B. The term *shear flow* has been introduced in this section, where *shear flow* $q =$ *shear stress* $*$ *wall thickness* $= \tau t$.

In certain cases, when the shear stress varies inversely with the wall thickness, so that the shear flow q is constant, it may be found convenient to carry out the calculation using q instead of τ.

2.9.1 EXAMPLE 2.7 SHEAR CENTRE POSITION FOR A CHANNEL SECTION

Determine the shear centre position for the channel section beam of Fig. 2.19.

2.9.2

From Fig. 2.20, it can be seen that the horizontal shearing stress distribution in the flange is linear, so that the average shear stress in the flange τ_{av}

$$= 11.19/2 = \underline{5.595\ \text{MN/m}^2}$$

Let,

$$F_F = \text{the resisting shearing force in the flange}$$

$$= 5.595 \times 0.01 \times 0.095 = 5.315\text{E-3 MN}$$

$$\underline{F_F = 5.315\ \text{kN}}$$

Now from vertical equilibrium considerations, the resisting shearing force in the web = $F = 30\,\text{kN}$.

Let the shear centre position be at "S" in the beam section, as shown in Fig. 2.23, and by taking moments about "S",

$$F * \Delta = F_F * 0.19$$

or,

$$30\Delta = 5.32 \times 0.19$$

$$\Delta = 0.0337\,\text{m} = 3.37\,\text{cm}$$

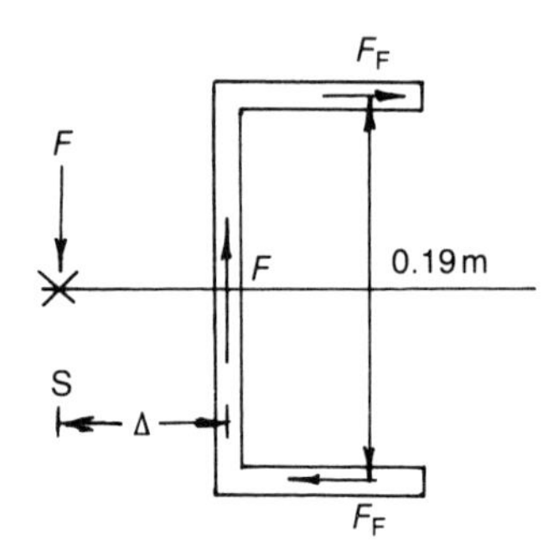

Fig. 2.23. Shearing forces on channel section.

2.10.1 EXAMPLE 2.8 SHEAR CENTRE POSITION FOR A THIN-WALLED CURVED SECTION

Determine the shear centre position for the thin-walled curved section shown in Fig. 2.24, which is of constant thickness t.

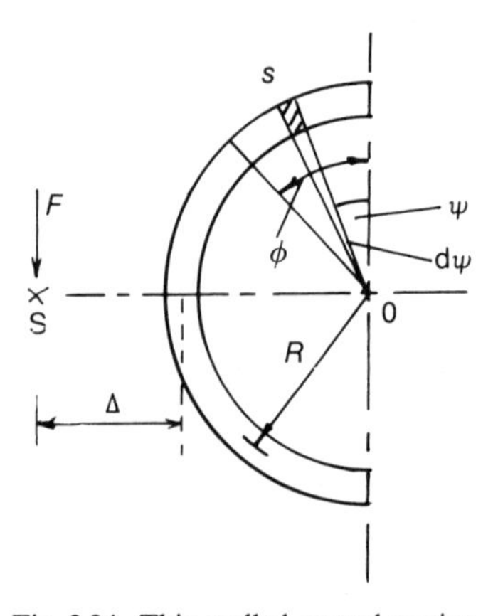

Fig. 2.24. Thin-walled curved section.

2.10.2

At any distance s, the shear stress due to bending =

$$\tau_s = \frac{F}{tI}\int y\cdot \mathrm{d}A$$

where,

$\int y\cdot \mathrm{d}A$ = first moment of area of the element, shaded in Fig. 2.24, about NA, i.e.

$$\tau_s = \frac{F}{tI}\int_0^{\phi}(R\cdot\cos\psi)(t\cdot R\cdot \mathrm{d}\psi) = \frac{FR^2\sin\phi}{I} \tag{2.8}$$

The shear flow in the section is shown in Fig. 2.25, where the magnitude and directions of the arrows are intended to give a measure of the magnitude and direction of the internal resisting shearing stresses.

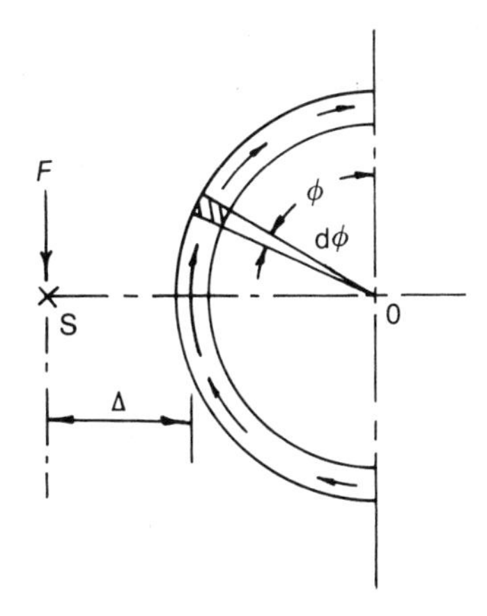

Fig. 2.25. Shear flow in thin-walled curved section.

To determine Δ, take moments about "O", i.e.

$$F(\Delta + R) = \int_0^{\pi}\tau_s * (R\cdot \mathrm{d}\phi) * t * R$$

$$= \frac{2FR^4 t}{I}$$

or,

$$\Delta = \frac{2R^4 t}{I} - R \tag{2.9}$$

2.10.3 To determine I

From reference [1],

$$I = \int y^2 \cdot dA$$

$$= \int_0^{\pi} (R\cos\phi)^2 t \cdot R \cdot d\phi$$

$$= tR^3 \int_0^{\pi} \frac{[1 + \cos 2\phi]}{2} \cdot d\phi$$

$$= \frac{tR^3}{2}\left[\phi + \frac{\sin 2\phi}{2}\right]_0^{\pi}$$

$$\underline{I = \pi R^3 t/2} \qquad (2.10)$$

Substituting equation (2.10) into (2.9),

$$\underline{\Delta = R(4/\pi - 1) = 0.273\,R}$$

2.11.1 EXAMPLE 2.9 SHEAR CENTRE POSITION FOR A THIN-WALLED CURVED SECTION WITH FLANGES

Determine the shear centre position for the thin-walled section of Fig. 2.26, which is of uniform thickness t.

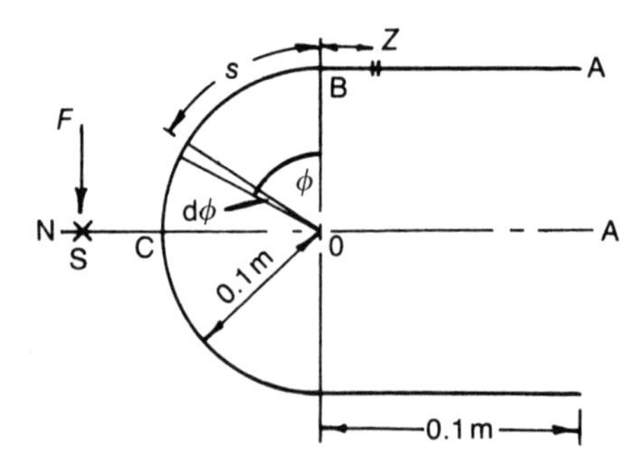

Fig. 2.26. Curved section with flanges.

2.11.2

Let,

I = second moment of area of section about NA, which can be obtained with the assistance of equation (2.10).

$$I = \frac{\pi R^3 t}{2} + 2\left(\frac{0.1t^3}{12} + t \times 0.1 \times 0.1^2\right)$$

but as the section is thin, higher order terms involving t can be ignored; therefore

$$I = \frac{\pi \times 0.1^3 t}{2} + 2 \times 0.001t$$

$$\underline{I = 3.57\text{E-}3t}$$

2.11.3 Consider the top flange

$$\tau_F = \text{shear stress in the top flange at } Z$$
$$= F(B - Z)\bar{y}/I$$

@ $Z = 0.1$,

$$\underline{\tau_A = 0} \quad (2.11)$$

@ $Z = 0$,

$$\tau_B = FB\bar{y}/I = F \times 0.1 \times 0.1/(3.57\text{E-}3t)$$

$$\underline{\tau_B = 2.8\, F/t} \quad (2.12)$$

2.11.4 Consider the curved section

@ $$s, \tau_s = \frac{F}{tI}\int y \cdot \mathrm{d}A$$

$$= \frac{F}{t * 3.57\text{E-}3t}\left\{\int_0^\phi (R\cos\psi \cdot tR\mathrm{d}\psi) + 0.01 * t * R\right\}$$

$$= \frac{F}{3.57\text{E-}3t}(R^2 \sin\phi + 0.1\, R)$$

$$\underline{\tau_s = \frac{F}{0.357t}(1 + \sin\phi)} \quad (2.13)$$

@ $\phi = 0$,

$\tau_B = 2.8\, F/t$ — as required (see equation (2.12))

@ $\phi = 90°$,

$\tau_C = 5.6\, F/t$ = maximum shear stress due to bending.

2.11.5 To calculate the shear centre position

Let,

$$F_F = \text{resisting shearing force in the flange}$$

$$= \frac{(\tau_A + \tau_B)}{2} * 0.1t$$

$$\underline{F_F = 0.14\, F}$$

To obtain Δ, consider rotational equilibrium about the point O of the section, as shown in Fig. 2.27.

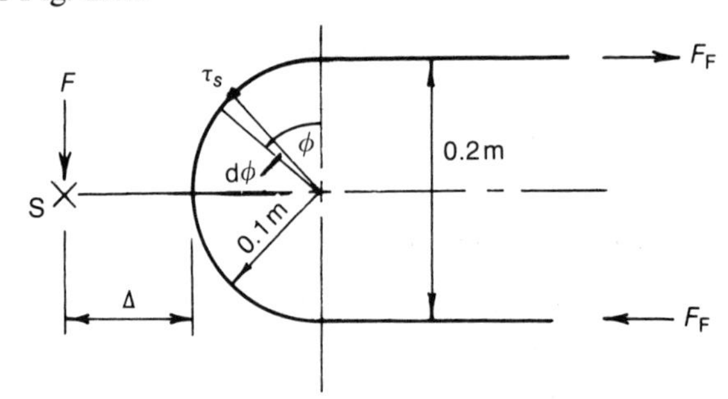

Fig. 2.27. Thin-walled open section.

$$F(\Delta + 0.1) = F_F \times 0.2 + \int_0^{\pi} \tau_s * R * d\phi * t * R$$

$$= 0.2 \times 0.14\, F + \int_0^{\pi} \frac{F(1 + \sin\phi)}{0.357t} * 0.1^2 t \cdot d\phi$$

$$= 0.028\, F + 0.028\, F[\phi - \cos\phi]_0^{\pi}$$

$$= 0.028\, F + 0.028\, F\{(\pi + 1) - (0 - 1)\}$$

$$= 0.172\, F$$

therefore

$$\underline{\Delta = 0.072\,\text{m}}$$

2.12.1 SHEAR CENTRE POSITIONS FOR CLOSED THIN-WALLED TUBES

The main problem of determining the shear stress due to bending in closed thin-walled tubes is that, initially, the shear stress is not known at any point. To overcome this difficulty, the assumption is made that the shear stress due to bending at a certain point in the section has an unknown value of τ_0, as shown in Fig. 2.28.

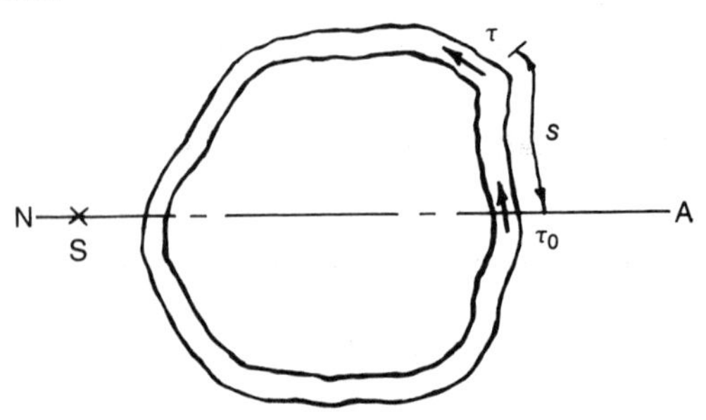

Fig. 2.28. Thin-walled closed tube.

2.12.2

Megson [8] has shown that for a thin-walled closed tube, the relationship between the twist and the shear stress is given by:

$$\oint \frac{\tau \cdot \mathrm{d}s}{G} = 2A \cdot \frac{\mathrm{d}\theta}{\mathrm{d}z}$$

where,

A = enclosed area of the cross-section of the tube

$\frac{\mathrm{d}\theta}{\mathrm{d}z}$ = twist/unit length

θ = angle of twist

z = distance along the axis of the tube

τ = shearing stress due to bending at any distance s from the "starting" point

$= \tau_0 + \tau_s$

τ_0 = shearing stress due to bending at the "starting" point

τ_s = shearing stress due to bending at any distance s for an equivalent OPEN tube

$\mathrm{d}s$ = elemental length

G = rigidity modulus

If F = the shearing force applied through the shear centre "S", then there will be no twist, i.e.

$$\frac{d\theta}{\mathrm{d}z} = 0 = \oint \frac{\tau \cdot \mathrm{d}s}{G}$$

or,

$$\oint (\tau_0 + \tau_s)\mathrm{d}s = 0$$

but,

$$\tau_0 = \text{constant}$$

therefore

$$\tau_0 \oint \mathrm{d}s = -\oint \tau_s \cdot \mathrm{d}s$$

or

$$\tau_0 = -\frac{\oint \tau_s \cdot \mathrm{d}s}{\oint \mathrm{d}s} \tag{2.14}$$

Once τ_0 is determined from equation (2.14), τ can be found.

2.13.1 EXAMPLE 2.10 SHEAR CENTRE FOR A THIN-WALLED CLOSED TUBE OF COMPLEX SHAPE

Determine the shear centre position for the thin-walled closed tube of Fig. 2.29, which is of uniform thickness t.

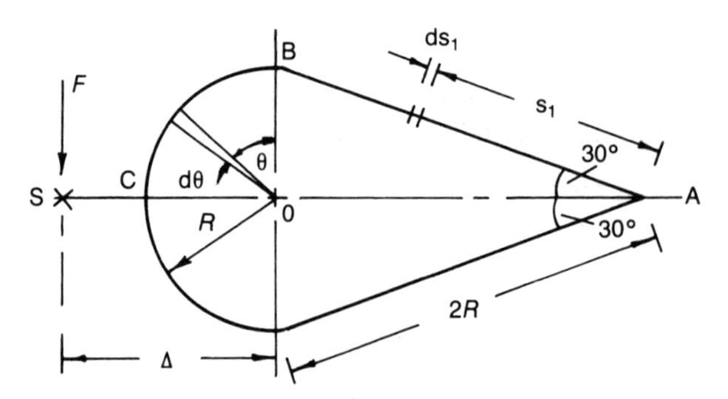

Fig. 2.29. Thin-walled closed tube.

2.13.2

Let,

$$\tau_0 = \text{shear stress due to bending at the point "A"}$$

$$I = 2\int_0^{2R} (s_1 \sin 30)^2 * (t \cdot ds_1) + \int_0^{\pi} (R \cdot \cos\phi)^2 * (t \cdot R \cdot d\phi)$$

$$= 0.5t \int_0^{2R} s_1^2 \cdot ds_1 + tR^3 \int_0^{\pi} \cos^2\phi \cdot d\phi$$

$$= \frac{0.5t}{3}[s_1^3]_0^{2R} + tR^3 \int_0^{\pi} \frac{(1+\cos 2\phi)}{2} d\phi$$

$$= \frac{4R^3 t}{3} + \frac{tR^3}{2}\left[\phi + \frac{\sin 2\phi}{2}\right]_0^{\pi}$$

$$\underline{I = R^3 t\left(\frac{4}{3} + \frac{\pi}{2}\right) = 2.904\, R^3 t}$$

2.13.3 Consider AB

@ any distance s_1,

$$\tau_{s1} = \frac{F}{tI}\int (s_1 \cdot \sin 30) * (t \cdot ds_1)$$

$$= \frac{0.5\,F}{I}\left[\frac{s_1^2}{2}\right]_0^{s_1}$$

$$\underline{\tau_{s1} = 0.25\,F s_1^2/I} \qquad (2.15)$$

2.13.4 Consider BC

@ any distance ϕ

$$\tau_\phi = \frac{F}{tI}\left\{\int_0^{\phi} (R \cdot \cos\phi) * (t \cdot R\, d\phi) + (2Rt) * (R/2)\right\}$$

$$= \frac{F}{I} \cdot R^2 \{ [\sin \phi]_0^\phi + 1 \}$$

$$\underline{\tau_\phi = FR^2(1 + \sin \phi)/I} \quad (2.16)$$

2.13.5

Now from equation (2.14),

$$\tau_0 = -\frac{\oint \tau_s \cdot \mathrm{d}s}{\oint \mathrm{d}s}$$

but,

$$\oint \mathrm{d}s = 2R * 2 + \pi R = \underline{7.14\,R} \quad (2.17)$$

and,

$$\oint \tau_s \,\mathrm{d}s = \oint \tau_{s1} \cdot \mathrm{d}s_1 + \oint \tau_\phi R \cdot \mathrm{d}\phi$$

$$= 2 \int_0^{2R} \frac{0.25\,Fs_1^2}{I} \cdot \mathrm{d}s_1 + \int \frac{FR^2}{I} (1 + \sin \phi) \cdot R \cdot \mathrm{d}\phi$$

$$= \frac{0.5F}{I} \left[\frac{s_1^3}{3} \right]_0^{2R} + \frac{FR^3}{I} [\phi - \cos\phi]_0^\pi$$

$$= \frac{F}{I} \left\{ \frac{4R^3}{3} + R^3 [(\pi + 1) - (0 - 1)] \right\}$$

$$= \frac{FR^3}{I} [\tfrac{4}{3} + \pi + 2]$$

or,

$$= \frac{6.475\,FR^3}{2.904\,R^3 t}$$

$$\oint \tau_s \cdot \mathrm{d}s = 2.23\,F/t \quad (2.18)$$

Hence, from equations (2.15) to (2.18),

$$\tau_0 = -\frac{2.23\,F}{t} * \frac{1}{7.14\,R}$$

$$\underline{\tau_0 = -0.312\,F/(Rt)} \quad (2.19)$$

so that,

$$\tau_{\mathrm{AB}} = \text{shear stress at any point between A and B}$$

$$= -\frac{0.312\,F}{Rt} + \frac{0.25\,Fs_1^2}{I}$$

$$= -\frac{0.312\,F}{Rt} + \frac{0.25\,Fs_1^2}{2.904\,R^3 t}$$

$$\underline{\tau_{\mathrm{AB}} = -\frac{0.312\,F}{Rt} + \frac{0.0861\,Fs_1^2}{R^3 t}} \quad (2.20)$$

τ_{BC} = shear stress at any point between B and C

$$= -\frac{0.312\,F}{Rt} + \frac{FR^2(1+\sin\phi)}{2.904\,R^3 t}$$

$$= -\frac{0.312\,F}{Rt} + \frac{F(1+\sin\phi)}{2.904\,Rt}$$

$$\tau_{BC} = \frac{F}{Rt}\left(0.0324 + \frac{\sin\phi}{2.904}\right) \tag{2.21}$$

2.13.6

Taking moments about "O",

$$F\Delta = \int_0^{\pi} \tau_{BC} * (t \cdot R \cdot \mathrm{d}\phi) * R$$

$$+ 2\int_0^{2R} \tau_{AB} * (t \cdot \mathrm{d}s_1) * R\cos 30$$

$$= \frac{F \cdot R^2 \cdot t}{2.904\,Rt}\int_0^{\pi} (0.0941 + \sin\phi) \cdot \mathrm{d}\phi$$

$$+ 1.732\,Rt\int_0^{2R}\left[-\frac{0.312\,F}{Rt} + \frac{0.0861\,Fs_1^2}{R^3 t}\right] \cdot \mathrm{d}s_1$$

$$= \frac{FR}{2.904}[0.0941\phi - \cos\phi]_0^{\pi}$$

$$+ 1.732\,Rt\left[-\frac{0.312\,Fs_1}{Rt} + \frac{0.0861 Fs_1^3}{3R^3 t}\right]_0^{2R}$$

$$= \frac{FR}{2.904}[(0.0941\pi + 1) - (-1)]$$

$$+ 1.732\,FRt\left[-\frac{0.312 \times 2R}{Rt} + \frac{8 \times 0.0861\,R^3}{3R^3 t}\right]$$

$$= FR(0.79 - 0.683)$$

therefore

$$\Delta = 0.107R$$

2.14.1 SHEAR DEFLECTIONS

Deflections of beams usually consist of two components, namely deflections due to bending and deflections due to shear. If a beam is long and slender, then deflections due to shear are small compared with deflections due to bending. If, however, the beam is short and stout, then deflections due to shear cannot be neglected.

Deflections due to shear are caused by shearing action alone, where each element of the beam tends to change shape, as shown in Fig. 2.30, where F is the shearing force acting on a typical element.

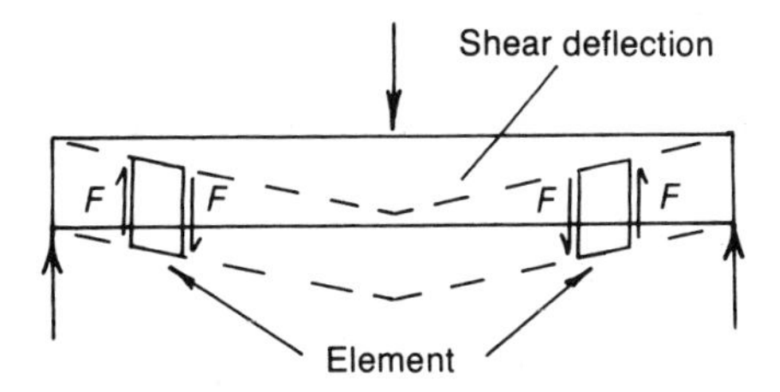

Fig. 2.30. Shear deflection of a beam.

2.15.1 EXAMPLE 2.11 SHEAR DEFLECTION OF A CANTILEVER

Determine the value of the maximum deflection due to shear, for the end-loaded cantilever of Fig. 2.31. The cantilever is of uniform rectangular section, of width b and depth d.

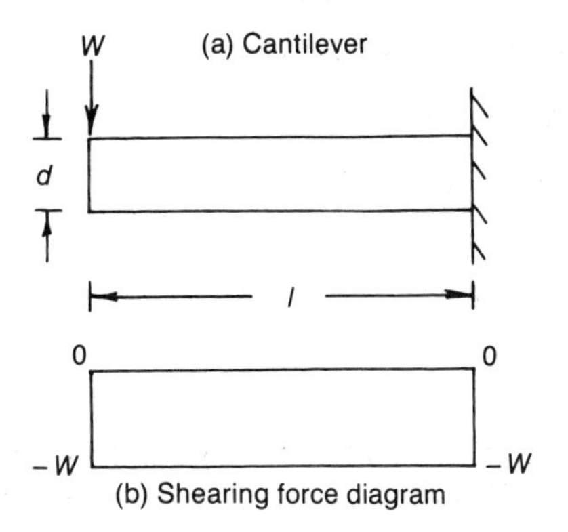

Fig. 2.31. Shearing force distribution in an end-loaded cantilever.

2.15.2

The shearing forces acting on the cantilever are of a constant value W, causing the shear deflected form shown in Fig. 2.32.

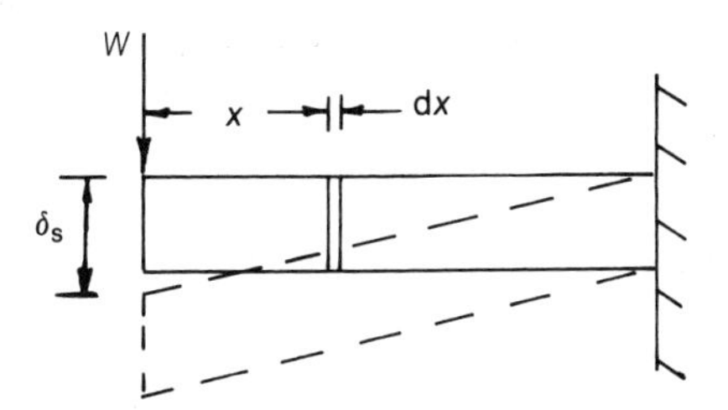

Fig. 2.32. Shear deflected form of cantilever.

From equation (2.3), the value of shear stress at any distance y from the neutral axis =

$$\tau = \frac{6W}{bd^3}\left(\frac{d^2}{4} - y^2\right)$$

From reference [1], the total shear strain energy of the cantilever =

$$\text{SSE} = \int \frac{\tau^2}{2G}\cdot \mathrm{d(vol)}$$

$$= \tfrac{2}{2}\int_0^{d/2} \frac{36W^2}{Gb^2d^6}\left(\frac{d^2}{4} - y^2\right)^2 l\cdot b\cdot \mathrm{d}y$$

$$= \frac{36\,W^2 l}{Gbd^6}\int_0^{d/2}\left(\frac{d^4}{16} - \frac{d^2y^2}{2} + y^4\right)\cdot \mathrm{d}y$$

$$\text{SSE} = \frac{3W^2 l}{5Gbd} \tag{2.22}$$

If,

δ_s = the maximum deflection due to shear

then the work done by the load W =

$$\text{WD} = \tfrac{1}{2}W\delta_s \tag{2.23}$$

Equating (2.22) and (2.23),

$$\delta_s = \frac{6Wl}{5Gbd} \tag{2.24}$$

2.16.1 EXAMPLE 2.12 SHEAR DEFLECTION OF A BEAM WITH A UNIFORMLY DISTRIBUTED LOAD

Determine the value of the maximum deflection due to shear for a cantilever, assuming that it is subjected to a uniformly distributed load w, as shown in Fig. 2.33.

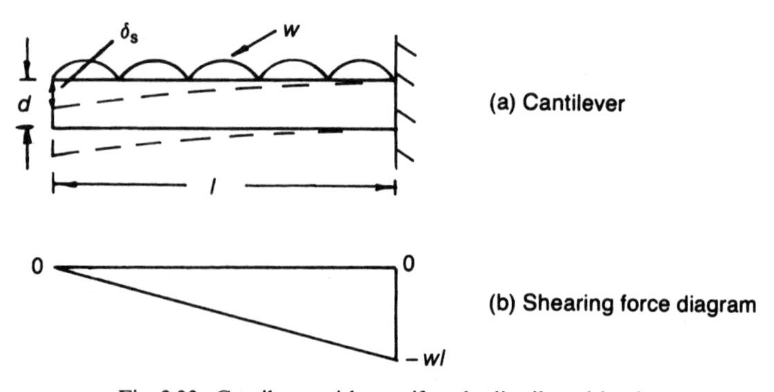

Fig. 2.33. Cantilever with a uniformly distributed load.

2.16.2

In this case, as the shearing force varies linearly along the length of the cantilever, the shear deflection will vary parabolically.

At any distance x from the free end, the shearing force on an element of the beam of length "dx" =

$$F = wx$$

From equation (2.24), the shear deflection of this element

$$= \frac{6wx \cdot \mathrm{d}x}{5Gbd}$$

therefore total shear deflection of the cantilever =

$$\delta_s = \int_0^l \frac{6wx}{5Gbd} \cdot \mathrm{d}x$$

$$\delta_s = \frac{3wl^2}{5Gbd}$$

2.17.1 TOTAL DEFLECTION OF A CANTILEVER

From reference [1], the maximum deflection, due to bending, of an end-loaded cantilever =

$$\delta_b = \frac{Wl^3}{3EI}$$

which for a rectangular section of width b and depth d becomes:

$$\delta_b = \frac{Wl^3}{3E} * \frac{12}{bd^3} = \frac{4Wl^3}{Ebd^3}$$

Now, the total deflection =

$$\delta = \delta_b + \delta_s = \frac{4Wl^3}{Ebd^3} + \frac{6Wl}{5Gbd}$$

$$= \frac{4Wl^3}{Ebd^3}\left[1 + \tfrac{3}{4}\left(\frac{d}{l}\right)^2\right] \tag{2.25}$$

where the second term in the square brackets represents the component of deflection due to shear, and it is assumed that $E = 2.5G$.

From equation (2.25), it can be seen that the deflection due to shear is important when (d/l) becomes relatively large.

2.18.1 Warping

The effects of warping have not been included in this chapter, as the subject of the warping of thin-walled sections is beyond the scope of this book.

Megson [8] describes warping as out-of-plane deformation of a cross-section, particularly when an unsymmetrical section is not loaded through its shear centre, as shown in Fig. 2.34.

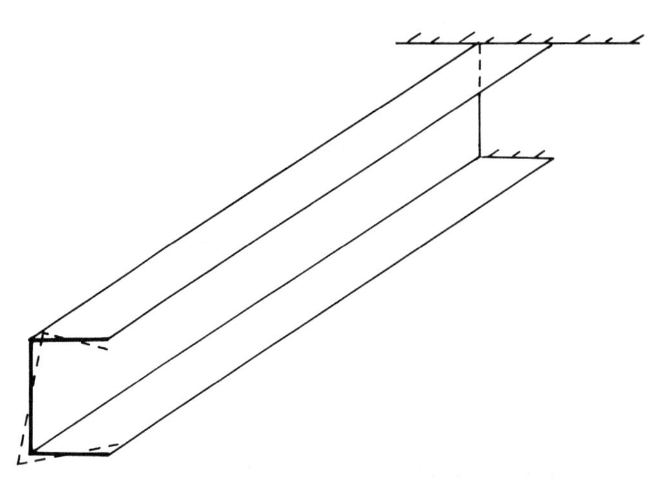

Fig. 2.34. Warping of a cross-section.

The longitudinal direct stresses caused by warping are of particular importance when the beam is restrained from axial movement. Warping is not of importance in solid circular section beams and in thin-walled tubes of circular and square cross-section.

EXAMPLES FOR PRACTICE 2

1. A beam of length 3 m is simply-supported at its ends, and it is subjected to a uniformly distributed load of 200 kN/m, spread over its entire length. If the beam has a uniform cross-section of depth 0.2 m and of width 0.1 m, determine the position and value of the maximum shearing stress due to bending. What will be the value of the maximum shear stress at mid-span?

   {22.5 MPa @ NA @ the ends; 0}

2. Determine the maximum values of shear stress due to bending, in the web and flanges of the sections of Fig. Q.2.2, when they are subjected to vertical shearing forces of 100 kN.

   {(*a*) 97.83 MPa, 35.87 MPa; (b) 127.4 MPa, 52.5 MPa}

Fig. Q.2.2. Symmetrical sections subjected to vertical shearing forces.

3. Determine an expression for the maximum shearing stress due to bending, for the section of Fig. Q.2.3, assuming that it is subjected to a shearing force of 0.5 MN acting through its centroid and in a perpendicular direction to NA.

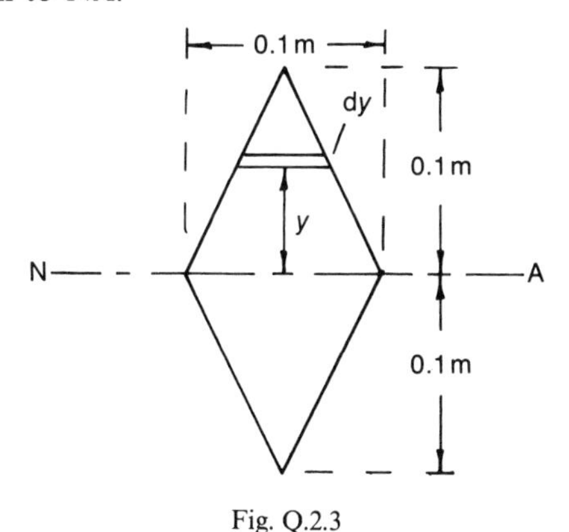

Fig. Q.2.3

$\{\tau = 5000[0.01 - (3y^2 - 20y^3)]/(1 - 10y);$
$\hat{\tau} = 56.25\text{ MPa} \text{ @ } y = \pm 0.025\text{ m}\}$

4. Determine the value of the maximum shear stress for the cross-section of Fig. Q.2.4, assuming that it is subjected to a shearing force of magnitude 0.5 MN acting through its centroid and in a perpendicular direction to NA.

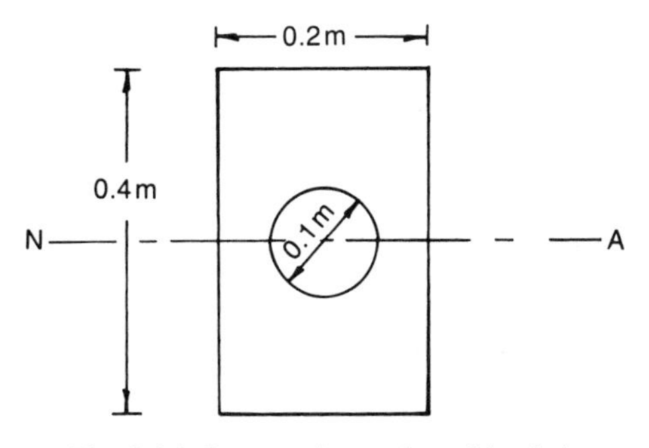

Fig. Q.2.4. Rectangular section with a hole.

{18.44 MPa}

5. A simply-supported beam, with a cross-section as shown in Fig. Q.2.5, is subjected to a centrally placed concentrated load of 100 MN, acting through its centroid and perpendicular to NA.

   Determine the values of the vertical shearing stress at intervals of 0.1 m from NA.

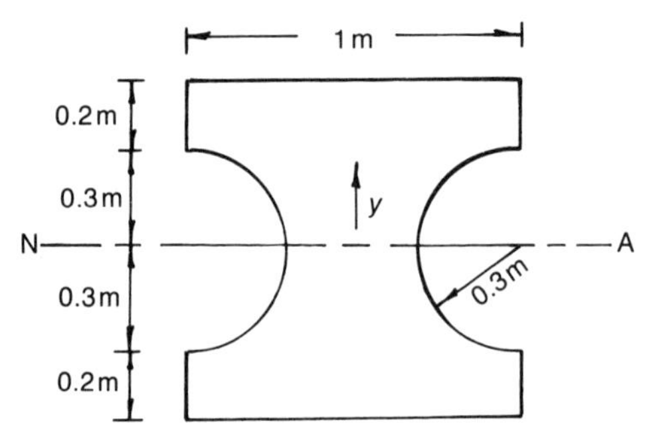

Fig. Q.2.5. Complex cross-section.

{@ $y=0$, $\tau_0 = 173.7$; $\tau_{0.1} = 156.95$; $\tau_{0.2} = 114.49$; $\tau_{0.3} = 51.95$; $\tau_{0.4} = 29.22$;
@ $y = 0.5$, $\tau_{0.5} = 0$—all in MPa}

6. Determine the positions of the shear centres for the thin-walled sections of Figs. Q.2.6(a) and (b).

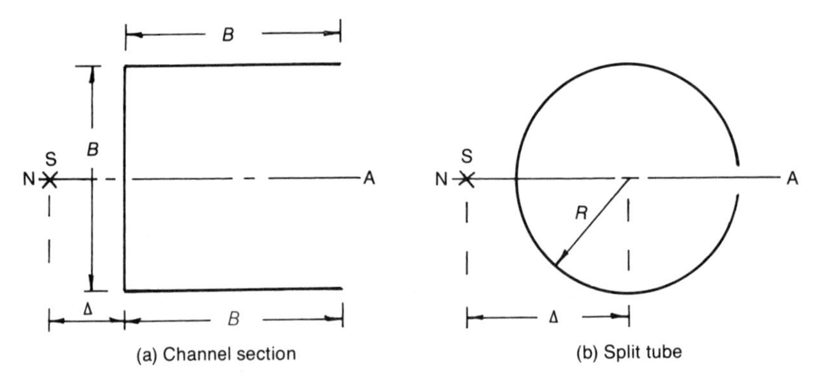

Fig. Q.2.6. Thin-walled open sections.

{(a) $\Delta = 0.429B$; (b) $\Delta = 2R$}

7. Determine the shear centre positions for the thin-walled sections of Figs. Q.2.7(*a*) and (*b*).

   {(a) $\Delta = 0.396$ m; $\Delta = 0.35$}

8. Determine the shear centre position for the thin-walled closed tube of Fig. Q.2.8, which is of uniform thickness.

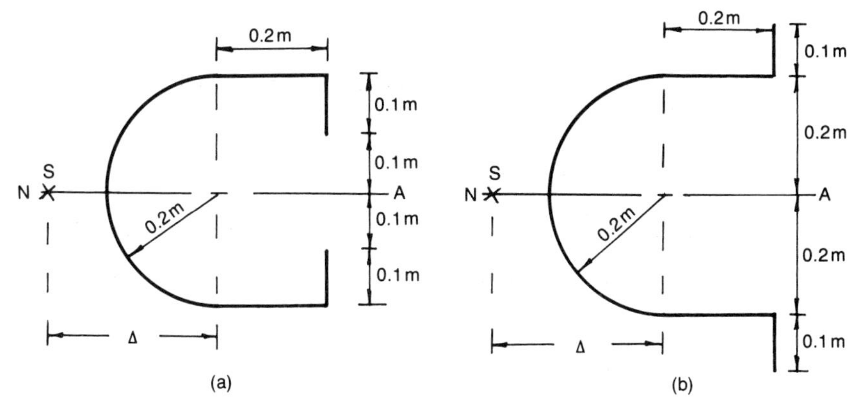

Fig. Q.2.7. Thin-walled complex open sections.

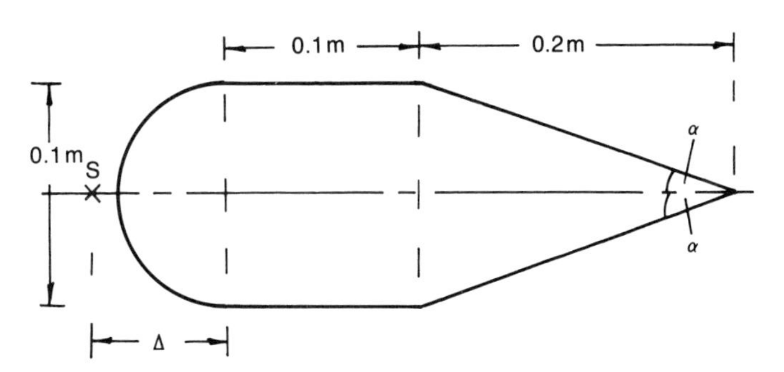

Fig. Q.2.8. Thin-walled closed tube.

{Portsmouth Polytechnic, 1984}
{$\Delta = -0.032\,\text{m}$}

9. Determine the shear centre position for the thin-walled closed tube of Fig. Q.2.9, which is of uniform thickness.

{Portsmouth Polytechnic, 1985}

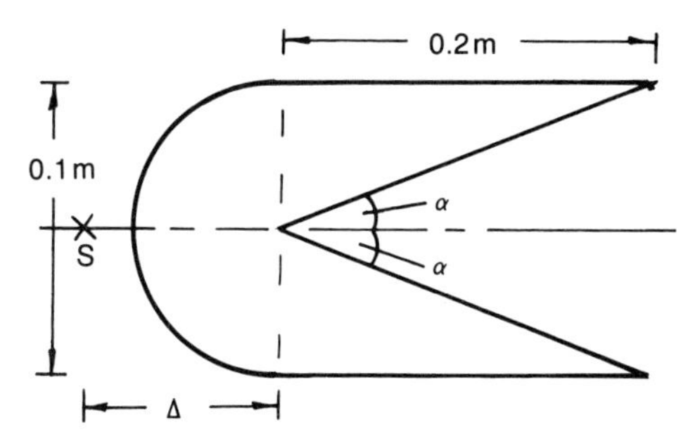

Fig. Q.2.9. Thin-walled closed tube.

{$\Delta = 0.079\,\text{m}$}

10. Determine the maximum deflections due to shear for the simply-supported beams of Figs. Q.2.10(a) to (c). In all cases, it may be assumed that the beam cross-sections are rectangular, of constant width b and of constant depth d.

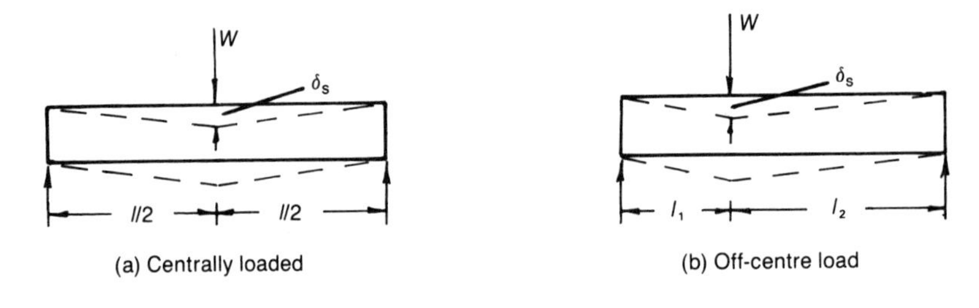

(a) Centrally loaded

(b) Off-centre load

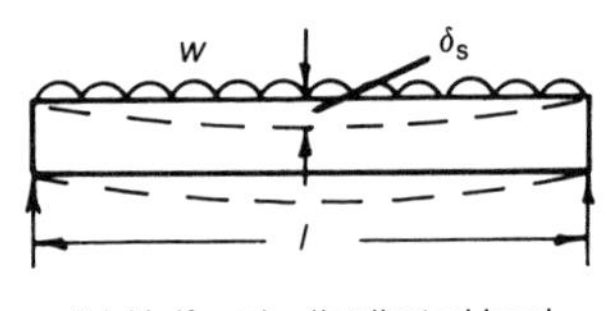

(c) Uniformly distributed load

Fig. Q.2.10. Shear deflections of simply-supported beams.

{(a) $\delta_s = 3Wl/(10\,Gbd)$; (b) $\delta_s = 6Wl_1l_2/(5\,Gbdl)$; (c) $\delta_s = 3wl^2/(20\,Gbd)$}

3

Theories of Elastic Failure

3.1.1 THE FIVE MAJOR THEORIES OF ELASTIC FAILURE

There are many theories as to which stress or strain, or any combination of these, causes the onset of yield, but in this chapter, we will restrict ourselves to a consideration of the five major theories, namely:

(a) Maximum principal stress theory.
(b) Maximum principal strain theory.
(c) Total strain energy theory.
(d) Maximum shear stress theory.
(e) Shear strain energy theory or octahedral shear stress theory.

The combination of stresses and strains that causes the onset of yield is of much importance in stress analysis, as the onset of yield is very often related to the ultimate failure of the structure.

For convenience, the five major theories are related to a triaxial principal stress system, where,

$$\sigma_1 > \sigma_2 > \sigma_3$$
$$\sigma_1 = \text{maximum principal stress}$$
$$\sigma_3 = \text{minimum principal stress}$$
$$\sigma_2 = \text{minimax principal stress}$$

The reason for choosing the triaxial principal stress system to investigate yield criteria is that such a system describes the complete stress situation at a point, without involving the complexities caused by the shear stresses on six planes, which would have resulted if a different three-dimensional coordinate system were adopted. The five major theories of elastic failure, together with some of the reasons why they have gained popularity, are described below.

3.2.1 MAXIMUM PRINCIPAL STRESS THEORY (RANKINE)

This theory states that yield will occur when,

$$\underline{\sigma_1 = \sigma_{yp}} \tag{3.1}$$

or, if the material is in compression, when,

$$\underline{\sigma_3 = \sigma_{ypc}} \tag{3.2}$$

where,

σ_{yp} = yield stress in tension, obtained in the simple uni-axial tensile test

σ_{ypc} = yield stress in compression, obtained in the simple uni-axial compression test

3.3.1 MAXIMUM PRINCIPAL STRAIN THEORY (ST. VENANT)

Experimental tests have revealed that comparison between the maximum principal stress theory and experiment was very often found to be poor, and St. Venant suggested that perhaps yield occurred owing to the maximum principal strain, rather than the maximum principal stress, as the former involved all three principal stresses and Poisson's ratio, whilst the latter did not.

The maximum principal strain theory states that yield will occur when,

$$\varepsilon_1 = \frac{1}{E}[\sigma_1 - \nu(\sigma_2 + \sigma_3)] = \sigma_{yp}/E$$

or,

$$\underline{\sigma_1 - \nu(\sigma_2 + \sigma_3) = \sigma_{yp}} \tag{3.3}$$

or if the material is in compression,

$$\underline{\sigma_3 - \nu(\sigma_1 + \sigma_2) = \sigma_{ypc}} \tag{3.4}$$

where,

ε_1 = maximum principal strain

3.3.2

In *two dimensions*, $\sigma_3 = 0$; therefore equations (3.3) and (3.4) become:

$$\sigma_1 - \nu\sigma_2 = \sigma_{yp} \tag{3.5}$$

$$\sigma_2 - \nu\sigma_1 = \sigma_{ypc} \tag{3.6}$$

For convenience, for the remaining three theories, the assumption will be made that σ_{yp} is of the same magnitude as σ_{ypc}. If the magnitude of σ_{ypc} is less than that of σ_{yp}, the theories can be easily modified to take this into account.

3.4.1 TOTAL STRAIN ENERGY THEORY (BELTRAMI AND HAIGH)

This theory states that elastic failure will occur when the total strain energy per unit volume, at a point, reaches the total strain energy per unit volume of a specimen made from the same material, when it is subjected to a simple uni-axial test.

Now for a three-dimensional stress system, the total strain energy per unit volume, in terms of the three principal stresses =

$$U_T/\text{vol} = \tfrac{1}{2}\sigma_1\varepsilon_1 + \tfrac{1}{2}\sigma_2\varepsilon_2 + \tfrac{1}{2}\sigma_3\varepsilon_3$$

but from reference [1],

$$\varepsilon_1 = \frac{1}{E}[\sigma_1 - \nu(\sigma_2 + \sigma_3) = \text{maximum principal strain}$$

$$\varepsilon_2 = \frac{1}{E}[\sigma_2 - \nu(\sigma_1 + \sigma_3) = \text{minimax principal strain}$$

$$\varepsilon_3 = \frac{1}{E}[\sigma_3 - \nu(\sigma_1 + \sigma_2)] = \text{minimum principal strain}$$

i.e.

$$U_T/\text{vol} = \frac{1}{2E}(\sigma_1^2 + \sigma_2^2 + \sigma_3^2) - \frac{\nu}{E}(\sigma_1\sigma_2 + \sigma_1\sigma_3 + \sigma_2\sigma_3) \tag{3.7}$$

In the simple uni-axial test,

$$\sigma_1 = \sigma_{yp} \quad \text{and} \quad \sigma_2 = \sigma_3 = 0$$

therefore

$$U_T/\text{vol} = \sigma_{yp}^2/2E \tag{3.8}$$

Equating (3.7) and (3.8),

$$\underline{\underline{\sigma_1^2 + \sigma_2^2 + \sigma_3^2 - 2\nu(\sigma_1\sigma_2 + \sigma_1\sigma_3 + \sigma_2\sigma_3) = \sigma_{yp}^2}} \tag{3.9}$$

This theory, like the principal strain theory, also involves all three principal stresses and Poisson's ratio.

3.4.2

In *two dimensions*, $\sigma_3 = 0$; therefore equation (3.9) becomes,

$$\underline{\sigma_1^2 + \sigma_2^2 - 2\nu\sigma_1\sigma_2 = \sigma_{yp}^2} \tag{3.10}$$

3.5.1 MAXIMUM SHEAR STRESS THEORY (TRESCA)

The problem with the three previous theories is that they all fail in the case of hydrostatic stress. Experiments have shown that, whether or not a solid piece of material is a soft and ductile or hard and brittle, when it is subjected to a large uniform external pressure, then despite the fact that the yield stress is grossly exceeded, the materials do not suffer elastic breakdown in this condition. For example, lumps of chalk or similar low strength substances can survive intact at great depths in the oceans.

In such cases, all the principal stresses are equal to the water pressure P, so that,

$$\sigma_1 = \sigma_2 = \sigma_3 = -P \tag{3.11}$$

From equation (3.11), it can be seen that there are no shear stresses in a hydrostatic stress condition, and this is why low strength materials survive intact under large values of water pressure.

An argument held against this hypothesis is that in a normal hydrostatic stress condition, the stresses are all compressive, and this is the reason why failure does not take place. However, a Russian scientist carried out tests on a piece of glass, which by a process of heating and cooling was believed to be subjected to a hydrostatic tensile stress of about 6.895E9 Pa (1E6 $\mathrm{lbf/in^2}$) at a certain point in the material, but inspection revealed that there were no signs of cracking at this point.

Thus, it can be concluded that for elastic failure to take place, the material must distort (change shape), and for this to occur, it is necessary for shear stress to exist.

The maximum shear stress theory states that elastic failure will take place when the maximum shear stress at a point equals the maximum shear stress obtained in a specimen, made from the same material, in the simple uni-axial test, i.e.

$$\underline{\sigma_1 - \sigma_3 = \sigma_{yp}} \tag{3.12}$$

3.5.2

In *two dimensions*, $\sigma_3 = 0$, and equation (3.12) becomes,

$$\sigma_1 - \sigma_2 = \sigma_{yp} \tag{3.13}$$

3.6.1 MAXIMUM SHEAR STRAIN ENERGY THEORY (HENCKY AND VON MISES)

The problem with the maximum shear stress theory is that it states that,

$$\underline{\tau_{yp} = 0.5\sigma_{yp}}$$

where,

$$\tau_{yp} = \text{shear stress at yield}$$

However, torsional tests on mild steel specimens have found that

$$\underline{\tau_{yp} \doteqdot 0.577\,\sigma_{yp}}$$

and this implies that the maximum shear stress theory is not always suitable, and the reason for this may be due to the fact that it ignores the effects of σ_2 and v.

A theory, therefore, that takes into consideration all these factors, and does not fail under the hydrostatic stress condition, is the shear strain energy theory, which states that elastic failure takes place when the shear strain energy per unit volume, at a point, equals the shear strain energy per unit volume in a specimen of the same material, in the simple uni-axial test.

Now, the shear strain energy/vol =

$$\text{Total strain energy/vol} - \text{Hydrostatic strain energy/vol}$$

i.e.

$$\text{SSE} = U_{\text{T}} - U_{\text{H}}$$

where,

$$U_{\text{T}} = \text{total strain energy} = \frac{1}{2E}[(\sigma_1^2 + \sigma_2^2 + \sigma_3^2) - 2v(\sigma_1\sigma_2 + \sigma_1\sigma_3 + \sigma_2\sigma_3)] * \text{volume (see equation (3.7))}$$

$$U_{\text{H}} = \text{hydrostatic strain energy} = \frac{1}{2E}(3 * P^2 - 2v * 3P^2) * \text{volume}$$

$$P = \text{hydrostatic stress} = (\sigma_1 + \sigma_2 + \sigma_3)/3$$

and U_{H} is obtained by substituting P for σ_1, σ_2 and σ_3 into U_{T}. Therefore

$$\text{SSE/vol} = \frac{1}{2E}[\sigma_1^2 + \sigma_2^2 + \sigma_3^2 - 2v(\sigma_1\sigma_2 + \sigma_1\sigma_3 + \sigma_2\sigma_3) + (2v - 1) * 3 * (\sigma_1 + \sigma_2 + \sigma_3)^2/9]$$

$$= \frac{(1 + v)}{6E}[(\sigma_1 - \sigma_2)^2 + (\sigma_1 - \sigma_3)^2 + (\sigma_2 - \sigma_3)^2]$$

but,

$$G = E/[2(1 + v)]$$

therefore

$$\underline{\text{SSE/vol} = \frac{1}{12G}[(\sigma_1 - \sigma_2)^2 + (\sigma_1 - \sigma_3)^2 + (\sigma_2 - \sigma_3)^2]} \tag{3.14}$$

The shear strain energy/vol for a specimen in the uni-axial tensile test can be obtained from equation (3.14) by substituting

$$\sigma_1 = \sigma_{\text{yp}} \quad \text{and} \quad \sigma_2 = \sigma_3 = 0$$

i.e.

$$\text{SSE/vol} = \frac{\sigma_{\text{yp}}^2}{6G} \tag{3.15}$$

Equating (3.14) and (3.15), the criterion for yielding, according to the shear strain energy theory, is as in equation (3.16):

$$\underline{(\sigma_1 - \sigma_2)^2 + (\sigma_1 - \sigma_3)^2 + (\sigma_2 - \sigma_3)^2 = 2\sigma_{\text{yp}}^2} \tag{3.16}$$

3.6.2

In *two dimensions*, $\sigma_3 = 0$, and equation (3.16) reduces to:

$$\underline{\sigma_1^2 + \sigma_2^2 - \sigma_1\sigma_2 = \sigma_{yp}^2} \tag{3.17}$$

Another interpretation of equation (3.16) is that elastic failure occurs in a structure when the octahedral shear stress at a point in it reaches the octahedral shear stress at yield, in a specimen made from the same material as the structure, when the specimen undergoes the simple uni-axial test. The octahedral shear stress (3 to 7), is that shear stress that lies on a regular octahedron, composed of eight tetrahedrons, one of which is shown in Fig. 3.1.

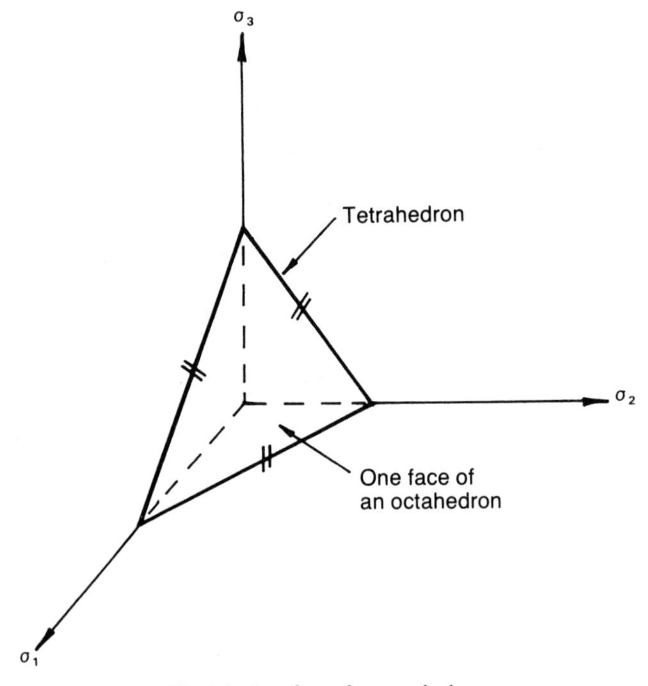

Fig. 3.1. One face of an octahedron.

3.7.1 YIELD LOCI

In two dimensions, equations (3.1), (3.2), (3.5), (3.6), (3.10), (3.13) and (3.17) can be expressed graphically, as shown in Fig. 3.2.

The figures are obtained by plotting the above equations, using σ_1 as the horizontal co-ordinate and σ_2 as the vertical co-ordinate, and the interpretation of each figure is that yield will not occur according to the theory under consideration if the point described by the values of σ_1 and σ_2 does not fall outside the appropriate figure.

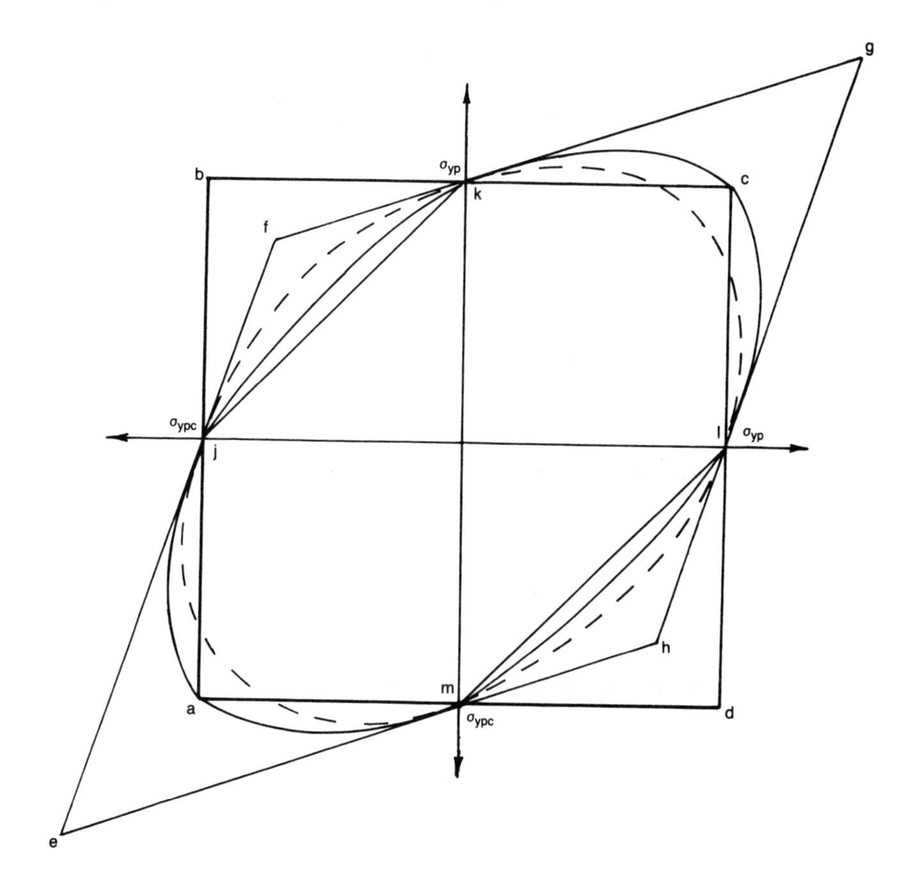

Figure	Theory enclosed by figure
abcd	Maximum principal stress theory
efgh	Maximum principal strain theory
ajkclm	Maximum shear stress theory
– – – ellipse	Total strain energy theory
______ ellipse	Distortion or shear strain energy theory

Fig. 3.2. Yield loci for a two-dimensional stress system.

3.8.1 EXAMPLE 3.1 THIN-WALLED CYLINDER UNDER UNIFORM PRESSURE

A long thin-walled cylinder of wall thickness 2 cm and of internal diameter 10 m is subjected to a uniform internal pressure. Determine the pressure that will cause yield, based on the five major theories of elastic failure, and given the following:

$$\sigma_{yp} = \sigma_{ypc} = 300\ \text{MN/m}^2$$
$$\nu = 0.3$$

From reference [1],

$$\sigma_1 = \text{hoop stress} = \frac{pR}{t} = \frac{p \times 5}{2E-2} = 250\,p$$

$$\sigma_2 = \text{longitudinal stress} = \frac{pR}{2t} = 125\,p$$

$$\sigma_3 = \text{radial stress on inside surface of cylinder wall} = -p$$

where,

R = internal radius of cylinder
t = wall thickness
p = internal pressure—to be determined

3.8.2 Maximum Principal Stress Theory

$$\sigma_1 = \sigma_{yp}$$
$$250\,p = 300\,\text{MPa}$$
$$\underline{p = 1.2\,\text{MPa}}$$

3.8.3 Maximum Principal Strain Theory

$$\sigma_1 - \nu(\sigma_2 + \sigma_3) = \sigma_{yp}$$
$$250\,p - 0.3(125\,p - p) = 300$$

therefore

$$p = \frac{300}{212.8} = \underline{1.41\,\text{MPa}}$$

3.8.4 Total Strain Energy Theory

$$\sigma_1^2 + \sigma_2^2 + \sigma_3^2 - 2\nu(\sigma_1\sigma_2 + \sigma_1\sigma_3 + \sigma_2\sigma_3) = \sigma_{yp}^2$$
$$[62\,500 + 15\,625 + 1 - 0.6(31\,250 - 250 - 125)]p^2 = 90\,000$$

therefore

$$p^2 = 90\,000/59\,601$$

or,

$$\underline{p = 1.229\,\text{MPa}}$$

3.8.5 Maximum Shear Stress Theory

$$\sigma_1 - \sigma_3 = \sigma_{yp}$$

or,

$$251\,p = 300$$
$$\underline{p = 1.195\,\text{MPa}}$$

3.8.6 Shear Strain Energy Theory

$$(\sigma_1 - \sigma_2)^2 + (\sigma_1 - \sigma_3)^2 + (\sigma_2 - \sigma_3)^2 = 2\sigma_{yp}^2$$
$$(15\,625 + 63\,001 + 15\,876)p^2 = 18\,000$$

therefore

$$\underline{p = 1.38\,\text{MPa}}$$

3.8.7

From the calculations in this section, it can be seen that the maximum principal strain theory is the most optimistic, and the maximum shear stress theory is the most pessimistic. As expected, the shear strain energy theory is more optimistic than the maximum shear stress theory.

3.9.1 EXAMPLE 3.2 SHAFT UNDER COMBINED BENDING AND TORSION

A solid circular section shaft of diameter 0.1 m is subjected to a bending moment of 15 kN m. Determine the required torque to cause yield, based on the five major theories of elastic failure, given the following:

$$\sigma_{yp} = 300\,\text{MN/m}^2 \quad \nu = 0.3$$

This is a two-dimensional system of stress; hence, it is necessary to use equations (3.1), (3.5), (3.10), (3.13) and (3.17).

Now,

$$I = \frac{\pi \times 0.1^4}{64} = 4.909\text{E-6}\,\text{m}^4$$
$$\bar{y} = 0.05\,\text{m}$$

therefore

$$\sigma_x = \frac{M\bar{y}}{I} = \text{maximum stress in axial direction}$$
$$= \frac{15\text{E3} \times 0.05}{4.909\text{E-6}}$$
$$\underline{\sigma_x = 152.8\,\text{MPa}}$$

By inspection,

$$\underline{\sigma_y = 0}$$

Now from reference [1],

$$T = \frac{\tau_{xy} * J}{r} = \frac{\tau_{xy} * 9.818\text{E-6}}{0.05}$$

therefore

$$\underline{T = 1.964\text{E-}4\,\tau_{xy}}$$

Also, from reference [1],

$$\sigma_1 = \frac{(\sigma_x + \sigma_y)}{2} + \tfrac{1}{2}\sqrt{[(\sigma_x - \sigma_y)^2 + 4\tau_{xy}^2]}$$

$$= \frac{152.8}{2} + \tfrac{1}{2}\sqrt{[152.8^2 + 4\tau_{xy}^2]}$$

therefore

$$\underline{\sigma_1 = 76.4 + k}$$

and,

$$\underline{\sigma_2 = 76.4 - k}$$

where,

$$\underline{k = \tfrac{1}{2}\sqrt{[152.8^2 + 4\tau_{xy}^2]}}$$

3.9.2 Maximum Principal Stress Theory

$$\sigma_1 = \sigma_{yp}$$

or,

$$76.4 + k = 300$$

therefore

$$k = 223.6 = \tfrac{1}{2}\sqrt{[152.8^2 + 4\tau_{xy}^2]}$$

$$447.2^2 = 152.8^2 + 4\tau_{xy}^2$$

$$\underline{\tau_{xy} = 210.1\,\text{MPa}}$$

but,

$$T = 1.964\text{E-}4 * \tau_{xy}$$

therefore

$$\underline{T = 41.3\,\text{kN m}}$$

3.9.3 Maximum Principal Strain Theory

$$\sigma_1 - \nu\sigma_2 = \sigma_{yp}$$

$$76.4 + k - 0.3 \times 76.4 + 0.3k = 300$$

or,

$$1.3k = 246.52$$

or,

$$k = 189.6 = \tfrac{1}{2}\sqrt{[152.8^2 + 4\tau_{xy}^2]}$$

therefore

$$\underline{\tau_{xy} = 173.5\,\text{MPa}}$$

and,

$$\underline{T = 34.08\,\text{kN m}}$$

3.9.4 Total Strain Energy Theory

$$\sigma_1^2 + \sigma_2^2 - 2\nu\sigma_1\sigma_2 = \sigma_{yp}^2$$

$$(76.4 + k)^2 + (76.4 - k)^2 - 0.6 * (76.4 + k) * (76.4 - k) = 90\,000$$

or,

$$k = 177.4 = \tfrac{1}{2}\sqrt{[152.8^2 + 4\tau_{xy}^2]}$$

therefore

$$\underline{\tau_{xy} = 160.1}$$

and,

$$\underline{T = 31.45\,\text{kN m}}$$

3.9.5 Maximum Shear Stress Theory

$$\sigma_1 - \sigma_2 = \sigma_{yp}$$

$$76.4 + k - 76.4 + k = 300$$

or,

$$k = 150 = \tfrac{1}{2}\sqrt{[152.8^2 + 4\tau_{xy}^2]}$$

therefore

$$\underline{\tau_{xy} = 129.1\,\text{MPa}}$$

and,

$$\underline{T = 25.35\,\text{kN m}}$$

3.9.6 Shear Strain Energy Theory

$$\sigma_1^2 + \sigma_2^2 - \sigma_1\sigma_2 = \sigma_{yp}^2$$

$$(76.4 + k)^2 + (76.4 - k)^2 - (76.4 + k) * (76.4 - k) = 300^2$$

or,

$$k = 167.5 = \tfrac{1}{2}\sqrt{[152.8^2 + 4\tau_{xy}^2]}$$

therefore

$$\underline{\tau_{xy} = 149\,\text{MPa}}$$

and,

$$\underline{T = 29.27\,\text{kN m}}$$

3.9.7

The calculations in this section have shown that for this example, the maximum principal stress theory is the most optimistic, and the maximum shear stress is the most pessimistic, where the ratio of the former to the latter is about 1.6:1.

3.10.1 CONCLUSIONS

The examples have shown that the predictions by the various yield criteria can be very different. Furthermore, from the heuristic arguments of Section 3.5.1, it would appear that the only two theories that do not fail the hydrostatic stress condition are the maximum shear stress theory and the shear strain energy theory, and in any case, when materials such as mild steel are tested to destruction, in tension, the characteristic "cup and cone" failure mode (Fig. 3.3) indicates the importance that shear stress plays in elastic failure. Because of this, many structural designers often prefer to use the maximum shear stress and the shear strain energy theories, when designing structures involving two- and three-dimensional stress systems, where the former often lends itself to neat mathematical computations.

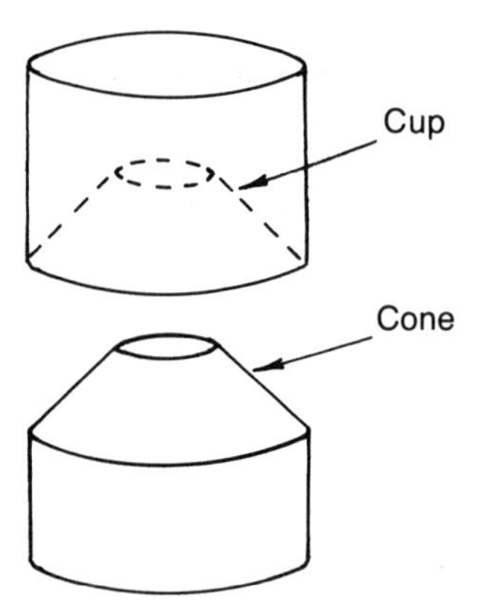

Fig. 3.3. Cup and cone failure, indicating the importance of shear stress.

Finally, it can be concluded from the calculations for both examples in this chapter, that in certain two- and three-dimensional stress systems, some of the theories of elastic failure can be dangerous, when applied to practical cases.

EXAMPLES FOR PRACTICE 3

1. A submarine pressure hull, which may be assumed to be a long thin-walled circular cylinder, of external diameter 10 m and of wall thickness 5 cm, is constructed from high tensile steel.

Assuming that buckling does not occur, determine the maximum permissible diving depths that the submarine can achieve, without suffering elastic failure, based on the five major theories of yield, and given the following:

$$\sigma_{yp} = -\sigma_{ypc} = 400\,\mathrm{MN/m^2} \quad \nu = 0.3$$

Density of water $= 1020\,\mathrm{kg/m^3}$

$g = 9.81\,\mathrm{m/s^2}$

{400 m; 470 m; 412.3 m; 404 m; 466.6 m}

2. A circular section torsion specimen, of diameter 2 cm, yields under a pure torque of 0.25 kN m. What is the shear stress due to yield? What is the yield stress according to (a) Tresca, (b) Hencky–von Mises? What is the ratio τ_{yp}/σ_{yp} according to these two theories?

   {159.1 MPa; 318.3 MPa; 275.6 MPa; 0.5; 0.577}

3. A shaft of diameter 0.1 m is found to yield under a torque of 30 kN m.

   Determine the pure bending moment that will cause a similar shaft, with no torque applied to it, to yield, assuming that the Tresca theory applies. What would be the bending moment to cause yield if the Hencky–von Mises theory applied?

   {30 kN m; 26 kN m}

4

Plasticity

4.1.1 PLASTIC AND ELASTIC DESIGN

The design of most structures is based on the small deflection theory of elasticity and Hooke's law, which states that load is proportional to extension. From this fundamental assumption, calculations are made of stresses that occur due to the applied loads, and the structure is designed so that its maximum stresses will not exceed a certain permissible stress for the material of construction. This permissible stress is usually several times lower than the material's yield stress.

However, the design of such a structure, based on Hookean elasticity, is somewhat illogical, as the designer has little or no idea of the initial stresses that occur in the structure due to welding, riveting, etc.

In this context, plastic theory is somewhat superior to elastic theory, as it depends a lot less on what built-in stresses there are in the structure. For example, in the plastic design of beams, plastic theory assumes that the beam collapses when the beam changes into a mechanism, the hinges of the mechanism being called plastic hinges, and these occur in regions of the beam where plastic flow takes place.

Although the concept of the plastic hinge was made as early as 1914 [9], its application to structural design did not take place until 1938 [10]. Since then, considerable research has taken place [11–14], although there are still some areas which require investigation.

In this chapter, considerations will be made of the plastic analysis and design of beams, frames and shafts, but no attempt will be made to investigate plastic buckling due to compressive stresses.

4.2.1 LOAD–EXTENSION RELATIONSHIP

The load–extension relationship for a material such as mild steel is shown in Fig. 4.1, where it can be seen that the material behaves in a lienar elastic manner from "A" to "B". From the points "B" to "C", the material becomes plastic, where the extension δ_p is about forty times the extension δ_{yp}.

After reaching the point "C", the material strain hardens, where α is approximately $\theta/50$.

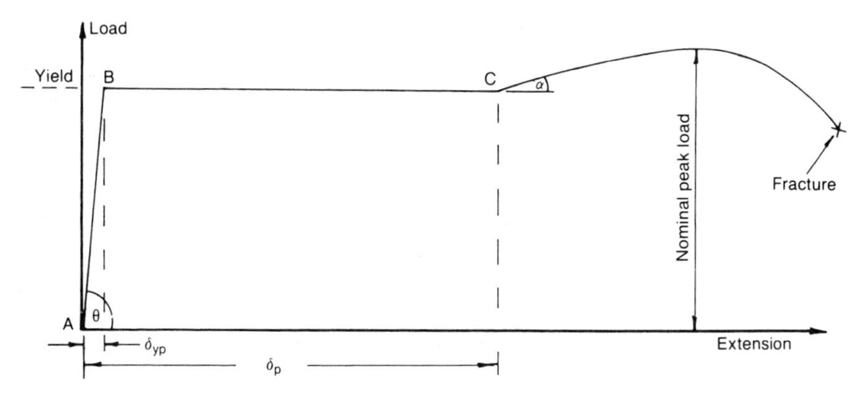

Fig. 4.1. Load–extension relationship for mild steel.

4.3.1 PLASTIC HINGE

To demonstrate the concept of the plastic hinge, and also plastic failure, consider the encastré beam of Fig. 4.2, and assume that it is of a constant rectangular cross-section.

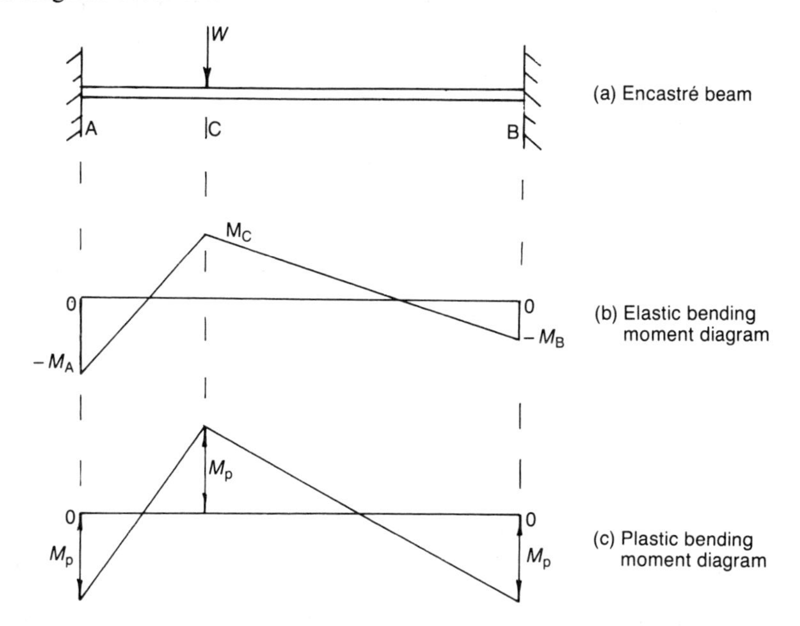

Fig. 4.2. Encastré beam and bending moment diagrams.

Now when the applied load W is of such a magnitude that the beam is entirely elastic, the stress distribution at the support "A", up to the yield stress (σ_{yp}), will be as shown in Fig. 4.3(a).

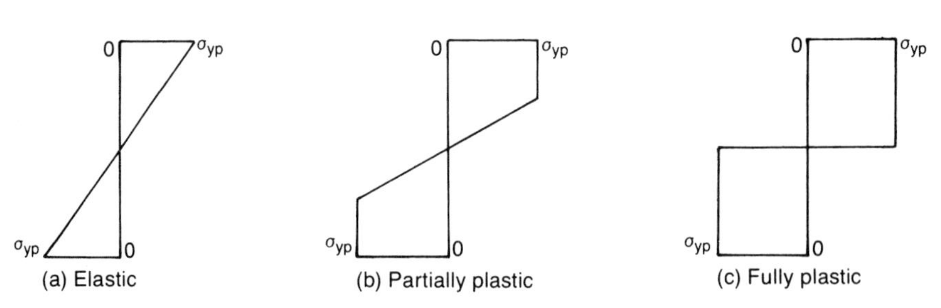

Fig. 4.3. Stress distributions across the beam's sections.

Further increase of the load W will cause the outermost fibres at the support "A" to first become plastic, as this section is the point with the maximum bending moment. The stress distribution in the partially plastic state is shown in Fig. 4.3(b), where some of the stresses are in the elastic zone, whilst others are in the plastic zone, "B" to "C" of Fig. 4.1.

A still further increase in the magnitude of W will cause more plastic penetration to take place, until eventually, the stress distribution at the support "A" of the beam becomes fully plastic, as shown in Fig. 4.3(c).

If W is now increased, the section at the support "A" will start to rotate like a hinge, because the stresses cannot increase at this section, so that the moment of resistance will not be able to increase either. The result of this will be that there is a redistribution of bending moment, so that the sections at "B" and "C", will start to become plastic. Eventually, when W reaches the collapse load W_c, the bending moment distribution will be as in Fig. 4.2(c), and all three sections at A, B and C will have developed fully plastic moments of resistance, as shown in Fig. 4.3(c).

At this stage, the slightest increase in load will cause catastrophic failure, as shown in Fig. 4.4, where the beam will become a mechanism, the hinges of the mechanism being at the points A, B and C. These hinges are called *plastic hinges*, and they can be imagined to behave somewhat like rusty hinges.

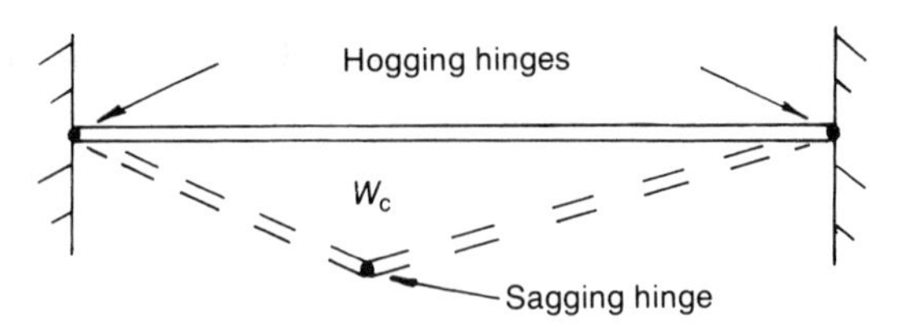

Fig. 4.4. Plastic hinges.

In general, the number of hinges required to cause the failure of a beam or a frame will be one more than that required to make the structure statically determinate. For example, the encastré beam of Fig. 4.2(a) is statically indeterminate to the second degree (i.e. it has two redundancies), so that three hinges are required to make the beam into a mechanism. If the beam of Fig. 4.2(a) were simply-supported at its ends, then only one hinge would be required to make it into a mechanism.

For the design of beams and frames, it is very often easier to use plastic theory in place of elastic theory, as the *plastic moment of resistance*, (M_p), is known at the hinges.

N.B. It should be pointed out that as δ_p is about forty times larger than δ_{yp}, for materials such as mild steel, there is usually sufficient plasticity in the material for the plastic hinges to occur, and for this reason, the load–extension distribution is usually assumed to be of the ideally elastic–plastic form of Fig. 4.5.

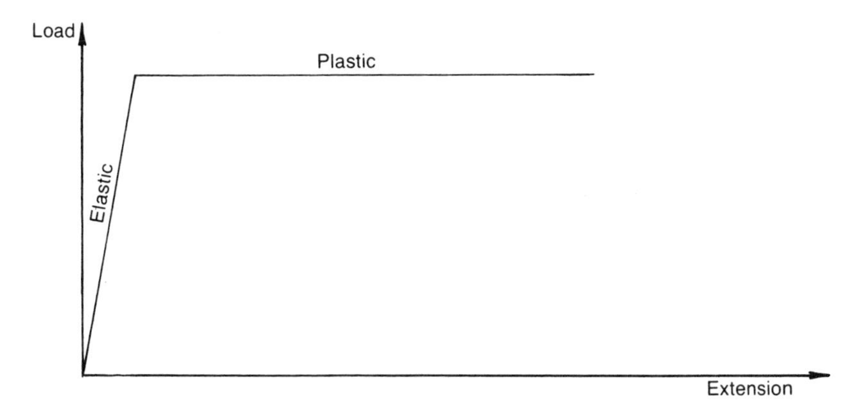

Fig. 4.5. Ideally elastic–plastic material

4.4.1 NEUTRAL AXES IN BENDING

The elastic neutral axis of a section lies on its centroidal axis, whilst the plastic neutral axis of the section lies on its central axis. The *centroidal axis* of a section can be defined as the centre of its *moment of area*, and the *central axis* can be defined as its *centre of area* [1]. These positions are obtained by considering equilibrium along the axis of the beam, where in elastic theory, the stress distribution varies linearly across the section, as shown in Fig. 4.3(a), and in plastic theory, the stress distribution is as shown in Fig. 4.3(c). The centroidal and central axes will lie at the same positions for sections which are symmetrical about both axes, such as for rectangles, circles, ellipses and RSJs, etc.

4.5.1 PLASTIC MOMENT OF RESISTANCE (M_p)

From Fig. 4.3(c), it can be seen that for a rectangular section, of width b and depth d,

$$M_p = \sigma_{yp} * \frac{bd}{2} * \frac{d}{4} * 2$$

$$\underline{M_p = \frac{\sigma_{yp} * bd^2}{4}} \tag{4.1}$$

Similarly, from Fig. 4.3(a), it can be seen that the elastic moment of resistance of this section at first yield,

$$M_{yp} = \sigma_{yp} * \frac{bd}{4} * \frac{d}{3} * 2$$

$$\underline{M_{yp} = \frac{\sigma_{yp} * bd^2}{6}} \tag{4.2}$$

Thus, it can be seen from equations (4.1) and (4.2) that, for a rectangular section, its plastic moment of resistance (M_p) is 50% greater than its maximum elastic moment of resistance (M_{py}).

4.6.1 SHAPE FACTOR (S)

This is defined as in equation (4.3).

$$\text{Shape factor} = S = \frac{M_p}{M_{yp}} \tag{4.3}$$

For a *rectangular section,*

$$S = \sigma_{yp} * \frac{bd^2}{4} * \frac{6}{\sigma_{yp} * bd^2}$$

$$\underline{S = 1.5}$$

For a typical *RSJ*,

$$S = 1.15$$

4.7.1 LOAD FACTOR (λ)

This is the plastic equivalent of the safety factor that is normally used in elastic theory, where,

$$\text{load factor} = \lambda = \frac{W_c}{W} \tag{4.4}$$

W_c = plastic collapse load
W = working load

The two main methods of calculating W_c are called the *statical method* and the *kinematical method*, and a simple example will be used to describe each method.

4.8.1 EXAMPLE 4.1 STATICAL METHOD

Calculate W_c for the uniform section encastré beam of Fig. 4.6 by the statical method.

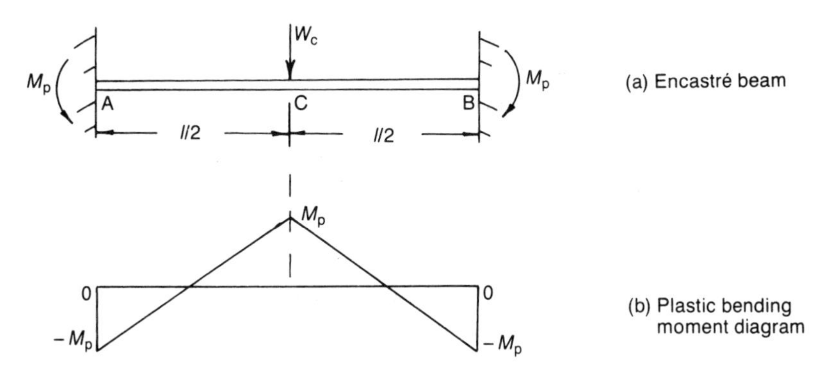

Fig. 4.6. Plastic collapse of a beam.

4.8.2

This method, which was developed by Baker [10], consists in examining the equilibrium of the beam, as follows.

The bending moment at C $= W_c/2 * l/2 - M_p$, but at collapse the bending moment at C $= M_p$ (sagging), so that

$$M_p = \frac{W_c l}{4} - M_p$$

or,

$$M_p = W_c l/8$$

and

$$\underline{\underline{W_c = 8M_p/l}} \qquad (4.5)$$

As M_p is effectively a sectional property, W_c can be found from equation (4.5).

4.9.1 EXAMPLE 4.2 KINEMATICAL METHOD

Calculate W_c for the uniform section encastré beam of Fig. 4.6, by the kinematical method.

4.9.2

This method, which was developed by Baker *et al.* [11], is based on the principle of virtual work [1].

At the failure load W_c, the beam of Fig. 4.6(a) will collapse in the manner shown in Fig. 4.7, where the beam elements are shown straight, as the beam's curvature will not change during the early stages of collpase. The reason why the curvature of the beam does not change during collapse is that the bending moment distribution remains constant during this period.

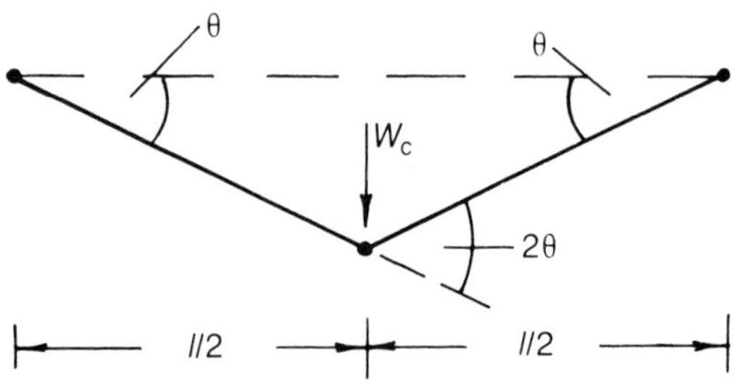

Fig. 4.7. Beam mechanism.

Assuming that the deflections are small, work done by $W_c =$

$$\mathrm{WD} = W_c * \frac{l}{2} * \theta \tag{4.6}$$

Work done by the plastic hinges

$$= M_p * \theta + M_p * 2\theta + M_p\theta \tag{4.7}$$

Equating (4.6) and (4.7),

$$4M_p = W_c * l/2$$

or

$$M_p = W_c l/8$$

and

$$\underline{W_c = 8M_p/l} \text{ (as before)} \tag{4.8}$$

In general, the kinematical method is often preferred to the statical method, especially if the positions of the hinges are known. It should, however, be pointed out that if incorrect assumptions are made, for the positions of the hinges, then the kinematical method will underestimate the value for M_p. Thus, when using the kinematical method, it should be ensured that all the possible mechanisms are investigated, and that the chosen value of M_p should be the maximum.

4.10.1 EXAMPLE 4.3 SHAPE FACTOR FOR A CIRCLE

Calculate the shape factor for a circular section.

4.10.2

The stress distribution for the solid circular section of Fig. 4.8(a) is shown in Fig. 4.8(b), where it can be seen that

$$M_p = \int_{-R}^{R} \sigma_{yp} * b\,dy * y$$

but,

$$b = 2R\cos\theta$$

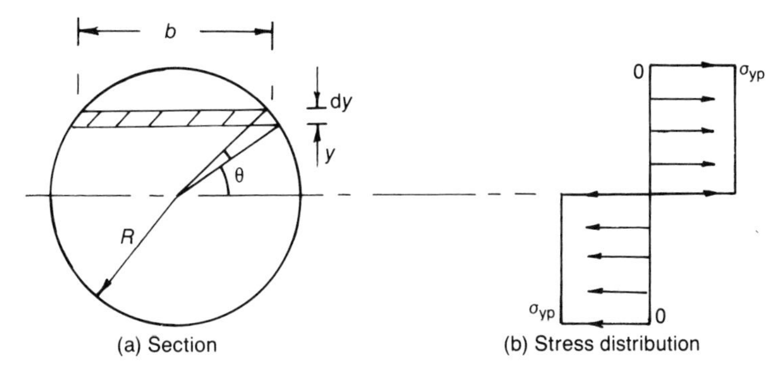

Fig. 4.8. Circular section.

and

$$y = R \sin \theta$$

and by differentiating y w.r.t. θ,

$$\mathrm{d}y = R \cos \theta . \mathrm{d}\theta$$

therefore

$$M_{\mathrm{p}} = \sigma_{\mathrm{yp}} \int_{\pi/2}^{\pi/2} 2R^3 \cos^2 \theta \sin \theta \, \mathrm{d}\theta$$

$$= -4R^3 \sigma_{\mathrm{yp}} \int_{0}^{\pi/2} \cos^2 \theta \, \mathrm{d}(\cos \theta)$$

$$= -4R^3 \sigma_{\mathrm{yp}} \left[\frac{\cos^3 \theta}{3} \right]_0^{\pi/2}$$

$$\underline{M_{\mathrm{p}} = 4\sigma_{\mathrm{yp}} \cdot R^3/3} \qquad (4.9)$$

From elementary elastic theory,

$$M_{\mathrm{yp}} = \frac{\sigma_{\mathrm{yp}}}{R} * I = \frac{\sigma_{\mathrm{yp}}}{R} * \frac{\pi R^4}{4}$$

$$\underline{M_{\mathrm{yp}} = \pi \sigma_{\mathrm{yp}} \cdot R^3/4} \qquad (4.10)$$

Now,

$$S = \frac{M_{\mathrm{p}}}{M_{\mathrm{yp}}}$$

therefore

$$\underline{S = 1.7}$$

4.11.1 EXAMPLE 4.4 SHAPE FACTORS FOR AN "I" SECTION AND A TEE SECTION

Determine S for the sections of Fig. 4.9(a) and (b).

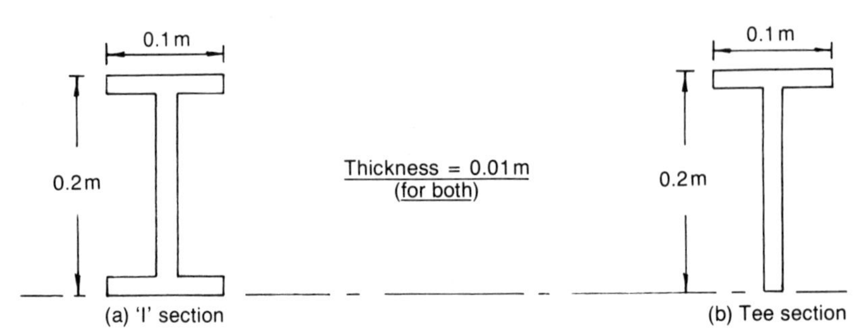

Fig. 4.9. Built-up sections.

4.11.2(a)

The plastic neutral axis of the "I" section will lie at its centre.

$$M_p = 2 * \sigma_{yp} * (0.1 * 0.01 * 0.095 + 0.09 * 0.01 * 0.09/2)$$

$$\underline{M_p = 2.71\text{E-}4\,\sigma_{yp}} \tag{4.11}$$

From elementary elastic theory,

$$M_{yp} = \frac{\sigma_{yp}}{0.1} * I$$

but

$$I = \frac{0.1 \times 0.2^3}{12} - \frac{0.09 \times 0.18^3}{12}$$

$$= 2.293\text{E-}5$$

therefore

$$\underline{M_{yp} = 2.293\text{E-}4\,\sigma_{yp}} \tag{4.12}$$

From (4.11) and (4.12),

$$\underline{S = \frac{2.71\text{E-}4}{2.293\text{E-}4} = 1.18}$$

4.11.3(b)

In this case it will be first necessary to determine the position of the plastic neutral axis which can be obtained by considering equilibrium along the axis of the beam.

From Fig. 4.10,

Force to "left" = Force to "right"

i.e.

$$\sigma_{yp} * 0.01\,\bar{Y} = \sigma_{yp} * [0.1 * 0.01 + 0.01 * (H - 0.01)]$$

but,

$$\bar{Y} = 0.2 - H$$

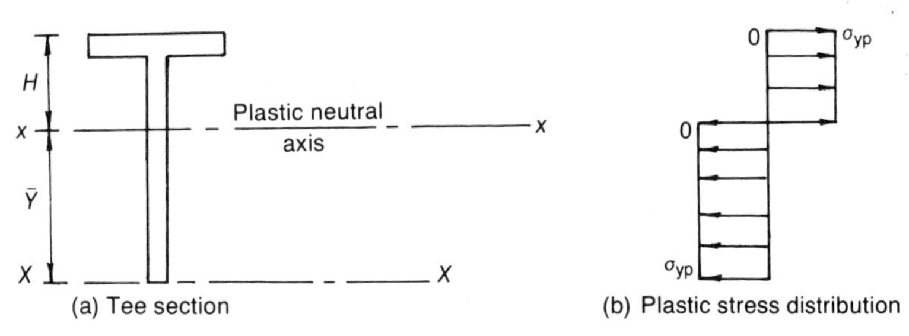

Fig. 4.10. Plastic stress distribution across tee section.

therefore

$$(0.2 - H) = 0.1 + (H - 0.01)$$

therefore

$$2H = 0.2 - 0.1 + 0.01 = 0.11$$

$$\underline{H = 0.055\,\text{m}}$$

and

$$\underline{\bar{Y} = 0.145\,\text{m}}$$

From Fig. 4.10(b),

$$\begin{aligned} M_p &= \sigma_{yp} * \{0.01 \times 0.1 \times (H - 0.005) \\ &\quad + 0.01 \times (H - 0.01) \times (H - 0.01)/2 \\ &\quad + 0.01 \times (0.2 - H) \times (0.2 - H)/2\} \\ &= \sigma_{yp}\{1\text{E-}3(0.05) + 0.005 \times 0.045^2 + 0.005 \times 0.145^2\} \\ &= \sigma_{yp}(5\text{E-}5 + 1.013\text{E-}5 + 1.051\text{E-}4) \end{aligned}$$

$$\underline{M_p = 1.6523\text{E-}4\,\sigma_{yp}} \tag{4.13}$$

To determine M_{yp}, it will be necessary to calculate I and the position of the elastic neutral axis, which can be achieved with the aid of Table 4.1. The symbols in Table 4.1 have the same meaning as in Chapter 1.

Table 4.1. Properties of the tee section

Section	a	y	ay	ay^2	i_H
①	1E-3	0.195	1.95E-4	3.803E-5	8.33E-9
②	1.9E-3	0.095	1.81E-4	1.715E-5	5.716E-6
$\sum$	2.9E-3	—	3.76E-4	5.518E-5	5.724E-6

$$\bar{y} = \frac{\sum ay}{\sum a} = \frac{3.76\text{E-}4}{2.9\text{E-}3} = \underline{0.1297\,\text{m}}$$

$$I_{xx} = \sum ay^2 + \sum i_H = \underline{6.09\text{E-}5\,\text{m}^4}$$

$$I_{NA} = \text{2nd moment of area about the elastic neutral axis}$$
$$= I_{xx} - \bar{y}^2 \sum a$$
$$= 6.09\text{E-5} - 0.1297^2 \times 2.9\text{E-3}$$
$$\underline{I_{NA} = 1.212\text{E-5 m}^4}$$

$$\underline{M_{yp} = \frac{\sigma_{yp}}{\bar{y}} * I_{NA} = 9.34\text{E-5}\,\sigma_{yp}} \tag{4.14}$$

From (4.13) and (4.14), therefore

$$\underline{\underline{S = \frac{1.652\text{E-4}}{9.34\text{E-5}} = 1.76}}$$

4.12.1 EXAMPLE 4.5 RESIDUAL STRESSES IN A BEAM

A steel beam of constant rectangular section, of depth 0.2 m and of width 0.1 m, is subjected to four-point loading, as shown in Fig. 4.11. Determine:

(a) the depth of plastic penetration at mid-span;
(b) the central deflection; and
(c) the length of the beam over which yield takes place.

If the above load is removed:

(d) determine the residual central deflection; and
(e) plot the residual stress distribution at mid-span.

The following may be assumed to apply:

$$\sigma_{yp} = 350\ \text{MN/m}^2 \qquad E = 2\text{E}11\ \text{N/m}^2$$

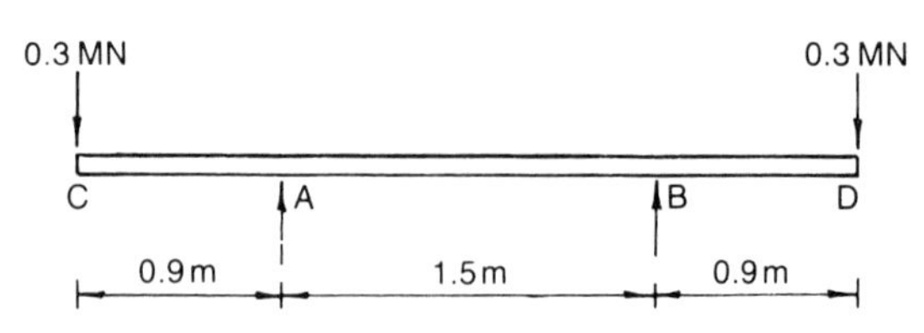

Fig. 4.11. Beam under four-point loading.

4.12.2(a) To Determine the Depth of Plastic Penetration

At mid-span and throughout the length AB, the bending moment =

$$M = 0.3\ \text{MN} * 0.9\ \text{m} = \underline{0.27\ \text{MN m}}$$

Owing to this bending moment, thre stress distribution will be as in Fig. 4.12.

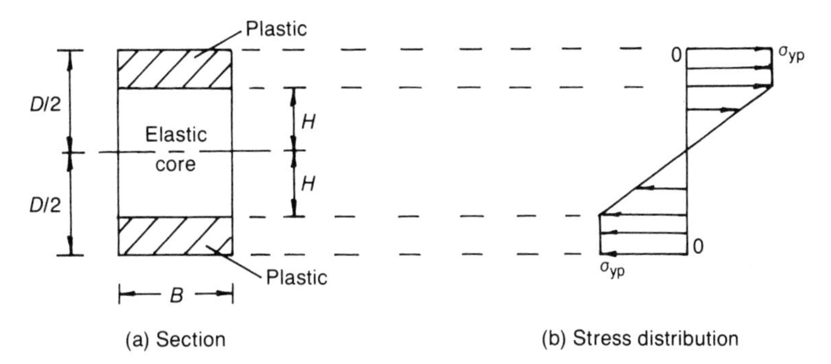

Fig. 4.12. Elastic–plastic deformation of beam section.

Let,

M_e = moment of resistance of the elastic portion of the beam's section

M'_p = moment of resistance of the plastic portion of the beam's section

I_e = 2nd moment of area of the elastic portion of the beam's section

$= B * (2H)^3/12 = 0.6667\, BH^3$

From elementary elastic theory [1],

$$\frac{\sigma}{y} = \frac{M}{I} = \frac{E}{R} \tag{4.15}$$

where,

σ = stress at any distance y from the neutral axis (NA)
M = bending moment
I = 2nd moment of area about NA
R = radius of curvature of NA

Applying equation (4.15) to the elastic portion of the beam's section,

$$M_e = \frac{\sigma_{yp} * I_e}{H}$$

$$= \frac{350 * 0.6667 * 0.1 * H^3}{H}$$

$$\underline{M_e = 23.33\, H^2} \tag{4.16}$$

From Fig. 4.12,

$$M'_p = 2 * \sigma_{yp} * B * (D/2 - H) * (D/2 + H)/2$$

$$= 35 * (0.1 - H) * (0.1 + H)$$

$$\underline{M'_p = 0.35 - 35\, H^2} \tag{4.17}$$

Now,

$$M = M_e + M'_p$$

or

$$0.27 = 23.33\,H^2 + 0.35 - 35\,H^2$$

$$11.667H^2 = 0.08$$

$$\underline{H = 0.0828\text{ m}} \qquad (4.18)$$

Therefore *depth of plastic penetration* at the top and the bottom of the beam

$$= 0.1 - 0.0828\text{ m}$$

$$= \underline{1.72\text{ cm}}$$

4.12.3(b) To Determine the Central Deflection (δ)

This can be calculated by substituting the value of M_e into equation (4.15), i.e.

$$\frac{M_e}{I_e} = \frac{E}{R}$$

Now,

$$I_e = 0.6667\,BH^3 = \underline{3.785\text{E-}5\text{ m}^4}$$

and

$$M_e = 23.33\,H^2 \quad = \underline{0.16\text{ MN m}}$$

therefore

$$R = \frac{EI_e}{M_e} \quad = \frac{2\text{E}11 \times 3.785\text{E-}5}{0.16\text{E}6}$$

$$\underline{R = 47.31\text{ m}}$$

The central deflection (δ) can be calculated from the properties of a circle, as shown in Fig. 4.13.

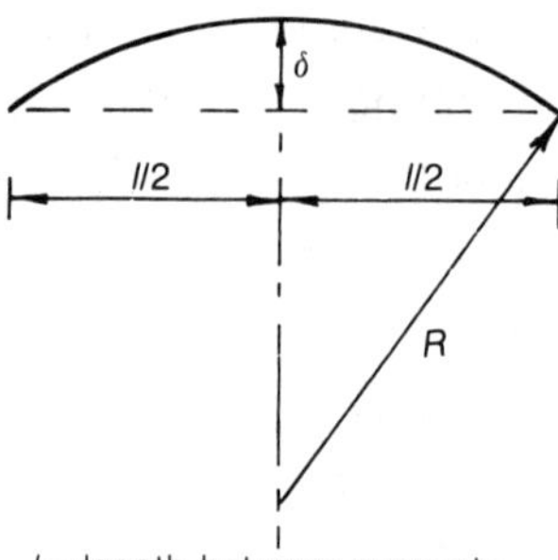

Fig. 4.13. Deflected form of beam.

$$\delta(2R - \delta) = (l/2)^2 \qquad (4.19)$$

$$\delta(94.62 - \delta) = 0.75^2$$

$$-\delta^2 + 94.62\,\delta - 0.5626 = 0$$

$$\delta = \frac{94.62 - \sqrt{[94.62^2 - 4 \times 0.5626]}}{2}$$

$$\underline{\delta = 5.945\text{E-3 m} = 5.945 \text{ mm}}$$

4.12.4(c) To Determine the Length of the Beam Over Which Yield Takes Place

Let,

$$M_{yp} = \text{moment required to just cause first yield}$$

$$= \frac{\sigma_{yp} * I}{(D/2)}$$

$$= \frac{\sigma_{yp} * BD^2}{6} = \frac{350 \times 0.1 \times 0.2^2}{6}$$

$$\underline{M_{yp} = 0.233 \text{ MN m}}$$

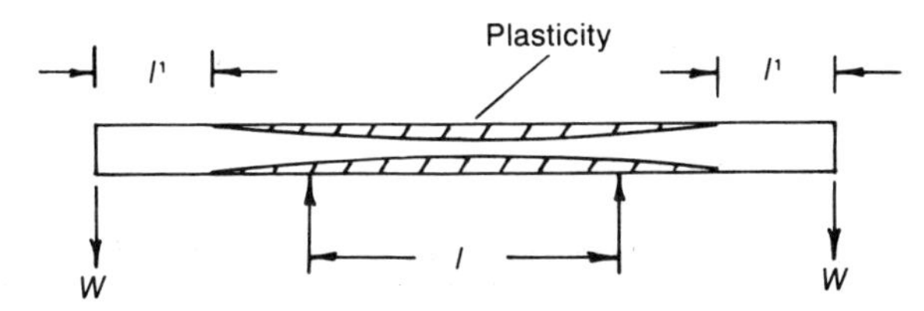

Fig. 4.14. Plastic region on beam.

From Fig. 4.14,

$$Wl' = M_{yp}$$

or

$$0.3\, l' = 0.2333$$

therefore

$$\underline{l' = 0.778 \text{ m}}$$

Therefore the length of the beam over which plasticity occurs

$$= 1.8 + 1.5 - 0.778 \times 2 = \underline{1.744 \text{ m}}$$

4.12.5(d) To Determine the Residual Central Deflection (δ_R)

On unloading the beam, it behaves elastically, as shown by Fig. 4.15, i.e.

$$R = \frac{EI}{M} = \frac{2\text{E}11 \times 0.1 \times 0.2^3}{12 \times 0.27\text{E}6}$$

$$\underline{R = 49.383 \text{ m}}$$

From equation (4.19),

$$\delta_1(98.765 - \delta_1) = 0.75^2$$

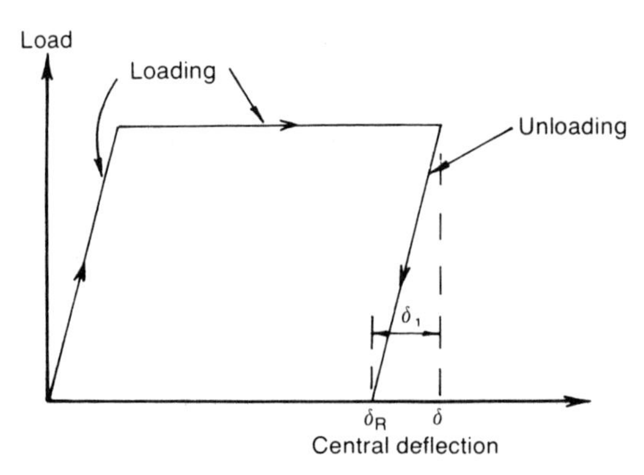

Fig. 4.15. Load–deflection relationship.

or

$$-\delta_1^2 + 98.765\,\delta_1 - 0.5625 = 0$$

therefore

$$\delta_1 = \frac{98.765 - \sqrt{[98.765^2 - 4 \times 0.5625]}}{2}$$

$$= 5.48\text{E-3 m}$$

Therefore residual deflection $= \delta_R = 5.95 - 5.48 = 0.47\,\text{mm}$

4.12.6(e) To Determine the Residual Stress Distribution

This can be obtained by superimposing the elastic–plastic stress distribution on loading, with the elastic stress distribution, on unloading, as shown in Figs. 4.16(a) to (c).

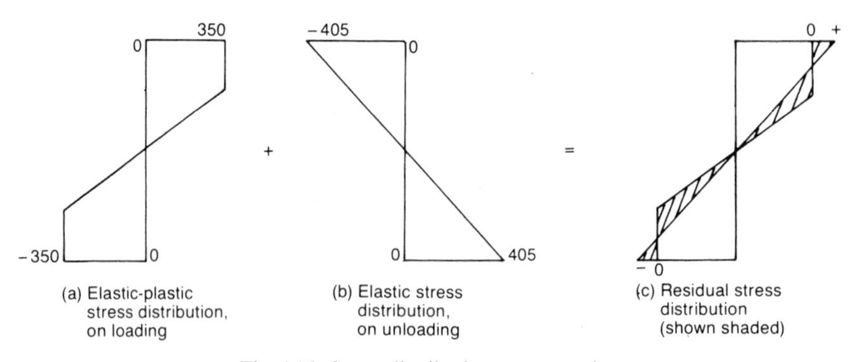

Fig. 4.16. Stress distribution across section.

4.13.1 EXAMPLE 4.6 PLASTIC DESIGN OF A PORTAL FRAME

Determine a suitable sectional modulus for the section shown in Fig. 4.17, assuming that the plastic moment of resistance of the top member is of twice the value of the plastic moment of resistance of the vertical columns.

The following are assumed to apply:

$$\lambda = 5, \quad S = 1.15, \quad \sigma_{yp} = 350\,\text{MN/m}^2$$

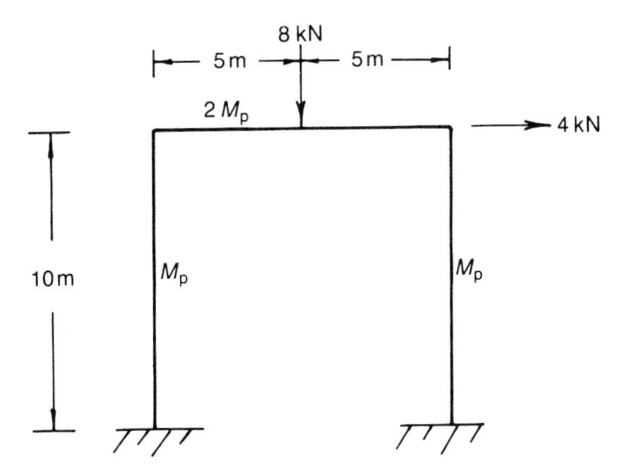

Fig. 4.17. Single-storey portal frame.

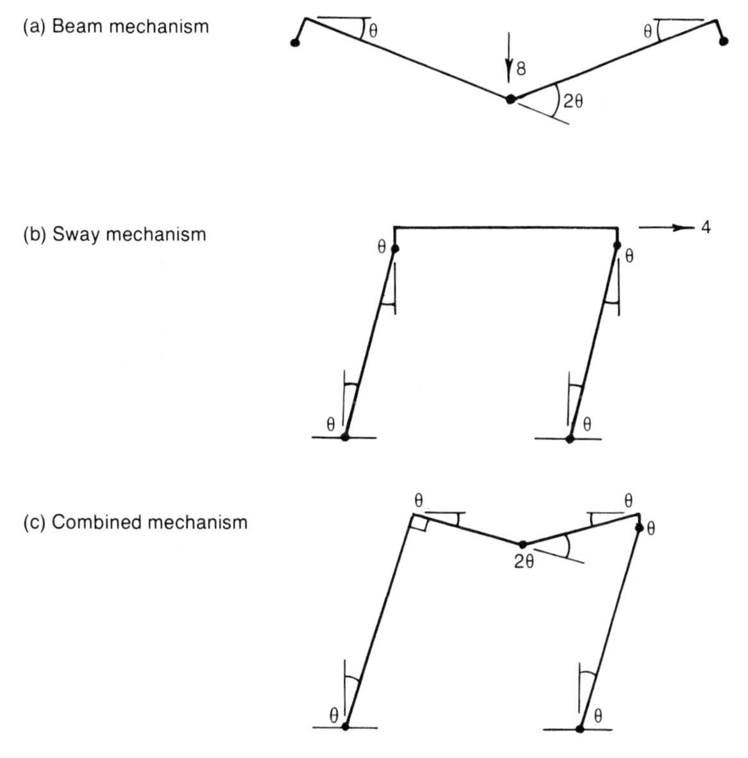

Fig. 4.18. Various mechanisms for the portal frame.

4.13.2

Experiments have shown that the frame of Fig. 4.17 can fail by three different mechanisms, namely

(a) beam mechanism;
(b) sway mechanism; and
(c) combined mechanism.

These three mechanisms are shown in Fig. 4.18, where it can be seen that the hogging hinges for the beam mechanism occur in the vertical columns, because these members are the weaker at these joints. Similar arguments apply to the sway and combined mechanisms.

4.13.3 Beam Mechanism

Three hinges are required for this mechanism, as the "beam" is statically indeterminate to the second degree.

Equating the work done by the 8 kN load to the work done by the plastic hinges, the following is obtained:

$$8 * 5\theta = M_p\theta + 2M_p * 2\theta + M_p\theta$$

therefore

$$\underline{M_p = \tfrac{40}{6} = 6.67\,\text{kN m}}$$

4.13.4 Sway Mechanism

This is a frame mechanism, and as the frame is statically indeterminate to the third degree, four hinges are required, as shown in Fig. 4.18(b).

Equating internal and external work done,

$$4 * 10\theta = M_p\theta * 4$$

therefore

$$\underline{M_p = 10\,\text{kN m}}$$

4.13.5 Combined Mechanism

This mechanism, which is also a frame mechanism, is a combination of the beam mechanism and the sway mechanism. Four plastic hinges are required, as the frame is statically indeterminate to the third degree.

Equating internal and external work done,

$$8 * 5\theta + 4 * 10\theta = M_p\theta + 2M_p * 2\theta + M_p\theta + M_p\theta + M_p\theta$$

therefore

$$\underline{M_p = 10\,\text{kN m}}$$

4.13.6

i.e.

$$\text{Design } M_p = 10\,\text{kN m} * \lambda$$
$$= \underline{50\,\text{kN m}}$$

Now,

$$M_{yp} = \frac{M_p}{S}$$
$$= \frac{50}{1.15} = \underline{43.48\,\text{kN m}}$$

and

$$Z = \text{sectional modulus}$$
$$= \frac{M_{yp}}{\sigma_{yp}} = \frac{43.48\text{E3 Nm}}{350\text{E6(N/m}^2)}$$
$$\underline{\underline{Z = 1.242\text{E-4}\,\text{m}^3\,\text{(verticals)}}}$$
$$\underline{Z = 2.484\text{E-4}\,\text{m}^3\,\text{(top beam)}}$$

4.14.1 EXAMPLE 4.7 PROPPED CANTILEVER

Determine a suitable section for the propped cantilever of Fig. 4.19, given that:

$$\lambda = 5, \quad S = 1.15, \quad \sigma_{yp} = 300\,\text{MN/m}^2$$

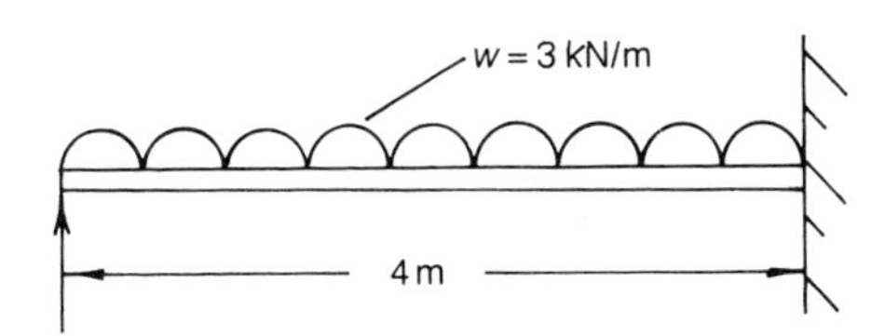

Fig. 4.19. Propped cantilever.

4.14.2

For this case, the beam is statically indeterminate to the first degree, so that it will be necessary for two plastic hinges to occur. One hinge will occur at the built-in end on the right, and the sagging hinge can be assumed to occur at a distance X from mid-span, as shown in Fig. 4.20.

Equating internal and external work done,

$$w * 4 * \frac{(2 + X)}{2}\theta = M_p * \alpha + M_p\beta \tag{4.20}$$

Now,

$$(2 + X)\theta = (2 - X)\alpha$$

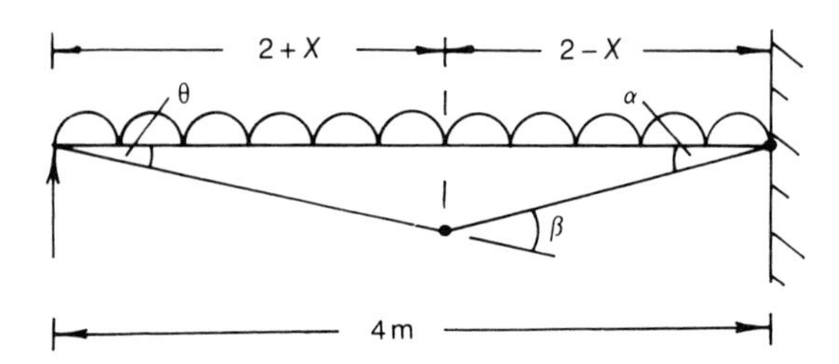

Fig. 4.20. Plastic hinges in propped cantilever.

therefore

$$\alpha = \left(\frac{2+X}{2-X}\right)\theta \tag{4.21}$$

and,

$$\beta = \alpha + \theta = \left(\frac{2-X}{2-X}\right)\theta + \left(\frac{2+X}{2-X}\right)\theta$$

$$= \frac{4}{(2-X)}\theta \tag{4.22}$$

Substituting (4.21) and (4.22) into (4.20),

$$2w(2+X)\theta = M_p\left(\frac{2+X}{2-X}\right)\theta + M_p * \frac{4\theta}{(2-X)}$$

therefore

$$2w(2+X) = M_p\frac{(2+X+4)}{(2-X)}$$

$$= \left(\frac{6+X}{2-X}\right)M_p$$

$$M_p = \frac{2*3*(2+X)*(2-X)}{(6+X)}$$

$$M_p = \frac{6*(4-X^2)}{(6+X)} \tag{4.23}$$

The maximum value of M_p occurs, when the condition,

$$\frac{dM_p}{dX} = 0$$

or

$$0 = \frac{1*6*(4-X^2) - 6*(6+X)(-2X)}{(6+X)^2}$$

or

$$4 - X^2 + 12X + 2X^2 = 0$$

$$X^2 + 12X - 4 = 0$$

$$X = -\frac{12 \pm \sqrt{[12^2+16]}}{2}$$

$$= -\frac{12 + 12.649}{2}$$

$$\underline{X = 0.325\,\text{m}} \qquad (4.24)$$

Substituting (4.24) into (4.23),

$$\underline{M_p = 3.694\,\text{kN m}}$$

4.14.3

$$\text{Design } M_p = 3.694\,\lambda$$
$$= 18.47\,\text{kN m}$$
$$M_{yp} = \frac{M_p}{S} = 16.06\,\text{kN m}$$

and,

$$Z = \frac{M_{yp}}{\sigma_{yp}} = \frac{16.06\text{E}3}{300\text{E}6}$$

$$\underline{Z = 5.353\text{E-}5\,\text{m}^3}$$

4.15.1 EXAMPLE 4.8 HYDROSTATICALLY LOADED BEAM

Determine a suitable sectional modulus for the ship's bulkhead stiffener of Fig. 4.21, assuming that it is hydrostatically loaded.

$$\lambda = 5, \quad S = 1.15, \quad \sigma_{yp} = 300\,\text{MN/m}^2$$

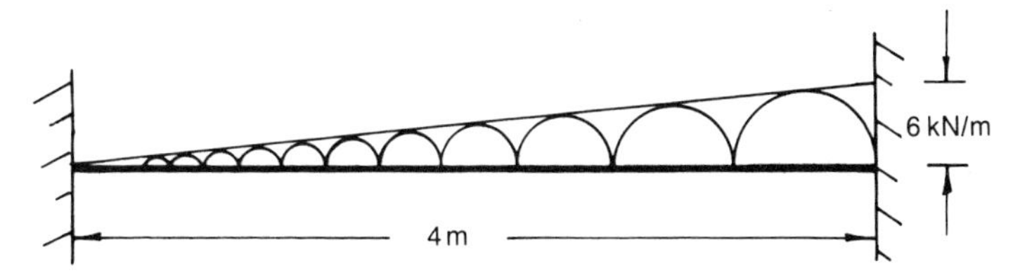

Fig. 4.21. Hydrostatically loaded beam.

4.15.2

As this beam is statically indeterminate to the second degree, three hinges are required to make it into a mechanism. The hogging hinges occur at the ends, and the sagging hinge may be assumed to occur at a distance X from mid-span, as shown in Fig. 4.22.

At a distance of $(2 + X)$ from the left end, the intensity of load =

$$w = \frac{(2 + X) * w_1}{4} = \frac{(2 + X) * 6}{4}$$

$$\underline{w = 1.5(2 + X)}$$

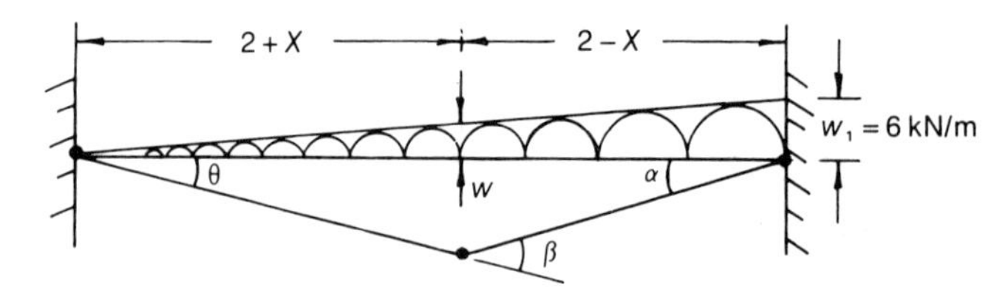

Fig. 4.22. Mechanism for hydrostatically loaded beam.

To determine α and β

$$(2+X)\theta = (2-X)\alpha$$

$$\alpha = \left(\frac{2+X}{2-X}\right)\theta \quad \text{and} \quad \beta = \left(\frac{4}{2-X}\right)\theta$$

The work done by the hinges $= M_p(\theta + \alpha + \beta)$

$$= M_p\theta\left[1 + \left(\frac{2+X}{2-X}\right) + \left(\frac{4}{2-X}\right)\right]$$

$$= M_p\theta\frac{2-X+2+X+4}{2-X} = \left(\frac{8M_p}{2-X}\right)\theta \tag{4.25}$$

The work done by the hydrostatic loads can be obtained by considering the separate effect of three components of distributed load, one from the "left" and two from the "right" of the beam, where in the case of the latter, the trapezoidal load can be divided into two triangular loads, as follows:

$$\begin{aligned}\text{WD} &= \tfrac{1}{2}*w*(2+X)*\tfrac{2}{3}(2+X)\theta \\ &\quad + \tfrac{1}{2}*w*(2-X)*\tfrac{2}{3}(2-X)\alpha \\ &\quad + \tfrac{1}{2}*w_1*(2-X)*\tfrac{1}{3}(2-X)\alpha \\ &= \tfrac{1}{6}\Bigg[1.5(2+X)*(2+X)*2*(2+X) \\ &\quad + 1.5(2+X)*(2-X)*2*(2-X)*\frac{(2+X)}{(2-X)} \\ &\quad + 6*(2-X)*(2-X)*\frac{(2+X)}{(2-X)}\Bigg]\theta\end{aligned} \tag{4.26}$$

Equating (4.25) and (4.26),

$$\frac{8M_p}{(2-X)} = \tfrac{1}{6}[3*(2+X)^3 + 3*(2+X)^2*(2-X) + 6*(2-X)*(2+X)]$$

$$M_p = \tfrac{1}{48}[3*(2+X)^3*(2-X) + 3*(2+X)^2*(2-X)^2 + 6*(2-X)^2*(2+X)] \tag{4.27}$$

For maximum M_p,

$$\frac{dM_p}{dX} = 0$$

therefore

$$0 = \tfrac{1}{48}[3*3*(2+X)^2*(2-X)+3*(2+X)^3*(-1)$$
$$+3*2*(2+X)*(2-X)^2+3*2*(2+X)^2*(2-X)*(-1)$$
$$+6*2*(2-X)*(2+X)*(-1)+6*(2-X)^2]$$

$$0 = 3*(4+X^2+4X)*(2-X)-(2+X)^3$$
$$+2*(2+X)*(4+X^2-4X)$$
$$-2*(4+X^2+4X)*(2-X)$$
$$-4*(4-X^2)+2*(4+X^2-4X)$$

$$0 = 3*(8+2X^2+8X-4X-X^3-4X^2)$$
$$-(8+12X+6X^2+X^3)$$
$$+2*(8+2X^2-8X+4X+X^3-4X^2)$$
$$-2*(8+2X^2+8X-4X-X^3-4X^2)$$
$$-16+4X^2+8+2X^2-8X$$

$$0 = 24-6X^2+12X-3X^3-8-6X^2-12X-X^3$$
$$+16-4X^2-8X+2X^3-16+4X^2-8X$$
$$+2X^3-8+6X^2-8X$$
$$0 = 8-24X-6X^2$$

or

$$3X^2+12X-4=0$$

$$X = \frac{-12+\sqrt{[144+48]}}{6}$$

$$= \frac{-12+13.856}{6}$$

$$\underline{X = 0.31\ \mathrm{m}} \tag{4.28}$$

Substituting (4.28) into (4.27),

$$M_p = \tfrac{1}{48}[3*2.31^3*1.69+3*2.31^2*1.69^2+6*1.69^2*2.31]$$
$$\underline{M_p = 3.08\ \mathrm{kN\,m}}$$

Design $M_p = 15.4\ \mathrm{kN\,m}$

$$\underline{M_{yp} = 13.39\ \mathrm{kN\,m}}$$

$$\underline{Z = \text{sectional modulus} = 4.46\text{E-}5\ \mathrm{m}^3}$$

4.16.1 EXAMPLE 4.9 PORTAL FRAME WITH A DISTRIBUTED LOAD

Design a suitable section for the portal frame of Fig. 4.23, given that,

$$\lambda = 4, \quad S = 1.14, \quad \sigma_{yp} = 300\ \mathrm{MPa}$$

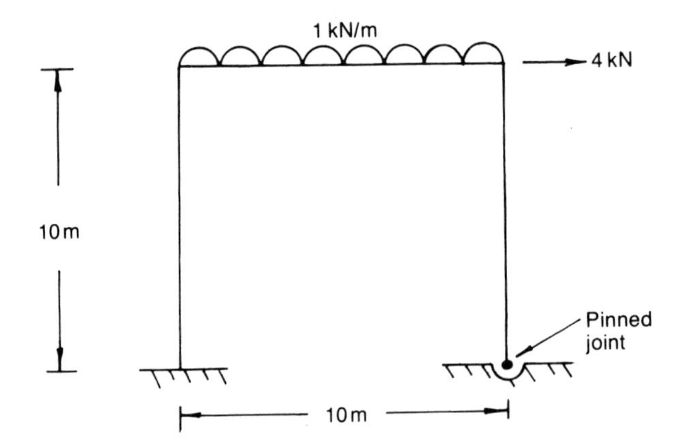

Fig. 4.23. Portal frame with a distributed load.

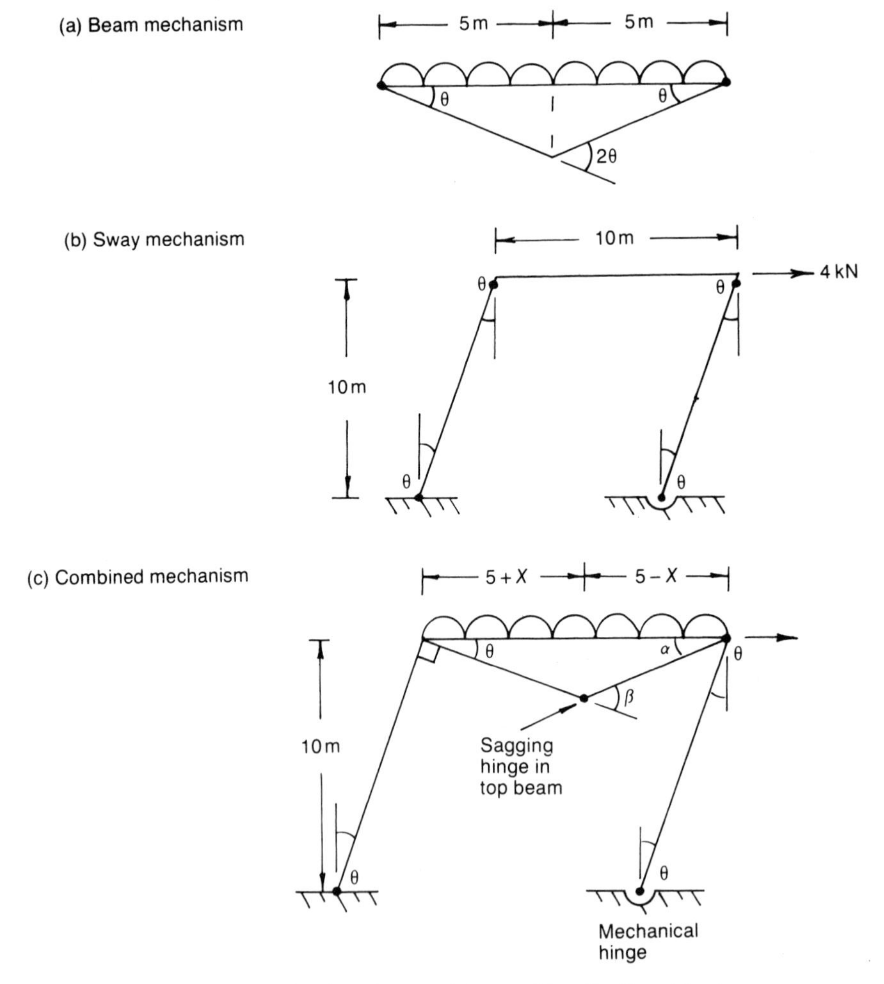

Fig. 4.24. Mechanisms for portal frame with a distributed load.

4.16.2

Experiments have shown that the frame of Fig. 4.23 can fail by three different mechanisms, namely

(a) beam mechanism;
(b) sway mechanism; and
(c) combined beam and sway mechanism.

These three mechanisms are shown in Fig. 4.24, where it can be seen that in the case of the combined mechanism, it is assumed that the sagging hinge in the top beam occurs at a distance X to the right of mid-span. The reason for this is that the value of the bending moment on the right of the top beam is M_p, and as a hinge does not occur on the left of the top beam, the bending moment at this point is less than M_p.

4.16.3 Beam Mechanism

Three hinges are required for this mechanism, as the beam is statically indeterminate to the second degree.

Equating internal and external work done, and with the aid of Fig. 4.24(a),

$$M_P(\theta + 2\theta + \theta) = 1\,\text{kN/m} * 10\,\text{m} * 2.5\theta\,\text{m}$$

therefore

$$\underline{M_p = 6.25\,\text{kN m}}$$

4.16.4 Sway Mechanism

This is a frame mechanism, and as the frame is statically indeterminate to the second degree, three plastic hinges are required to reduce the frame into a mechanism.

Equating internal and external work done in Fig. 4.24(b),

$$M_p(0 + \theta + \theta) = 4\,\text{kN} * 10\theta\,\text{m}$$

$$\underline{M_p = 13.33\,\text{kN m}}$$

4.16.5 Combined Mechanism

This is a frame mechanism, and like the sway mechanism, only three plastic hinges are required to change the frame into a mechanism.

Equating internal and external work done in Fig. 4.24(c),

$$M_p\theta + M_p\beta + M_p\alpha + M_p\theta = 1 * 10 * \frac{(5 + X)}{2}\theta + 4 * 10\theta$$

but from Section 4.14.2,

$$\alpha = \left(\frac{5 + X}{5 - X}\right)\theta$$

and

$$\beta = \alpha + \theta$$

$$= \left(\frac{10}{5 - X}\right)\theta$$

therefore

$$M_p\left[1 + \left(\frac{10}{5 - X}\right) + \left(\frac{5 + X}{5 - X}\right) + 1\right] = 5(5 + X) + 40$$

or

$$M_p\left(\frac{5 - X + 10 + 5 + X + 5 - X}{5 - X}\right) = 25 + 5X + 40$$

$$M_p(25 - X) = (65 + 5X)(5 - X)$$

$$M_p = \frac{(325 + 25X - 65X - 5X^2)}{25 - X}$$

$$M_p = \frac{(325 - 40X - 5X^2)}{(25 - X)} \qquad (4.29)$$

To determine maximum M_p,

$$\frac{dM_p}{dX} = 0$$

or

$$0 = \frac{(25 - X)(-40 - 10X) - (325 - 40X - 5X^2)(-1)}{(25 - X)^2}$$

or

$$0 = -(1000 - 40X + 250X - 10X^2) + 325 - 40X - 5X^2$$

$$0 = -675 - 250X + 5X^2$$

or

$$X^2 - 50X - 135 = 0$$

$$X = \frac{50 - \sqrt{50^2 + 4 \times 135}}{2}$$

$$\underline{X = -2.568\ \text{m}} \qquad (4.30)$$

Substituting (4.30) into (4.29),

$$M_p = \frac{325 + 102.72 - 32.97}{27.568}$$

$$\underline{M_p = 14.32\ \text{kN m}}$$

$$\text{Design } M_p = 14.32 * 4 = 57.28\ \text{kN m}$$

and,

$$M_{yp} = 50.25\ \text{kN m}$$

Hence,

$$\underline{Z = \frac{50.25\text{E}3}{300\text{E}6} = 1.675\text{E} - 4\ \text{m}^3}$$

4.17.1 EXAMPLE 4.10 TWO-STOREY PORTAL FRAME

Design a suitable section for the two-storey portal frame of Fig. 4.25, given that,

$$\lambda = 4.5, \quad S = 1.14, \quad \sigma_{yp} = 300\,\text{MPa}$$

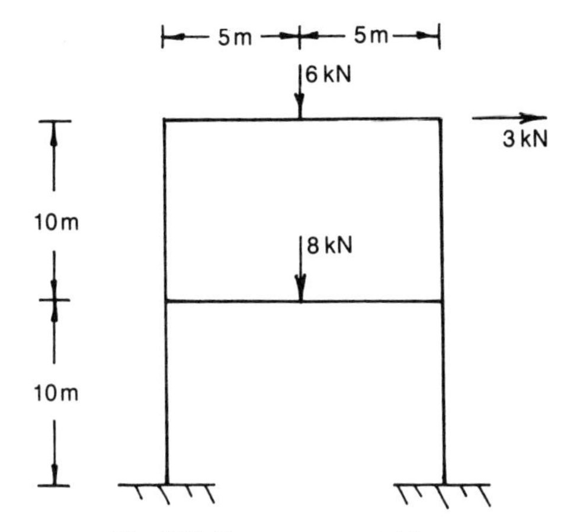

Fig. 4.25. Two-storey portal frame.

4.17.2

The possible mechanisms for this structure are shown in Fig. 4.26.

4.17.3 Top Beam Mechanism

$$M_p(\theta + 2\theta + \theta) = 6 * 5\theta$$

therefore

$$\underline{M_p = 7.5\,\text{kN m}}$$

4.17.4 Bottom Beam Mechanism

$$M_p(\theta + 2\theta + \theta) = 8 * 5\theta$$

$$\underline{M_p = 10\,\text{kN m}}$$

4.17.5 Top Sway Mechanism

$$M_p(\theta + \theta + \theta + \theta) = 3 * 10\theta$$

$$\underline{M_p = 7.5\,\text{kN m}}$$

4.17.6 Bottom Sway Mechanism

$$M_p\theta * 4 = 3 * 10\theta$$

$$\underline{M_p = 7.5\,\text{kN m}}$$

(a) Top beam

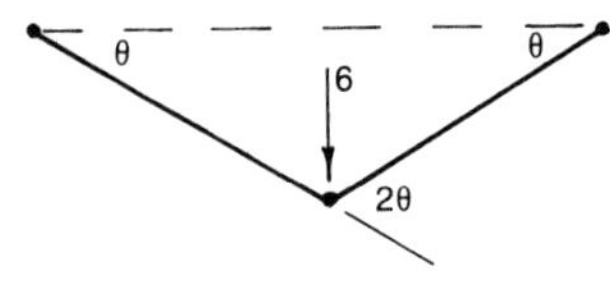

(b) Bottom beam

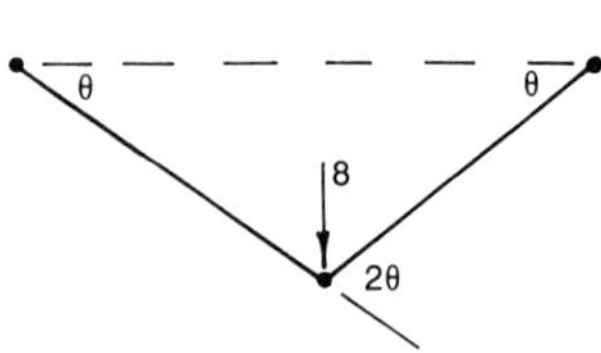

(c) Sway mechanisms (3 types)

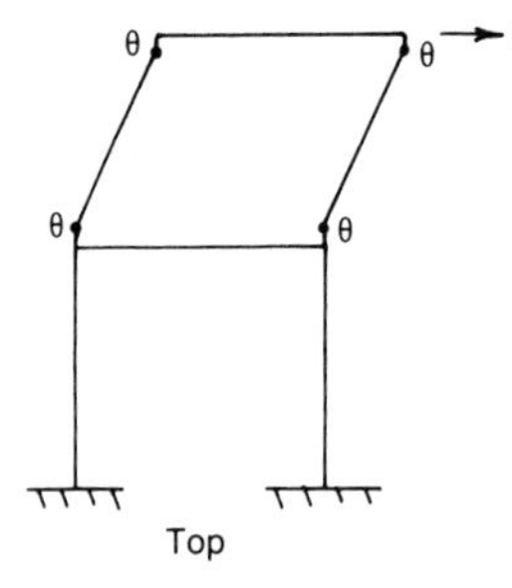

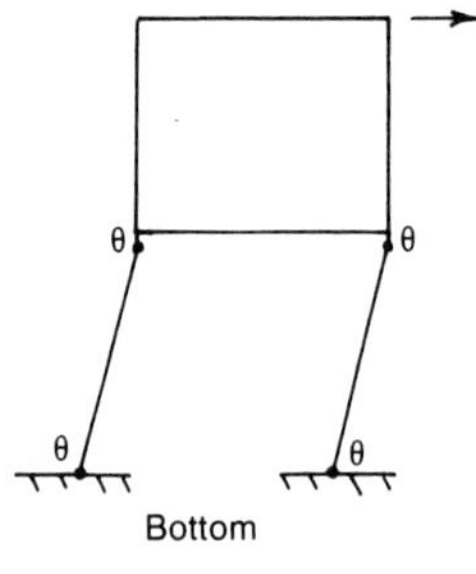

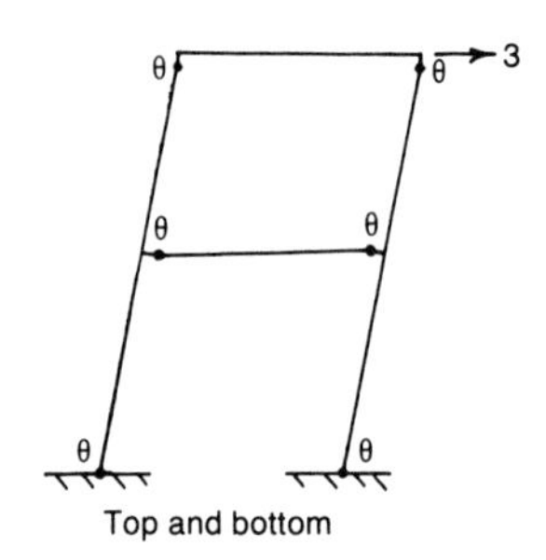

(d) Combined mechanisms (3 types)

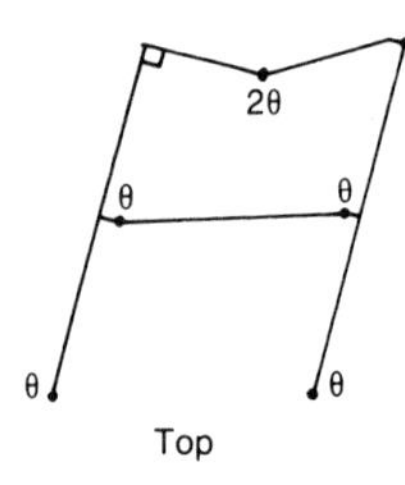

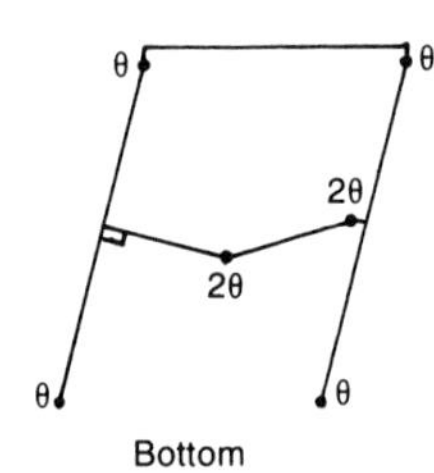

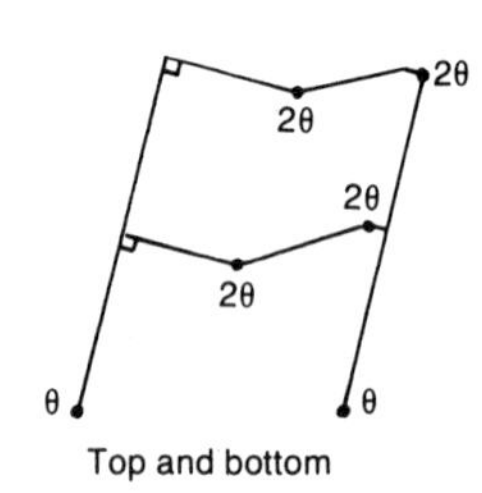

Fig. 4.26. Mechanisms for two-storey framework.

4.17.7 Top and Bottom Sway Mechanism

$$M_p(\theta + \theta + \theta + \theta + \theta + \theta) = 3 * 20\theta$$

$$\underline{M_p = 10\,\text{kN m}}$$

4.17.8 Combined Mechanism (Top)

$$M_p(\theta + \theta + 2\theta + 2\theta + \theta + \theta) = 6 * 5\theta + 3 * 20\theta$$

$$8M_p = 90$$

$$\underline{M_p = 11.25\,\text{kN m}}$$

4.17.9 Combined Mechanism (Bottom)

$$M_p(\theta + \theta + 2\theta + 2\theta + \theta + \theta) = 8 * 5\theta + 3 * 20\theta$$

$$8M_p = 100$$

$$M_p = 12.5\,\text{kN m}$$

4.17.10 Combined Mechanism (Top and Bottom)

$$M_p(\theta + 2\theta + 2\theta + 2\theta + 2\theta + \theta) = 30\theta + 40\theta + 60\theta$$

threfore

$$10M_p = 130$$

$$M_p = 13\,\text{kN m}$$

4.17.11

$$\text{Design } M_p = 13 \times 4.5 = 58.5\,\text{kN m}$$

$$M_{yp} = \frac{58.5}{1.14} = 51.32\,\text{kN m}$$

$$Z = \frac{51.32\text{E}3}{300\text{E}6} = 1.71\text{E-}4\,\text{m}^3$$

4.18.1 EXAMPLE 4.11 TWO-BAY PORTAL FRAMEWORK

Design a suitable section for the two-bay portal frame of Fig. 4.27, given that,

$$\lambda = 4.5, \quad S = 1.14, \quad \sigma_{yp} = 300\,\text{MPa}$$

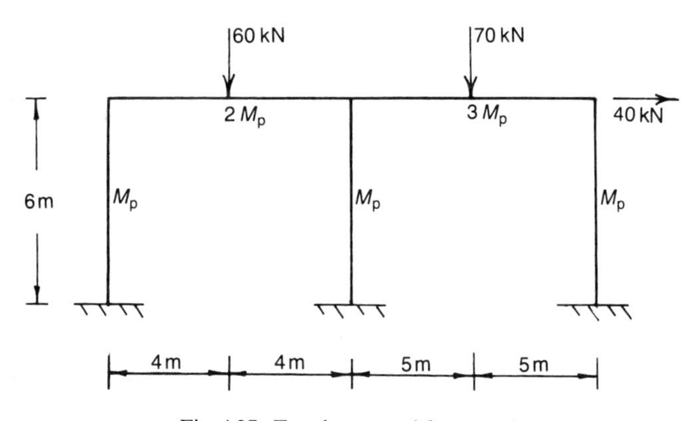

Fig. 4.27. Two-bay portal framework.

4.18.2

The mechanisms for this framework are shown in Fig. 4.28.

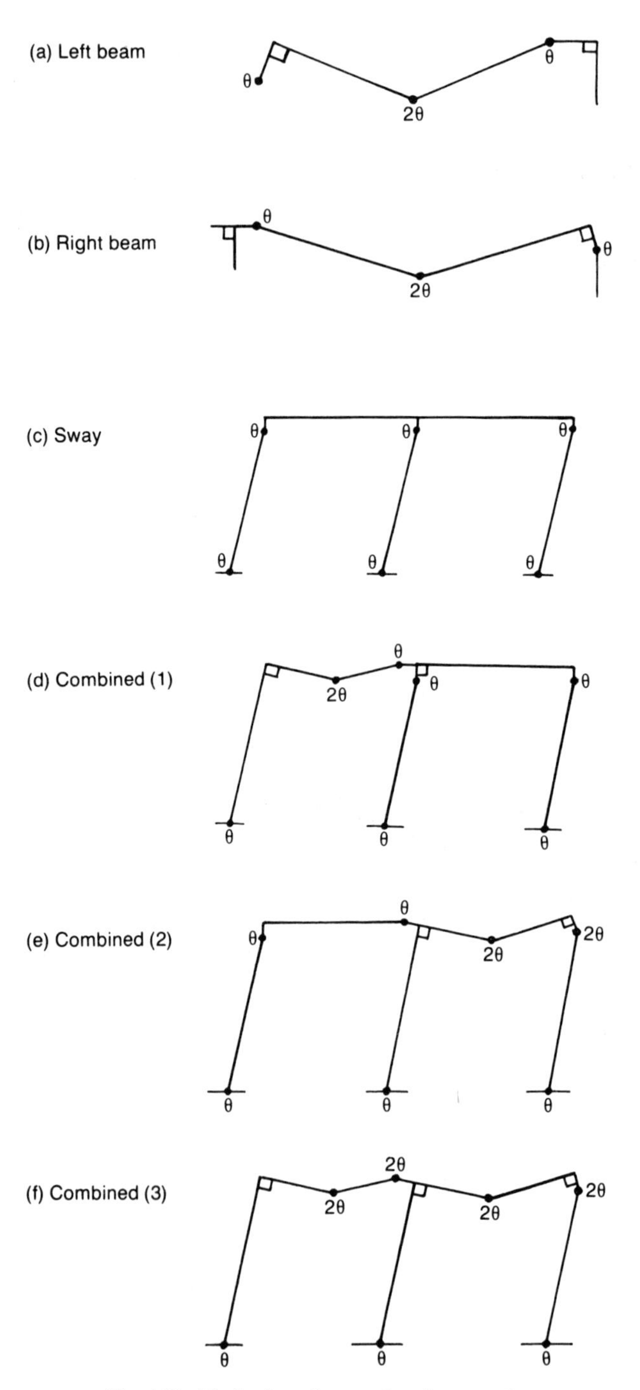

Fig. 4.28. Mechanisms for two-bay framework.

4.18.3 Left Beam

$$M_p(\theta + 4\theta + 2\theta) = 60 * 4\theta$$
$$\underline{M_p = 34.29\,\text{kN m}}$$

4.18.4 Right Beam

$$M_p(3\theta + 6\theta + \theta) = 70 * 5\theta$$
$$\underline{M_p = 35\,\text{kN m}}$$

4.18.5 Sway

$$M_p(\theta + \theta + \theta + \theta + \theta + \theta) = 40 * 6\theta$$
$$\underline{M_p = 40\,\text{kN m}}$$

4.18.6 Combined (1)

$$M_p(\theta + 4\theta + 2\theta + \theta + \theta + \theta + \theta) = 60 * 4\theta + 40 * 6\theta$$
$$11M_p = 480$$
$$\underline{M_p = 43.64\,\text{kN m}}$$

4.18.7 Combined (2)

$$M_p(\theta + \theta + 2\theta + \theta + 6\theta + 2\theta + \theta) = 70 * 5\theta + 40 * 6\theta$$
$$14M_p = 590$$
$$M_p = 42.14\,\text{kN m}$$

4.18.8 Combined (3)

$$M_p(\theta + 4\theta + 4\theta + \theta + 6\theta + 2\theta + \theta) = 60 * 4\theta + 70 * 5\theta + 40 * 6\theta$$
$$19M_p = 830$$
$$\underline{M_p = 43.68\,\text{kN m}}$$

4.18.9

$$\text{Design } M_p = 43.7 \times 4.5 = 196.7\,\text{kN m}$$
$$M_{yp} = \frac{196.7}{1.14} = 172.5\,\text{kN m}$$
$$\underline{Z = \frac{172.5\text{E3}}{300\text{E6}} = 5.75\,\text{E-4}\,\text{m}^3}$$

4.19.1 EXAMPLE 4.12 ELASTIC–PLASTIC TORSION OF CIRCULAR SECTION SHAFTS

A solid steel shaft of diameter 0.1 m and of length 0.7 m is subjected to a torque of 0.05 MN m, causing the shaft to suffer elastic–plastic deformation.

Determine:

(a) the depth of plastic penetration;
(b) the angle of twist on application of the torque;
(c) the residual angle of twist on release of the torque;
(d) the full plastic torsional resistance of a similar shaft (T_p); and
(e) the ratio of T_p to torque at first yield (T_{yp}).

The shaft may be assumed to have the following material properties:

$$\tau_{yp} = \text{shear yield stress} = 200\ \text{MN/m}^2$$
$$G = \text{rigidity modulus} = 7.7\text{E}10\ \text{N/m}^2$$

4.19.2

As the shaft section is partially elastic and partially plastic, the shear stress distribution will be as shown in Fig. 4.29, where it can be seen that there is an elastic core.

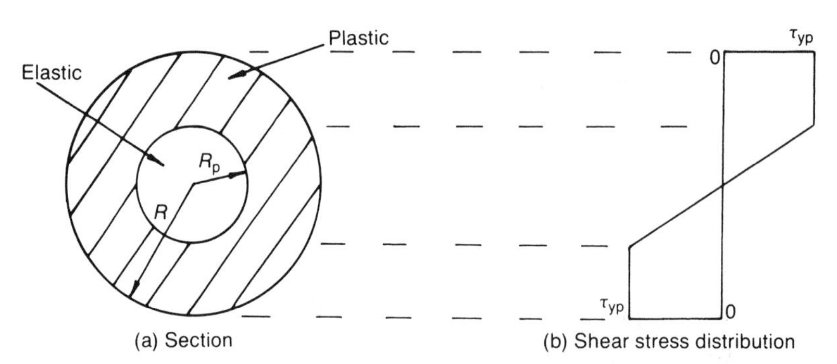

Fig. 4.29. Circular section shaft.

Let,

R = radius of shaft
R_p = outer radius of elastic core
T = applied torque = $T_e + T_{p'}$
T_e = torsional resistance of elastic core
$T_{p'}$ = torsional resistance of plastic portion of section (shaded in Fig. 4.29(a))
J_e = polar 2nd moment of area of elastic core
τ_{yp} = shear stress at yield

4.19.3(a) To Calculate Angle of Twist θ

First, it will be necessary to calculate T_e and $T_{p'}$.

From reference [1], the torsional equation for circular section shafts can be applied to the elastic core, i.e.

$$\frac{\tau_{yp}}{R_p} = \frac{T_e}{J_e} = \frac{G\theta}{l} \tag{4.31}$$

where,

G = rigidity or shear modulus
θ = angle of twist over the length l of the shaft
$J_e = \pi R_p^4/2$

therefore

$$T_e = \frac{\tau_{yp}}{R_p} * \frac{\pi R_p^4}{2} = \frac{200\,\pi R_p^3}{2}$$

or,

$$\underline{T_e = 314.16\,R_p^3} \tag{4.32}$$

To calculate $T_{p'}$, consider an annular element of radius r and the thickness "dr" in the plastic zone, where the shear stress is of constant value τ_{yp}, as shown in Fig. 4.30.

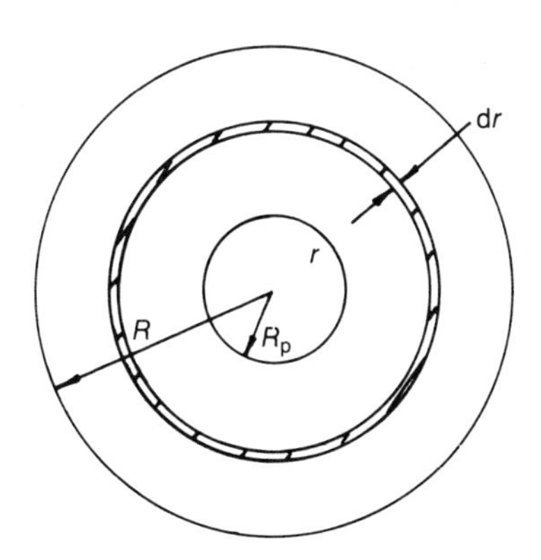

Fig. 4.30. Plastic section of shaft.

From Fig. 4.29,

$$T_{p'} = \int_{R_p}^{R} \tau_{yp} * (2\pi r * \mathrm{d}r) * r$$

$$= 2\pi\tau_{yp}\left[\frac{r^3}{3}\right]_{R_p}^{R}$$

$$= \frac{2\pi}{3}\tau_{yp}(R^3 - R_p^3) \tag{4.33}$$

$$\underline{T_{p'} = 418.9\,(1.25\text{E-}4 - R_p^3)}$$

Now,

$$T = T_e + T_{p'}$$

or,

$$0.05\,\text{MN m} = 314.16\,R_p^3 + 418.9\,(1.25\text{E-}4 - R_p^3)$$
$$= 0.0524 - 104.7\,R_p^3$$
$$\underline{R_p = 0.028\,\text{m}} \tag{4.34}$$

Therefore *depth of plastic penetration* $= 0.05 - 0.028$

$$= \underline{0.022\,\text{m}}$$

4.19.4(b) To Calculate θ_1, the Angle of Twist due to T

From (4.31),

$$\theta_1 = \frac{\tau_{yp} * l}{R_p * G} = \frac{200\text{E}6 \times 0.7}{0.028 \times 7.7\text{E}10} = 0.064 \text{ rads}$$
$$\underline{\theta_1 = 3.67^\circ}$$

4.19.5 To Calculate the Residual Angle of Twist (θ_R)

The T–θ relationship on loading and on unloading is shown in Fig. 4.31, where on removal of the applied torque, the shaft behaves elastically.

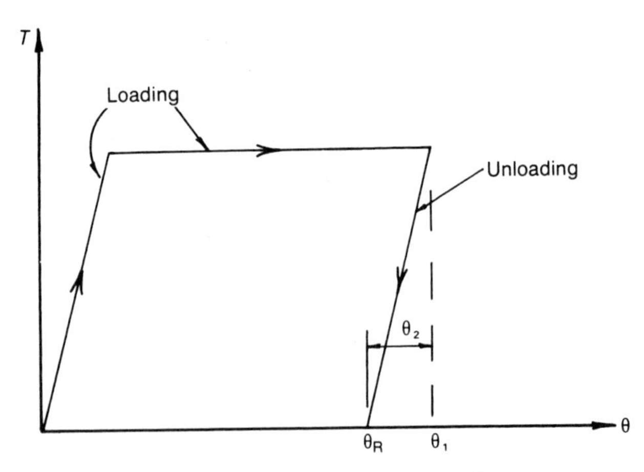

Fig. 4.31. T–θ relationship for shaft.

From Fig. 4.31, it can be seen that θ_2 can be calculated by applying the full torque T to the whole section, and assuming elastic behaviour on unloading, i.e.

$$\theta_2 = \frac{Tl}{GJ} = \frac{0.5\text{E}5 * 0.7}{7.7\text{E}10 * J}$$

but,

$$J = \pi \times R^4/2 = 9.817\text{E-6}\,\text{m}^4$$

therefore

$$\underline{\theta_2 = 0.046\,\text{rads} = 2.65^\circ}$$

Therefore residual angle of twist =

$$\underline{\theta_R = \theta_1 - \theta_2 = 1.02^\circ}$$

4.19.6 To Determine Fully Plastic Torsional Resistance (T_p)

From equation (4.33),

$$T_p = \frac{2\pi}{3} * \tau_{yp} * R^3$$

$$= \frac{2\pi \times 200 \times 0.05^3}{3}$$

$$\underline{T_p = 0.0524\,\text{MN m}}$$

4.19.7 To Determine the Ratio T_p/T_{yp}

$$T_{yp} = \text{maximum torsional resistance up to first yield}$$

$$= \frac{\tau_{yp} * J}{R} = \frac{200 * \pi * 0.1^4}{0.05 * 32}$$

$$\underline{T = 0.0393\,\text{MN m}}$$

therefore

$$\underline{\frac{T_p}{T_{yp}} = \frac{0.0524}{0.0393} = 1.334}$$

4.20.1 EXAMPLE 4.13 COMPOUND SHAFT UNDER ELASTIC–PLASTIC DEFORMATION

A compound shaft of length 0.7 m consists of an aluminium alloy core, of diameter 0.07 m, surrounded co-axially by a steel tube of external diameter 0.1 m. Determine:

(a) the torque that can be applied without causing yield;
(b) the resulting angle of twist on applying a torque of 36 000 N m;
(c) the residual angle of twist remaining on the removal of this torque;
(d) the fully plastic torsional resistance of this shaft; and
(e) the ratio of the torque obtained from (d) to that obtained from (c).

The following may be assumed to apply:

Steel

$$\tau_{yps} = \text{shear yield stress in steel} = 200\,\text{MN/m}^2$$
$$G_s = \text{rigidity modulus} = 7.7\text{E}10\,\text{N/m}^2$$

Aluminium alloy

$$\tau_{ypa} = \text{shear yield stress in aluminium alloy} = 100\,\text{MN/m}^2$$
$$G_a = \text{rigidity modulus} = 2.6\text{E}10\,\text{N/m}^2$$

The assumption can be made that radial lines remain straight on application or removal of the torque, i.e.

$$\theta = \text{angle of twist for the steel and the aluminium alloy} = \text{constant}$$

4.20.2(a) To Determine T_{yp}

Prior to determining T_{yp}, it will be necessary to determine whether the aluminium alloy or the steel will yield first.

Considering the angle of twist in the steel

$$\theta = \frac{\tau_{yps} * l}{R * G} = \frac{200\text{E}6 * 0.7}{0.05 * 7.7\text{E}10}$$

$$\underline{\theta = 0.0364\,\text{rads}} \tag{4.35}$$

Considering the angle of twist in the aluminium alloy

$$\theta = \frac{\tau_{ypa} * l}{G * r} = \frac{100\text{E}6 * 0.7}{2.6\text{E}10 * 0.035}$$

$$\underline{\theta = 0.0769\,\text{rads}} \tag{4.36}$$

i.e. if the aluminium alloy were allowed to reach yield, then the steel would become plastic; therefore, the yield stress in the steel is the design criterion.

$$\underline{\text{“}\textit{Design}\text{” } \theta = 0.0364\,\text{rads}} \tag{4.37}$$

Let,

J_s = polar 2nd moment of area of the steel tube
J_a = polar 2nd moment of the aluminium alloy shaft
T_s = elastic torque in steel, due to the application of T_{yp}
T_a = elastic torque in the aluminium alloy due to the application of T_{yp}

$$J_s = \frac{\pi \times (0.1^4 - 0.07^4)}{32} = \underline{7.46\text{E-}6\,\text{m}^4}$$

$$J_a = \frac{\pi \times 0.07^4}{32} = \underline{2.357\text{E-}6\,\text{m}^4}$$

From (4.31),

$$T_a = \frac{G_a\theta * J_a}{l} = \frac{2.6E10 * 0.0364 * 2.357E\text{-}6}{0.7}$$

$$\underline{T_a = 3187\ \text{N m}}$$

Similarly,

$$T_s = \frac{G_s\theta * J_s}{l} = \frac{7.7E10 * 0.0364 * 7.46E\text{-}6}{0.7}$$

$$\underline{T_s = 29\,871\ \text{N m}}$$

Now,

$$T_{yp} = T_a + T_s = 3187 + 29\,871$$

$$\underline{T_{yp} = 33\,058\ \text{N m}}$$

4.20.3(b) To Determine the Angle of Twist θ, Due to a Torque of 36 000 N m

On application of the torque, both the steel and the aluminium alloy may be assumed to go plastic, as shown in Fig. 4.32.

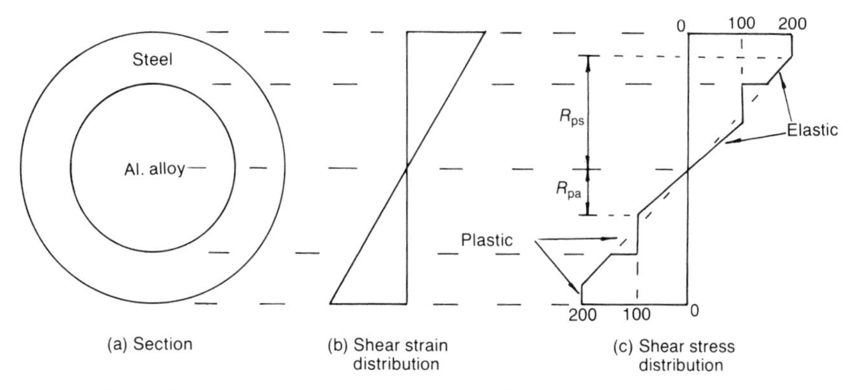

Fig. 4.32. Shear stress and shear strain distributions in compound shaft.

From equation (4.33),

$T_{p's}$ = torque contribution from the part of the steel that becomes plastic

$$= \frac{2\pi}{3} * \tau_{yps} * (0.05^3 - R_{ps}^3)$$

$$\underline{T_{p's} = 418.9 * (1.25E\text{-}4 - R_{ps}^3)} \quad (4.38)$$

Also, from equation (4.33),

$$T_{p'a} = \text{torque contribution from that part of the aluminium alloy that becomes plastic}$$

$$= \frac{2\pi}{3} * \tau_{ypa} * (0.035^3 - R_{pa}^3)$$

$$\underline{T_{p'a} = 209.4 * (4.288\text{E-5} - R_{pa}^3)} \tag{4.39}$$

The elastic components of torque in the steel (T_{es}) and in the aluminium alloy (T_{ea}) can be calculated from elementary elastic theory, as follows:

$$T_{es} = \frac{\tau_{yps}}{R_{ps}} * \frac{\pi * (R_{ps}^4 - 0.035^4)}{2} \tag{4.40}$$

and,

$$T_{ea} = \frac{\tau_{ypa}}{R_{pa}} * \frac{\pi * R_{pa}^4}{2}$$

giving,

$$\underline{T_{es} = 314.2 * (R_{ps}^4 - 1.5\text{E-6})/R_{ps}} \tag{4.41}$$

and,

$$\underline{T_{ea} = 157.1\, R_{pa}^3} \tag{4.42}$$

Now, the total torque =

$$T = T_{p's} + T_{p'a} + T_{es} + T_{ea}$$

therefore

$$\frac{36\,000}{1\text{E6}} = 418.9 * (1.25\text{E-4} - R_{ps}^3) + 209.4 * (4.288\text{E-5} - R_{pa}^3) + 314.2 * (R_{ps}^4 - 1.5\text{E-6})/R_{ps} + 157.1\, R_{pa}^3$$

therefore

$$0.036 = 0.0524 - 418.9\, R_{ps}^3 + 8.979\text{E-3} - 209.4\, R_{pa}^3 + 314.2\, R_{ps}^3 - 4.713\text{E-4}/R_{ps} + 157.1\, R_{pa}^3$$

$$0.036 = 0.0614 - 104.7 R_{ps}^3 - 52.3 R_{pa}^3 - 4.713\text{E-4}/R_{ps}$$

or

$$0.0254 = 104.7\, R_{ps}^3 + 52.3\, R_{pa}^3 + 4.713\text{E-4}/R_{ps} \tag{4.43}$$

This problem is statically indeterminate; hence, it will be necessary to consider compatibility, i.e.

$$\theta = \text{constant} = \frac{\tau}{R} * \frac{l}{G}$$

therefore

$$\theta = \frac{\tau_{yps}}{R_{ps}} * \frac{l}{G_s} = \frac{\tau_{ypa}}{R_{pa}} * \frac{l}{G_a}$$

or,

$$\frac{200E6}{R_{ps} * 7.7E10} = \frac{100E6}{R_{pa} * 2.6E10}$$

therefore

$$R_{ps} = R_{pa} * 2 * 2.6/7.7$$

$$\underline{R_{ps} = 0.675\, R_{pa}} \tag{4.44}$$

but R_{ps} cannot be less than R_{pa}; therefore the aluminium alloy must be completely elastic, so that,

$$\underline{R_{pa} = 0.035\,\text{m} \quad \text{and} \quad T_{p'a} = 0} \tag{4.45}$$

Furthermore, it can no longer be assumed that the maximum shear stress in the aluminium alloy will reach τ_{ypa}, so that it will be necessary to determine a new expression for T_{ea}.

T_{ea} can be obtained in terms of R_{ps} by considering the compatibility condition that,

$$\theta = \text{constant} = \frac{\tau * l}{G * r}$$

therefore

$$\frac{\tau_{yps} * l}{G_s * R_{ps}} = \frac{\tau_a * l}{G_a * 0.035}$$

where,

τ_a = the maximum shear stress in the aluminium alloy

therefore

$$\tau_a = \frac{0.035 * \tau_{yps} * G_a}{G_s * R_{ps}}$$

$$= \frac{0.035 \times 200 \times 2.6E10}{7.7E10 \times R_{ps}}$$

$$\underline{\tau_a = 2.364/R_{ps}} \tag{4.46}$$

and,

$$T_{ea} = \frac{\tau_a}{0.035} * \frac{\pi * 0.035^4}{2}$$

$$\underline{T_{ea} = 1.592E\text{-}4/R_{ps}} \tag{4.47}$$

Hence, from (4.38), (4.40) and (4.47),

$$0.036 = 418.9(1.25E\text{-}4 - R_{ps}^3) + \frac{\tau_{yps}}{R_{ps}} \times \frac{\pi(R_{ps}^4 - 0.035^4)}{2} + 1.592E\text{-}4/R_{ps}$$

$$0.036 = 0.0524 - 418.9\,R_{ps}^3 + 314.2\,R_{ps}^3$$
$$- 4.714E\text{-}4/R_{ps} + 1.592E\text{-}4/R_{ps}$$

therefore

$$0 = 0.0164 - 104.7\,R_{ps}^3 - 3.122\text{E-}4/R_{ps}$$

or,

$$104.7\,R_{ps}^4 - 0.0164\,R_{ps} + 3.122\text{E-}4 = 0 \tag{4.48}$$

Using the computer program of Appendix II, the four roots of the quartic were found to be as follows:

$$R_{ps} = 0.02\,\text{m}, \quad 0.0448\,\text{m}, \quad (-0.0325 \pm 0.0475\,\text{j})$$

The root of interest is $\underline{R_{ps} = 0.0448\,\text{m}}$.

Hence, from (4.47),

$$\underline{T_{ea} = 3.554\text{E-}3\ \text{MN m}}$$

therefore

$$\theta = \frac{3.554\text{E-}3 \times 0.7}{(\pi \times 0.035^4/2) \times 2.6\text{E}4} = 0.0406\,\text{rads}$$

$$\underline{\theta = 2.326^\circ}$$

4.20.4 To Determine the Residual Angle of Twist θ_R

On release of the torque of 36 000 N m, the shaft is assumed to behave elastically.

$$\theta_1 = \frac{Tl}{GJ}\text{(see Fig. 4.33)}$$

$$= \frac{T_s l}{G_s J_s} = \frac{T_a l}{G_a J_a}$$

therefore

$$T_s = \frac{T_a G_s J_s}{G_a J_a} = \frac{T_a \times 7.7\text{E}10 \times 7.46\text{E-}6}{2.6\text{E}10 \times 2.357\text{E-}6}$$

$$\underline{T_s = 9.373\,T_a}$$

therefore

$$T_a = \frac{36000}{10.373} = \underline{3470\ \text{N m}}$$

therefore

$$\theta_1 = \frac{3470 \times 0.7}{2.6\text{E}10 \times 2.357\text{E-}6} = 0.0396\,\text{rads}$$

$$\underline{\theta_1 = 2.27^\circ}$$

and,

$$\theta_R = \text{residual angle of twist}$$

$$\underline{\theta_R = \theta - \theta_1 = 0.056^\circ}$$

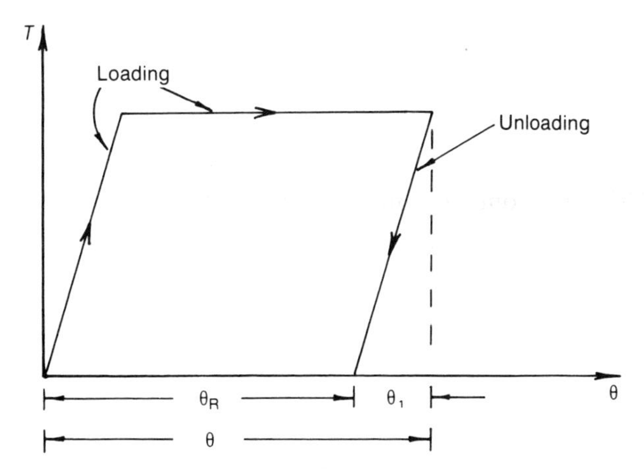

Fig. 4.33. T–θ relationship for the compound shaft.

4.20.5 To Determine the Fully Plastic Torsional Resistance of the Compound Shaft (T_p)

The steel tube will first become fully plastic and then it will rotate, until the aluminium alloy will become fully plastic.

From,

$$T_{pa} = \frac{2\pi}{3} * \tau_{ypa} * 0.035^3$$

$$= \frac{2\pi}{3} \times 100 \times 0.035^3$$

$$\underline{T_{pa} = 8980 \text{ N m}}$$

Similarly,

$$T_{ps} = \frac{2\pi}{3} * \tau_{yps} * (0.05^3 - 0.035^3)$$

$$\underline{T_{ps} = 34\,400 \text{ N m}}$$

Therefore total fully plastic moment of resistance =

$$\underline{T_p = 8980 + 34400 = 43380 \text{ N m}}$$

and,

$$\underline{\frac{T_p}{T_{yp}} = \frac{43\,380}{33\,058} = 1.312}$$

4.21.1 SHAKEDOWN

Shakedown is a form of incremental plastic collapse, where owing to repeated cyclic loading into the plastic region, the structure, or part of it, fails through low cycle fatigue.

After plastic flow takes place, on the first application of load, the material strain hardens, and the structure behaves elastically with residual stresses. Continued cyclic loading into the plastic region eventually causes the structure to fail through incremental collapse.

According to Calladine [15], a practical example of shakedown is that used in the manufacture of gun barrels. In this case, the gun barrels are subjected to an internal pressure to cause plastic flow, prior to machining, so that strain hardening takes place. Thus, as the material now has an apparent higher "yield stress", the gun barrel does not suffer further plastic deformation under normal use, so that the bore of the barrel remains true.

EXAMPLES FOR PRACTICE 4

1. Determine the shape factors for the sections shown in Figs. Q.4.1(a) to (e).

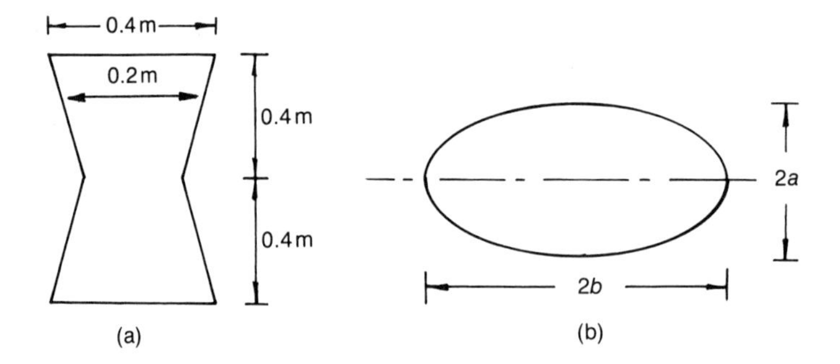

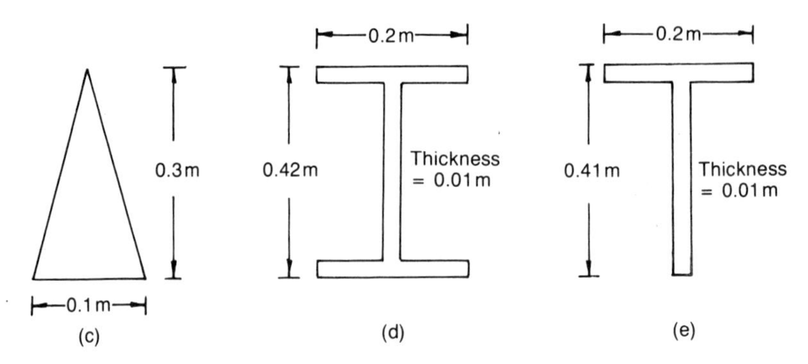

Fig. Q.4.1. Various sections.

{1.427; 1.7; 2.34; 1.16; 1.72}

2. Obtain suitable values for the sectional moduli (Z), for the beams of Figs. Q.4.2(a) to (h), given that,

$$\lambda = 4, \quad S = 1.15, \quad \sigma_{yp} = 300\,\text{MN/m}^2$$

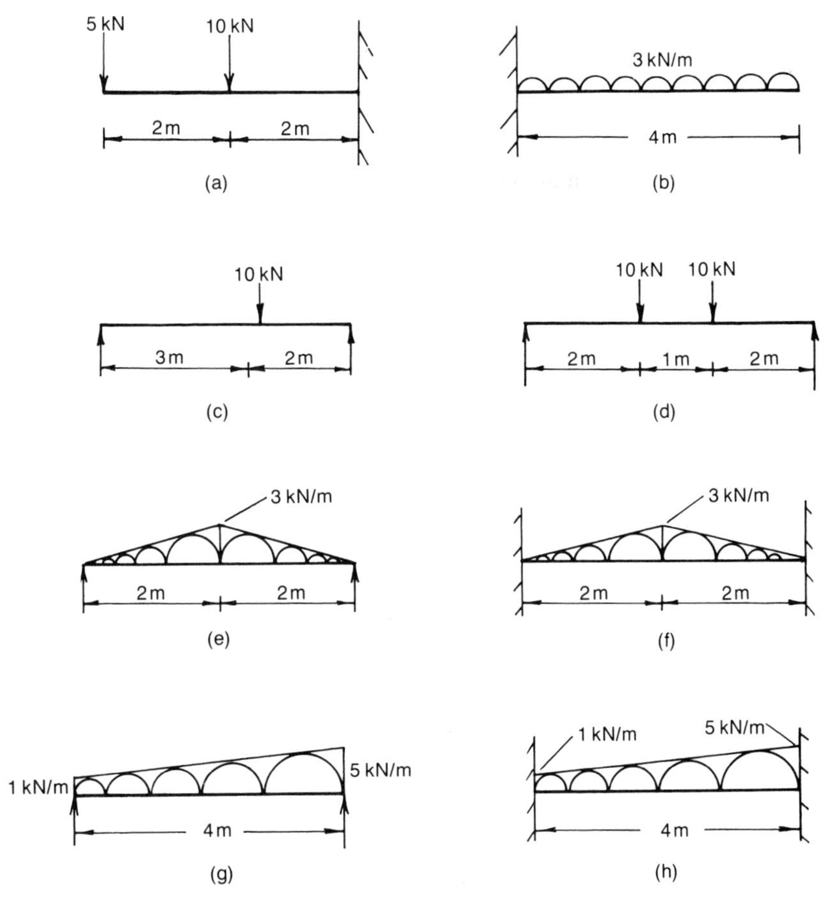

Fig. Q.4.2. Various beams.

{4.638E-4 m^3; 2.783E-4 m^3; 1.39E-4 m^3; 2.9E-4 m^3; 4.64E-5 m^3; 2.319E-5 m^3; 7.08E-5 m^3; 4.95E-5 m^3}

3. Determine suitable values for the sectional moduli for the frames shown in Figs. Q.4.3(a) to (d).

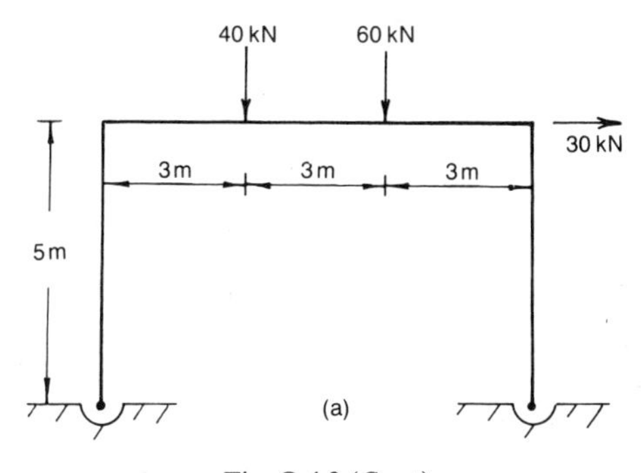

Fig. Q.4.3 (*Cont.*)

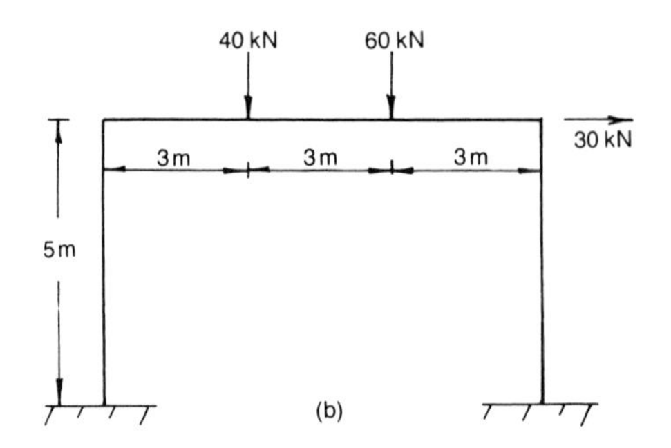

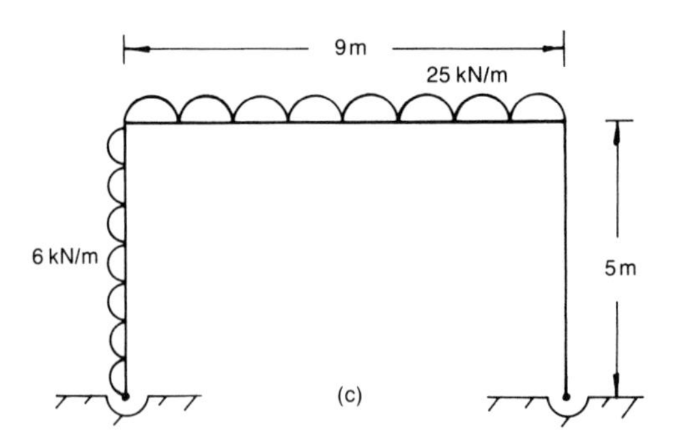

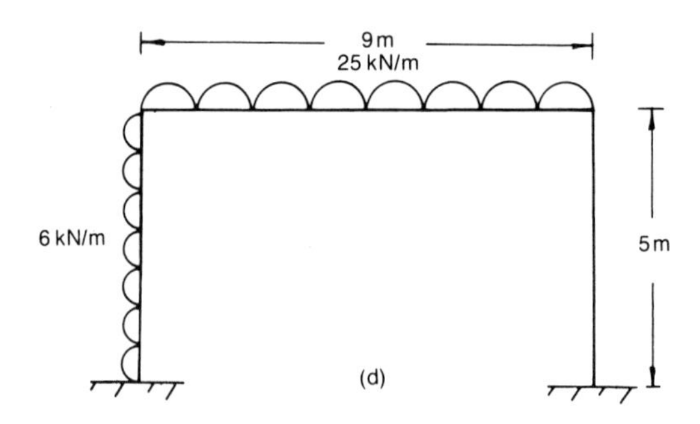

Fig. Q.4.3. Various frameworks.

{1.217E-3; 9.13E-4; 1.693E-3; 1.468E-3 (all in m^3)}

4. A portal frame of uniform section is subjected to the loading shown in Fig. Q.4.4.

   Using the plastic hinge theory, determine a suitable section for a load factor of 4, a shape factor of 1.15 and a yield stress of 300 MN/m^2.

   {Portsmouth Polytechnic, 1982}

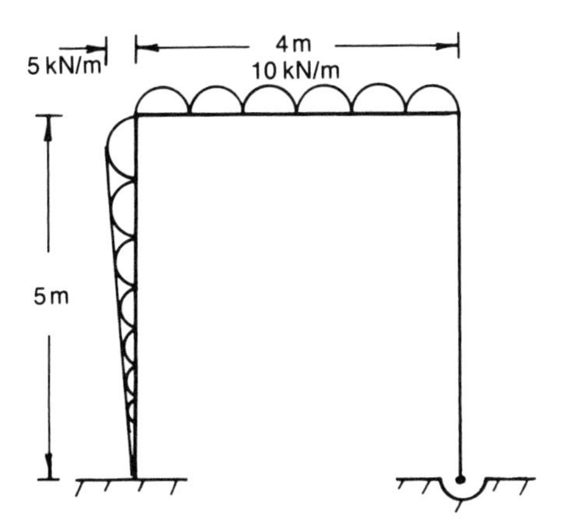

Fig. Q.4.4.

{1.98E-4 m^3}

5. Using the plastic hinge theory, determine a suitable section for the two-storey rectangular frame shown in Fig. Q.4.5, given the following:

$$\text{Yield stress} = 300\ \text{MN/m}^2$$
$$\text{Shape factor} = 1.14$$
$$\text{Load factor} = 5$$

{Portsmouth Polytechnic, 1986}

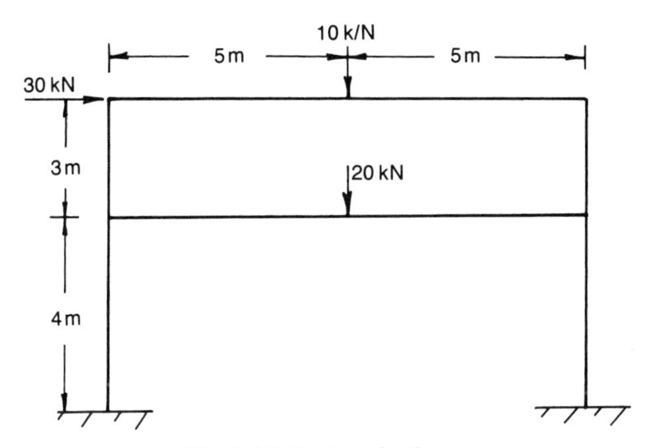

Fig. Q.4.5. Rectangular frame.

{5.66E-4 m^3}

6. Using the plastic hinge theory, determine a suitable section for the two-bay rectangular frame, shown in Fig. Q.4.6, given the following:

$$\text{Yield stress} = 300\ \text{MN/m}^2$$
$$\text{Shape factor} = 1.15$$
$$\text{Load factor} = 3$$

{Portsmouth Polytechnic, 1986}

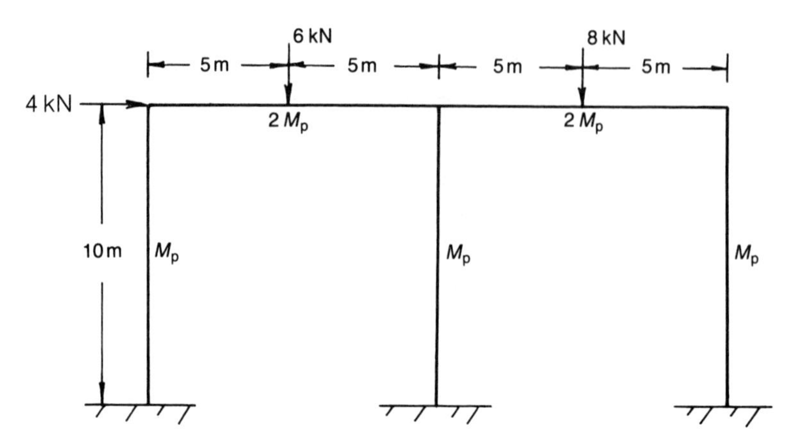

Fig. Q.4.6. Two-bay portal frame.

{Top: 1.159E-4 m^3; Verticals: 5.797E-5 m^3}

7. A uniform section tee beam, which has the same sectional properties as in Fig. Q.4.1(e), is subjected to four-point loading, as shown in Fig. Q.4.7.

   Determine the central deflection on application of the load, and the residual central deflection on its removal.

   $\sigma_{yp} = 300\,\mathrm{MN/m^2}$, $\mathrm{E} = 2\mathrm{E}11\,\mathrm{N/m^2}$

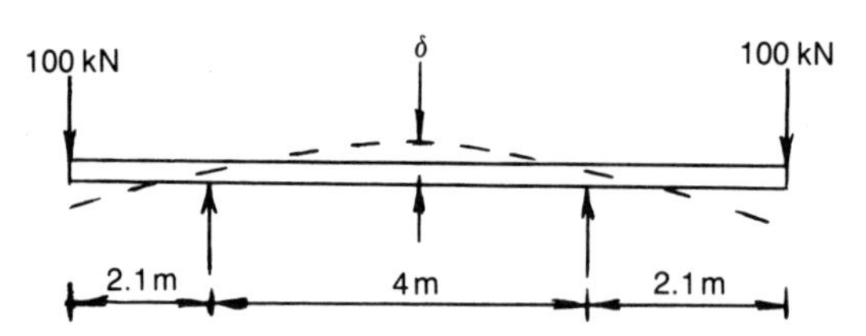

Fig. Q.4.7. Beam under four-point loading.

{0.11 m; 0.07 m}

8. A circular section shaft of diameter 0.2 m and of length 1 m is subjected to a torque that causes an angle of twist of 3.5°. Determine this torque, and the residual angle of twist on removal of this torque.

$$G = 7.7\mathrm{E}10\,\mathrm{N/m^2}, \qquad \tau_{yp} = 180\,\mathrm{MN/m^2}$$

{0.37 MN.m; 1.75°}

5

Torsion of Non-circular Sections

5.1.1 CIRCULAR AND NON-CIRCULAR SECTIONS

The torsional theory of circular sections cannot be applied to the torsion of non-circular sections, as the shear stresses for non-circular sections are no longer circumferential. Furthermore, plane cross-sections do not remain plane and undistorted on the application of torque, and, in fact, warping of the cross-section takes place.

As a result of this behaviour, the polar second moment of area of the section is no longer applicable for static stress analysis, and it has to be replaced by a torsional constant, whose magnitude is very often a small fraction of the magnitude of the polar second moment of area.

5.2.1 TO DETERMINE THE TORSIONAL EQUATION

Consider a prismatic bar of uniform non-circular section, subjected to twisting action, as shown in Fig. 5.1.

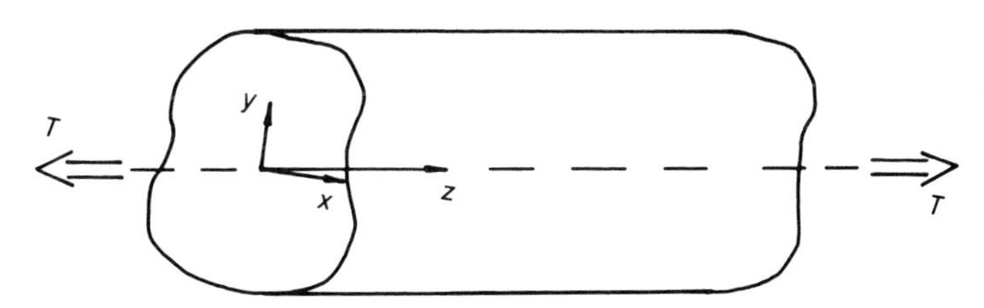

Fig. 5.1. Non-circular section under twist.

Let,

T = torque
u = displacement in the x direction
v = displacement in the y direction

$w =$ displacement in the z direction
$\quad =$ the warping function
$\theta =$ rotation/unit length
$x, y, z =$ Cartesian co-ordinates

Consider any point "P" in the section, which, owing to the application of T, will rotate and warp, as shown in Fig. 5.2.

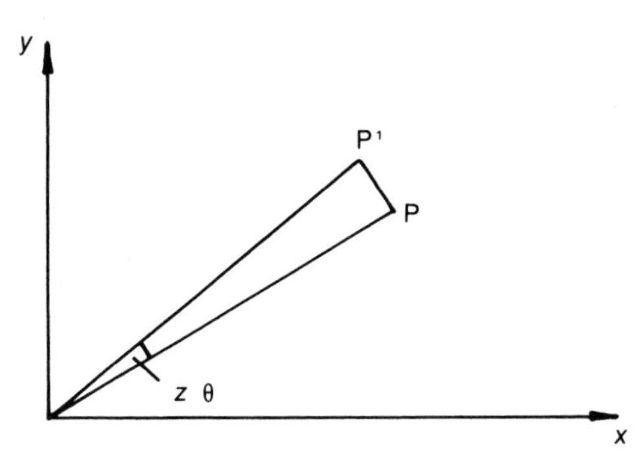

Fig. 5.2. Displacement of "P".

From Fig. 5.2,

$$\left.\begin{aligned} u &= -\,yz\theta \\ v &= xz\theta \end{aligned}\right\} \text{due to rotation} \tag{5.1}$$

and,

$$\begin{aligned} w &= \theta * \psi(x, y) \text{ (due to warping)} \\ &= \theta * \psi \end{aligned} \tag{5.2}$$

The theory assumes that,

$$\varepsilon_x = \varepsilon_y = \varepsilon_z = \gamma_{xy} = 0 \tag{5.3}$$

and therefore the only shearing strains that exist are γ_{xz} and γ_{yz}, which from references [3–7] are defined as follows:

$$\begin{aligned} \gamma_{xz} &= \text{shear strain in the } x\text{–}z \text{ plane} \\ &= \frac{\partial w}{\partial x} + \frac{\partial u}{\partial z} = \theta\left(\frac{\partial \psi}{\partial x} - y\right) \end{aligned} \tag{5.4}$$

$$\begin{aligned} \gamma_{yz} &= \text{shear strain in the } y\text{–}z \text{ plane} \\ &= \frac{\partial w}{\partial y} + \frac{\partial v}{\partial z} = \theta\left(\frac{\partial \psi}{\partial y} + x\right) \end{aligned} \tag{5.5}$$

5.2.2

The equations of equilibrium of an infinitesimal element of dimensions $\mathrm{d}x * \mathrm{d}y * \mathrm{d}z$ can be obtained with the aid of Fig. 5.3, where,

$$\tau_{xz} = \tau_{zx}$$

and,

$$\tau_{yz} = \tau_{zy}$$

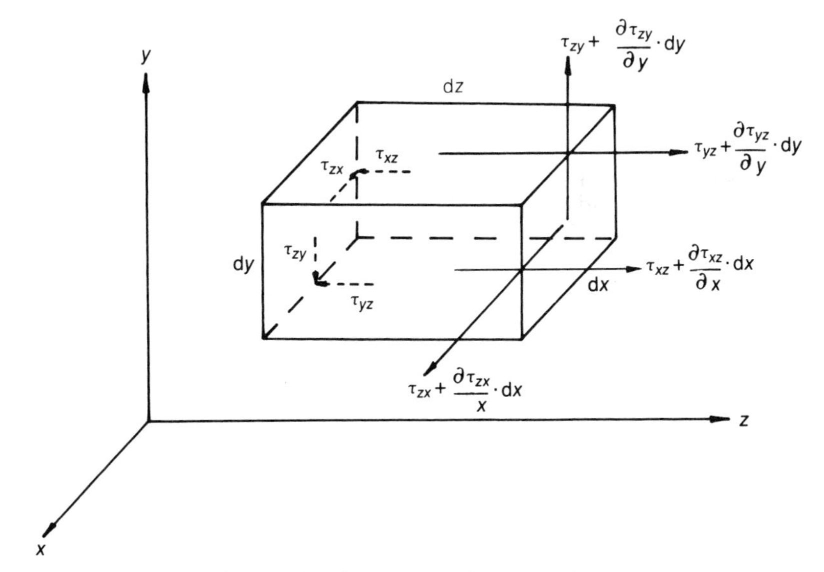

Fig. 5.3. Shearing stresses acting on an element.

Resolving in the z Direction

$$\frac{\partial \tau_{yz}}{\partial y} * \mathrm{d}y * \mathrm{d}x * \mathrm{d}z + \frac{\partial \tau_{xz}}{\partial x} * \mathrm{d}x * \mathrm{d}y * \mathrm{d}z = 0$$

or,

$$\frac{\partial \tau_{xz}}{\partial x} + \frac{\partial \tau_{yz}}{\partial y} = 0 \tag{5.6}$$

but from (5.4) and (5.5):

$$\tau_{xz} = G\gamma_{xz} = G\theta\left(\frac{\partial \psi}{\partial x} - y\right) \tag{5.7}$$

and,

$$\tau_{yz} = G\gamma_{yz} = G\theta\left(\frac{\partial \psi}{\partial y} + x\right) \tag{5.8}$$

Let,

$$\frac{\partial \chi}{\partial y} = \frac{\partial \psi}{\partial x} - y \tag{5.9}$$

and,

$$-\frac{\partial \chi}{\partial x} = \frac{\partial \psi}{\partial y} + x \tag{5.10}$$

where,

$\chi =$ a shear stress function

By differentiating (5.9) and (5.10) w.r.t. y and x, respectively, the following is obtained:

$$\frac{\partial^2 \chi}{\partial x^2} + \frac{\partial^2 \chi}{\partial y^2} = \frac{\partial^2 \psi}{\partial x \cdot \partial y} - 1 - \frac{\partial^2 \psi}{\partial x \cdot \partial y} - 1$$

or,

$$\frac{\partial^2 \chi}{\partial x^2} + \frac{\partial^2 \chi}{\partial y^2} = -2 \tag{5.11}$$

Equation (5.11) can be described as the *torsion equation for non-circular sections.*

From (5.7) and (5.8):

$$\tau_{xz} = G\theta \frac{\partial \chi}{\partial y} \tag{5.12}$$

and,

$$\tau_{yz} = -G\theta \frac{\partial \chi}{\partial x} \tag{5.13}$$

Equation (5.11), which is known as Poisson's equation, can be put into the alternative form of equation (5.14), which is known as Laplace's equation.

$$\frac{\partial^2 \psi}{\partial x^2} + \frac{\partial^2 \psi}{\partial y^2} = 0 \tag{5.14}$$

5.3.1 TO DETERMINE EXPRESSIONS FOR THE SHEAR STRESS τ AND THE TORQUE T

Consider the non-circular cross-section of Fig. 5.4.

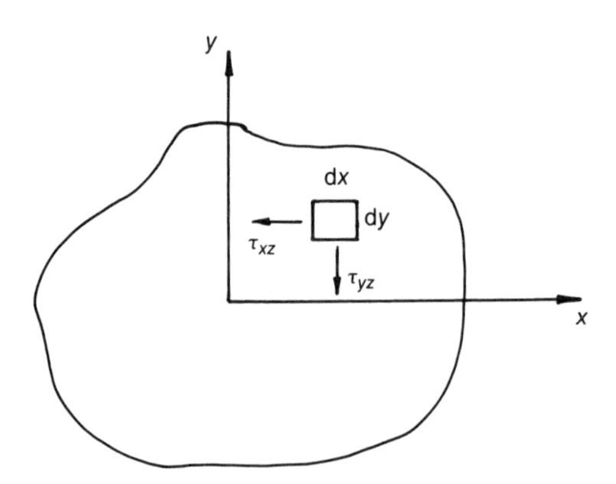

Fig. 5.4. Shearing stresses acting on an element.

From Pythagoras' theorem:

τ = shearing stress at any point (x, y) on the cross-section

$$= \sqrt{(\tau_{xz}^2 + \tau_{yz}^2)} \tag{5.15}$$

From Fig. 5.4, the torque =

$$T = \iint (\tau_{xz} * y - \tau_{yz} * x)\mathrm{d}x \cdot \mathrm{d}y \tag{5.16}$$

5.3.2

To determine the *boundary value for* χ, consider an element on the boundary of the section, as shown in Fig. 5.5, where the shear stress acts tangentially. Now, as the shear stress perpendicular to the boundary is zero,

$$\tau_{yz} \sin \phi + \tau_{xz} \cos \phi = 0$$

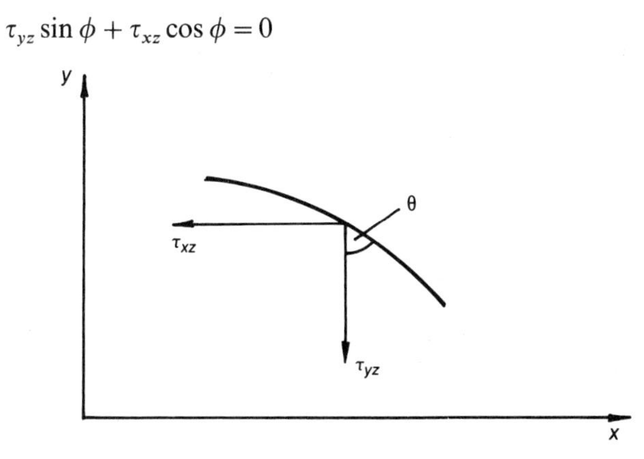

Fig. 5.5. Shearing stresses on boundary.

or,

$$-G\theta * \frac{\partial \chi}{\partial x}\left(-\frac{\mathrm{d}x}{\mathrm{d}s}\right) + G\theta * \frac{\partial \chi}{\partial y}\left(\frac{\mathrm{d}y}{\mathrm{d}s}\right) = 0$$

or,

$$G\theta \frac{\mathrm{d}\chi}{\mathrm{d}s} = 0$$

where, s is any distance along the boundary, i.e.

<u>χ is a constant along the boundary</u>

5.4.1 EXAMPLE 5.1 SHEARING STRESSES IN AN ELLIPTICAL SECTION

Determine the shear stress function χ for an elliptical section, and hence, or otherwise, determine expressions for the torque T, the warping function w and the torsional constant J.

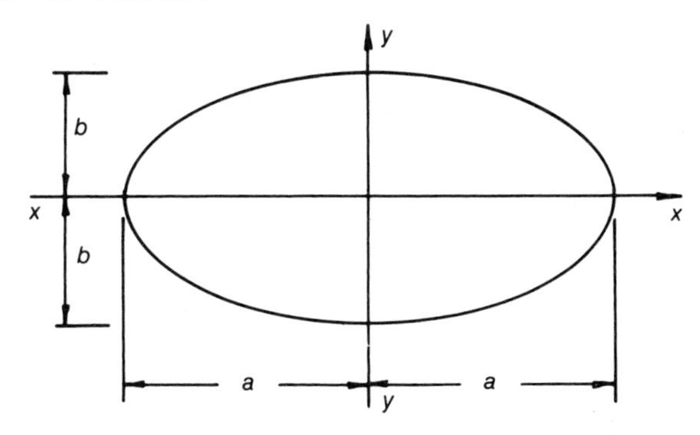

Fig. 5.6. Elliptical section.

5.4.2

The equation for the ellipse of Fig. 5.6 is given by:

$$\frac{x^2}{a^2}+\frac{y^2}{b^2}=1 \tag{5.17}$$

and this equation can be used for determining the shear stress function χ as follows:

$$\chi=C\left(\frac{x^2}{a^2}+\frac{y^2}{b^2}-1\right) \tag{5.18}$$

where,

$$C=\text{a constant, to be determined}$$

Equation (5.18) ensures that χ is constant along the boundary, as required.

The constant C can be determined by substituting (5.18) into (5.11), i.e.

$$C\left(\frac{2}{a^2}+\frac{2}{b^2}\right)=-2$$

therefore

$$C=\frac{-a^2b^2}{a^2+b^2}$$

and,

$$\chi=\frac{a^2b^2}{(a^2+b^2)}\left(1-\frac{x^2}{a^2}-\frac{y^2}{b^2}\right) \tag{5.19}$$

where χ is the required stress function for the elliptical section.

5.4.3

Now,

$$\tau_{xz}=G\theta\frac{\partial\chi}{\partial y}=-G\theta\cdot\frac{2ya^2}{a^2+b^2}$$

$$\tau_{yz}=-G\theta\frac{\partial\chi}{\partial x}=\frac{G\theta\cdot 2xb^2}{a^2+b^2}$$

and,

$$T=\int(\tau_{xz}y-\tau_{yz}x)\,\mathrm{d}A$$

$$=-G\theta\int\left(\frac{2x^2b^2}{a^2+b^2}+\frac{2y^2a^2}{a^2+b^2}\right)\mathrm{d}A$$

$$=-2G\theta\frac{a^2b^2}{a^2+b^2}\left[\int\frac{x^2}{a^2}\,\mathrm{d}A+\int\frac{y^2}{b^2}\,\mathrm{d}A\right]$$

but,

$$\int y^2\,\mathrm{d}A=I_{xx}=\frac{\pi ab^3}{4}=\text{second moment of area about } x\text{–}x$$

and,

$$\int x^2 \, dA = I_{yy} = \frac{\pi a^3 b}{4} = \text{second moment of area about } y\text{–}y$$

therefore

$$T = -2G\theta \frac{a^2 b^2}{a^2+b^2}\left(\frac{\pi ab}{4} + \frac{\pi ab}{4}\right)$$

$$\underline{\underline{T = \frac{-G\theta\pi a^3 b^3}{a^2+b^2}}} \tag{5.20}$$

therefore

$$\tau_{xz} = \frac{-2a^2 y}{(a^2+b^2)} \cdot \frac{-(a^2+b^2)T}{\pi a^3 b^3}$$

$$\underline{\underline{\tau_{xz} = \frac{2Ty}{\pi ab^3}}} \tag{5.21}$$

$$\underline{\underline{\tau_{yz} = \frac{-2Tx}{\pi a^3 b}}} \tag{5.22}$$

By inspection, it can be seen that $\hat{\tau}$ is obtained by substituting $y = b$ into (5.21), providing $a > b$.

$$\hat{\tau} = \text{maximum shear stress}$$

$$= \frac{2T}{\pi ab^2} \tag{5.23}$$

and occurs at the extremities of the minor axis.

5.4.4

The warping function can be obtained from equation (5.2).

Now,

$$\frac{\partial \chi}{\partial y} = \frac{\partial \psi}{\partial x} - y$$

or,

$$\frac{2ya^2b^2}{(a^2+b^2)b^2} = \frac{\partial \psi}{\partial x} - y$$

i.e.

$$\frac{\partial \psi}{\partial x} = \frac{(-2a^2 + a^2 + b^2)}{(a^2+b^2)} y$$

therefore

$$\underline{\psi = \left(\frac{b^2 - a^2}{a^2 + b^2}\right) xy} \tag{5.24}$$

Similarly, from the expression,

$$-\frac{\partial \chi}{\partial x} = \frac{\partial \psi}{\partial y} + x$$

the same equation for ψ, namely equation (5.24), can be obtained.

Now,

$$w = \text{warping function}$$
$$= \theta * \psi$$

therefore

$$\underline{w = \frac{(b^2 - a^2)}{(a^2 + b^2)}\theta xy} \tag{5.25}$$

5.4.5

From simple torsion theory,

$$\frac{T}{J} = G\theta \tag{5.26}$$

or,

$$T = G\theta J \tag{5.27}$$

Equating (5.20) and (5.27), and ignoring the negative sign in (5.20),

$$G\theta J = \frac{G\theta\pi a^3 b^3}{(a^2 + b^2)}$$

therefore

$$J = \text{torsional constant for an elliptical section}$$
$$\underline{J = \frac{\pi a^3 b^3}{(a^2 + b^2)}} \tag{5.28}$$

5.5.1 EXAMPLE 5.2 SHEAR STRESSES IN A TRIANGULAR CROSS-SECTION

Determine the shear stress function χ and the value of the maximum shear stress $\hat{\tau}$ for the equilateral triangle of Fig. 5.7.

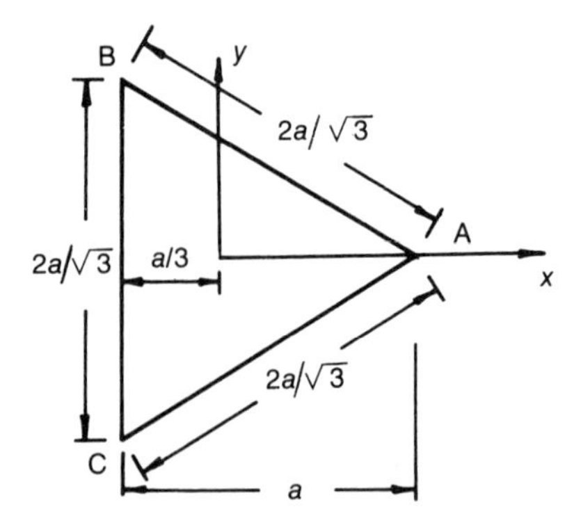

Fig. 5.7. Equilateral triangle.

5.5.2

The equations of the three straight lines representing the boundary can be used for determining χ, as it is necessary for χ to be a constant along the boundary.

Side BC
This side can be represented by the expression:

$$x = -\frac{a}{3} \quad \text{or} \quad x + \frac{a}{3} = 0 \tag{5.29}$$

Side AC
This side can be represented by the expression:

$$x - \sqrt{3}y - \frac{2a}{3} = 0 \tag{5.30}$$

Side AB
This side can be represented by the expression:

$$x + \sqrt{3}y - \frac{2a}{3} = 0 \tag{5.31}$$

The stress function χ can be obtained by multiplying together equations (5.29) to (5.31):

$$\begin{aligned} \chi &= C(x + a/3) * (x - \sqrt{3}y - 2a/3) * (x + \sqrt{3}y - 2a/3) \\ &= C\{(x^3 - 3xy^2) - a(x^2 + y^2) + 4a^3/27\} \end{aligned} \tag{5.32}$$

From equation (5.32), it can be seen that $\chi = 0$ (i.e. constant) along the external boundary, so that the boundary condition is satisfied.

Substituting χ into (5.11):

$$C(6x - 2a) + C(-6x - 2a) = -2$$

$$-4aC = -2$$

$$\underline{C = 1/(2a)}$$

therefore

$$\underline{\chi = \frac{1}{2a}(x^3 - 3xy^2) - \tfrac{1}{2}(x^2 + y^2) + \frac{2a^2}{27}} \tag{5.33}$$

Now,

$$\tau_{xz} = G\theta\frac{\partial\chi}{\partial y}$$

$$= G\theta\left\{\frac{1}{2a}(-6xy) - \tfrac{1}{2} * 2y\right\}$$

$$\underline{\tau_{xz} = -G\theta\left(\frac{3xy}{2a} + y\right)} \tag{5.34}$$

Along $y = 0$,

$$\tau_{xz} = 0$$

Now,

$$\tau_{yz} = -G\theta\frac{\partial \chi}{\partial x} = -G\theta\left\{\frac{1}{2a}(3x^2 - 3y^2) - \tfrac{1}{2} * 2x\right\}$$

therefore

$$\underline{\tau_{yz} = -\frac{3G\theta}{2a}\left\{(x^2 - y^2) - \frac{2ax}{3}\right\}} \tag{5.35}$$

Now, as the triangle is equilateral, the maximum shear stress $\hat{\tau}$ can be obtained by considering the variation of τ_{yz} along any edge. Consider the edge BC (i.e. $x = -a/3$):

$$\tau_{yz}\,(\text{edge BC}) = -\frac{3G\theta}{2a}\left(\frac{a^2}{9} - y^2 + \frac{2a^2}{9}\right)$$

$$= -\frac{3G\theta}{2a}\left(\frac{a^2}{3} - y^2\right) \tag{5.36}$$

where it can be seen from (5.36) that

$$\hat{\tau} \text{ occurs at } y = 0$$

therefore

$$\underline{\hat{\tau} = -G\theta a/2} \tag{5.37}$$

5.6.1 NUMERICAL SOLUTIONS OF THE TORSIONAL EQUATION

Equation (5.11) lends itself to satisfactory solution by either the finite element method [16–20] or the finite difference method (see Chapter 10), and Fig.

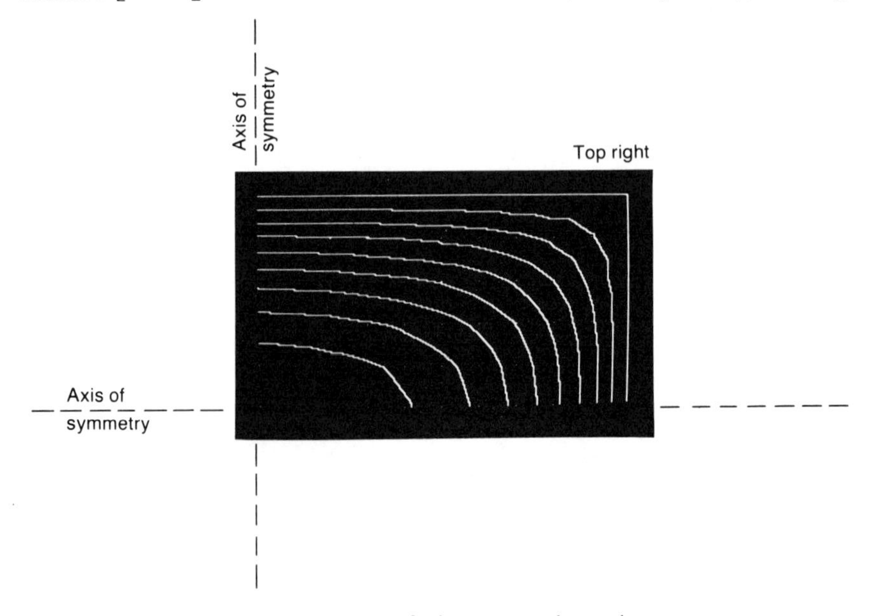

Fig. 5.8. Variation of χ in a rectangular section.

5.8 shows the variation of χ for a rectangular section, as obtained by the computer program "LAPLACE". (Solution was carried out on an Apple II + microcomputer, and the screen was then photographed.)

As the rectangular section had two axes of symmetry, it was only necessary to consider the top right-hand quadrant of the rectangle.

5.7.1 PRANDTL'S MEMBRANE ANALOGY

Prandtl noticed that the equations describing the deformation of a thin weightless membrane were similar to the torsion equation. Furthermore, he realised that as the behaviour of a thin weightless membrane under lateral pressure was more readily understood than that of the torsion of a non-circular section, the application of a membrane analogy to the torsion of non-circular sections considerably simplified the stress analysis of the latter.

Prior to using the membrane analogy, it will be necessary to develop the differential equation of a thin weightless membrane under lateral pressure. This can be done by considering the equilibrium of the element AA′BB′ in Fig. 5.9.

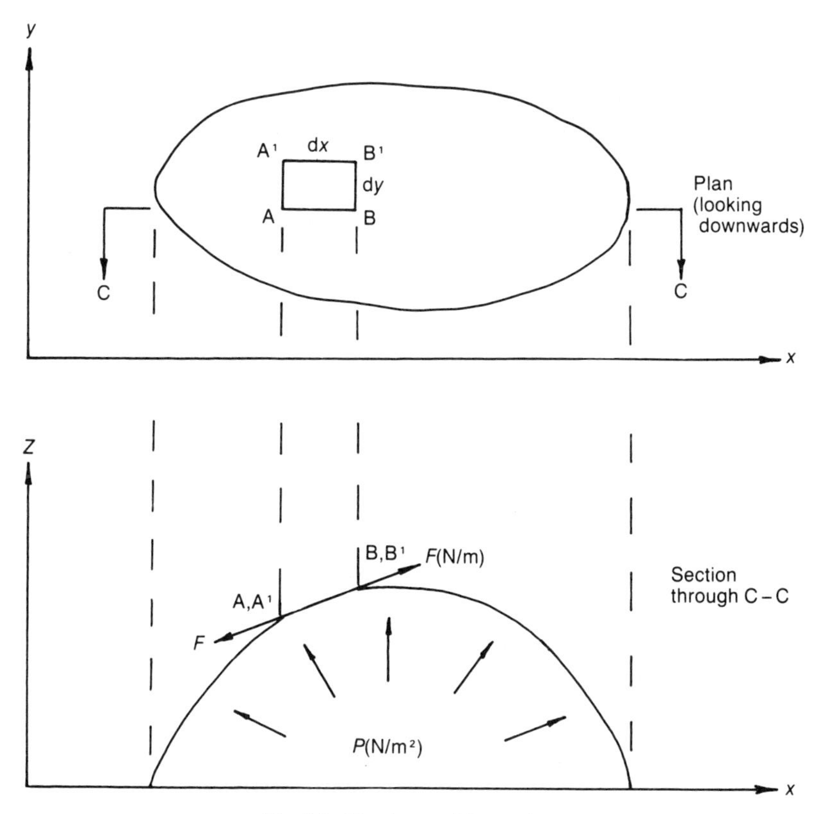

Fig. 5.9. Membrane deformation.

5.7.2

Let,

F = membrane tension per unit length (N/m)
Z = deflection of membrane (m)
P = pressure (N/m^2)

$$\text{Component of force on AA}' \text{ in } z \text{ direction} = F * \frac{\partial Z}{\partial x} * \mathrm{d}y \qquad \downarrow$$

$$\text{Component of force on BB}' \text{ in } z \text{ direction} = F\left(\frac{\partial Z}{\partial x} + \frac{\partial^2 Z}{\partial x^2} * \mathrm{d}x\right)\mathrm{d}y \qquad \uparrow$$

$$\text{Component of force on AB in } z \text{ direction} = F * \frac{\partial Z}{\partial y} * \mathrm{d}x \qquad \downarrow$$

$$\text{Component of force on A}'\text{B}' \text{ in } z \text{ direction} = F\left(\frac{\partial Z}{\partial y} + \frac{\partial^2 Z}{\partial y^2} * \mathrm{d}y\right)\mathrm{d}x \qquad \uparrow$$

Resolving vertically

$$F\left(\frac{\partial^2 Z}{\partial x^2} + \frac{\partial^2 Z}{\partial y^2}\right)\mathrm{d}x * \mathrm{d}y = -P * \mathrm{d}x * \mathrm{d}y$$

therefore

$$\frac{\partial^2 Z}{\partial x^2} + \frac{\partial^2 Z}{\partial y^2} = -\frac{P}{F} \tag{5.38}$$

If $Z = \chi$ in equation (5.38), and the pressure is so adjusted that $P/F = 2$, then it can be seen that equation (5.38) can be used as an analogy to equation (5.11).

From equations (5.12) and (5.13), it can be seen that:

$$\left.\begin{aligned} \tau_{xz} &= G\theta * \text{slope of the membrane in the } y \text{ direction} \\ \tau_{yz} &= G\theta * \text{slope of the membrane in the } x \text{ direction} \end{aligned}\right\} \tag{5.39}$$

Now, the torque =

$$T = \iint (\tau_{xz} * y - \tau_{yz} * x)\mathrm{d}x \cdot \mathrm{d}y$$

$$= G\theta \iint \left(\frac{\partial Z}{\partial y} * y + \frac{\partial Z}{\partial x} * x\right)\mathrm{d}x\,\mathrm{d}y \tag{5.40}$$

Consider the integral:

$$\iint \frac{\partial Z}{\partial y} * y * \mathrm{d}x\,\mathrm{d}y = \iint \partial Z * y * \mathrm{d}x$$

Now y and $\mathrm{d}x$ are shown in Fig. 5.10, where it can be seen that $\iint y * \mathrm{d}x$ = area of section. Therefore

$$\iint \frac{\partial Z}{\partial y} * y * \mathrm{d}x * \mathrm{d}y = \text{volume under membrane} \tag{5.41}$$

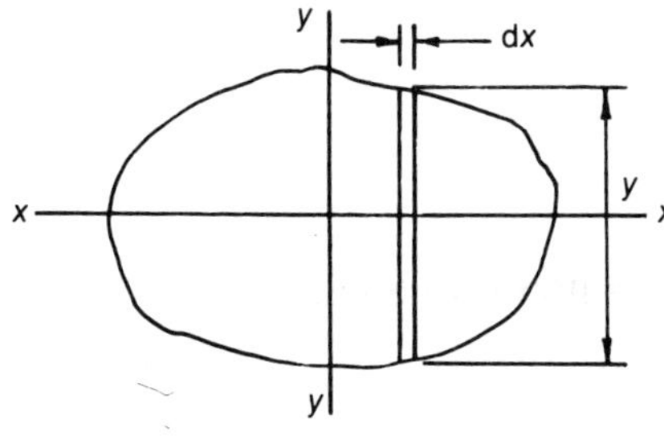

Fig. 5.10

Similarly, it can be shown that:

$$\iint \frac{\partial Z}{\partial x} * x * \mathrm{d}x * \mathrm{d}y = \text{volume under membrane} \tag{5.42}$$

Substituting (5.41) and (5.42) into (5.40):

$$\underline{T = 2G\theta * \text{ volume under membrane}} \tag{5.43}$$

Now,

$$\frac{T}{J} = G\theta$$

which, on comparison with (5.43), gives

$$\underline{J = \text{torsional constant}}$$

$$= \underline{2 * \text{volume under membrane}} \tag{5.44}$$

5.8.1 EXAMPLE 5.3 THE TORSION OF LONG THIN SECTIONS

Determine expressions for J and $\hat{\tau}$ for a thin rectangular section under a torque T.

5.8.2

If the section is long and thin, with an aspect ratio > 5, as shown in Fig. 5.11, then the long side does very little in resisting the pressure P, so that,

$$F\frac{\partial^2 Z}{\partial y^2} = 0$$

For this case, equation (5.38) reduces to:

$$\frac{\mathrm{d}^2 Z}{\mathrm{d}x^2} = -\frac{P}{F} \tag{5.45}$$

or,

$$\frac{\mathrm{d}Z}{\mathrm{d}x} = -\frac{P}{F}x + C_1$$

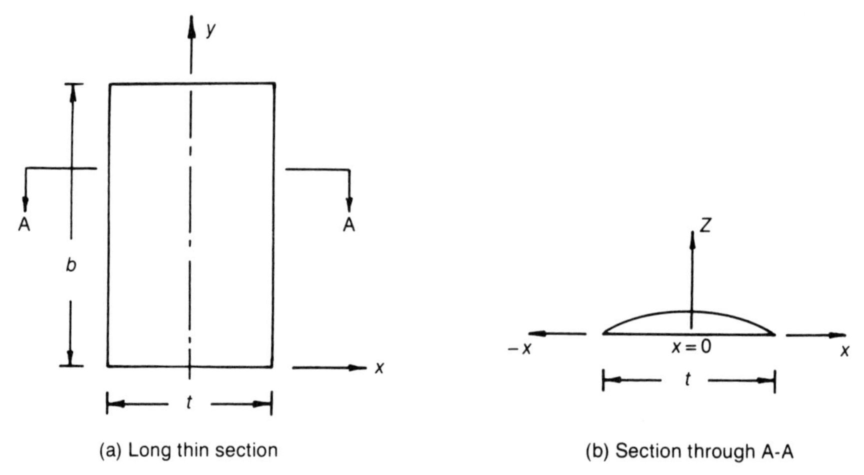

Fig. 5.11. Long thin membrane under lateral pressure.

and,

$$Z = -\frac{P}{2F}x^2 + C_1x + C_2$$

@ $x = 0$,

$$\frac{dZ}{dx} = 0$$

therefore

$$C_1 = 0$$

Now,

$$P/F = 2$$

therefore

$$Z = -x^2 + C_2$$

@ $x = t/2$,

$$Z = 0$$

therefore

$$C_2 = t^2/4$$

or,

$$\underline{Z = t^2/4 - x^2} \text{ (parabolic)}$$

@ $x = 0$,

$$\hat{Z} = t^2/4 \tag{5.46}$$

therefore

$$\text{volume under membrane} = b * t * \tfrac{2}{3} * t^2/4$$
$$= bt^3/6$$

But,

$$J = 2 * \text{volume under membrane}$$
$$= bt^3/3 \tag{5.47}$$

From (5.39),

$$\tau_{xz} = 0$$
$$\underline{\tau_{yz} = 2G\theta\, x}$$

i.e. the shear stress varies linearly across the smaller dimension, having a value of zero at the centre and a maximum value @ $x = t/2$.

$$\hat{\tau} = \text{maximum shear stress} = G\theta t$$

but,

$$T/J = G\theta$$

therefore

$$\underline{\hat{\tau} = Tt/J = G\theta t} \tag{5.48}$$

From equation (5.48), it can be seen that for a thin-walled open section under torsion, the *maximum shear stress* occurs on the web or flange with the *largest* value of t.

5.8.3

Some typical values for J for thin-walled open sections are given in Table 5.1.

Table 5.1. J and $\hat{\tau}$ for some thin-walled open sections

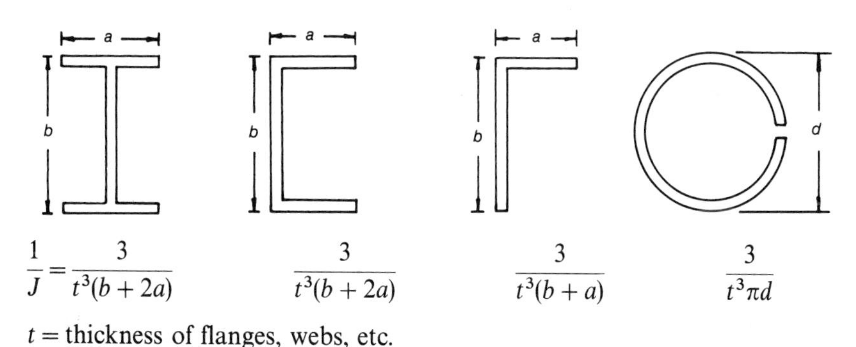

$$\frac{1}{J} = \frac{3}{t^3(b+2a)} \qquad \frac{3}{t^3(b+2a)} \qquad \frac{3}{t^3(b+a)} \qquad \frac{3}{t^3\pi d}$$

t = thickness of flanges, webs, etc.

5.9.1 EXAMPLE 5.4 TORSION OF AN "I" SECTION

Determine J for the "I" section shown in Fig. 5.12. If this section is subjected to a torque of 1000 N m, determine the maximum shear stress due to torsion and the resulting angle of twist over a length of 1.25 m. $G = 7.7\text{E}10$ Pa

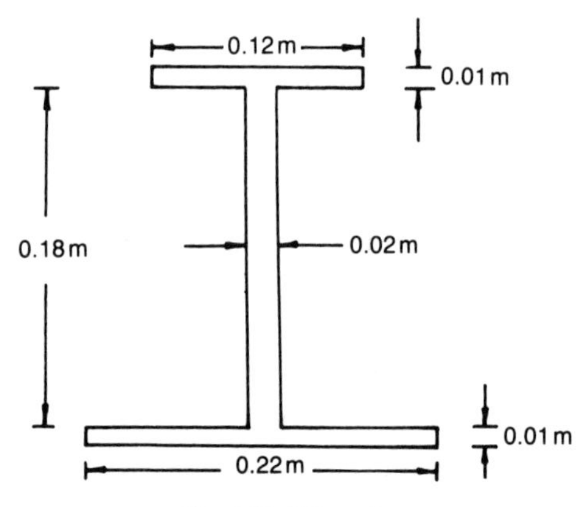

Fig. 5.12. "I" section.

5.9.2

Now, for a built-up open section,

$$J = \sum bt^3/3 \tag{5.49}$$

$$= (0.1 \times 0.01^3 + 0.2 \times 0.01^3 + 0.18 \times 0.02^3)/3$$

$$\underline{J = 5.8\text{E-}7\,\text{m}^4}$$

From (5.48), it can be seen that as J is a constant, $\hat{\tau}$ occurs on the flange or web with the largest value of t, i.e.

$$\hat{\tau} = \frac{T * 0.02}{5.8\text{E-}7}$$

$$= 34.48\,\text{MPa—on the outer surfaces of the web}$$

From (5.27),
Angle of twist over a length of 1.5 m

$$= 1.5\theta = 1.5 * T/(GJ)$$

$$= \frac{1.5 \times 1000}{7.7\text{E}10 \times 5.8\text{E-}7} = 0.033\,\text{rads} = \underline{1.92^\circ}$$

5.10.1 THE TORSION OF LONG THIN SOLID AEROFOIL SECTIONS

This is of much interest in aeronautical engineering and naval architecture, where for the latter, such sections are often met in the design of propellers.

For such sections, an approximate formula for J is as follows:

$$J = \tfrac{1}{3} \sum_{i=1}^{N} ht_i^3 \tag{5.50}$$

where,

h = spacing between ordinates (assumed equal), as shown in Fig. 5.13

and

t_i = thickness of the aerofoil of the ith ordinate
$\hat{t}$ = maximum thickness of the aerofoil
N = number of elements

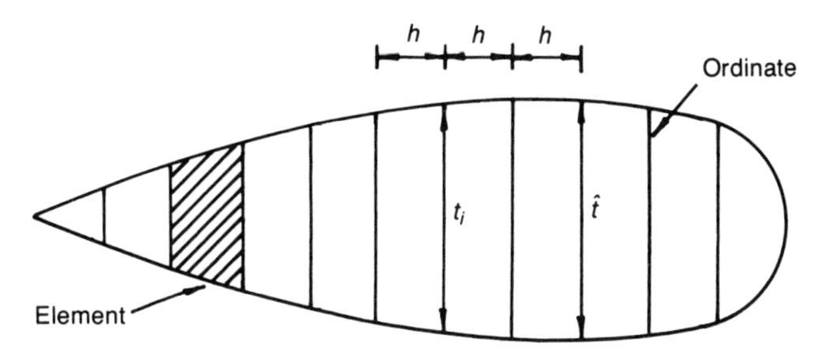

Fig. 5.13. Aerofoil section.

5.10.2

However, this method of calculation is incorrect, as it does not allow for the fact that t may vary between adjacent ordinates.

A method [21] of allowing for this can be obtained by considering the variation of t over two adjacent elements, as follows.

Let,

$$t = a + bx + cx^2 \tag{5.51}$$

where,

a, b and c are arbitrary constants, and
t varies parabolically with x, as shown in Fig. 5.14.

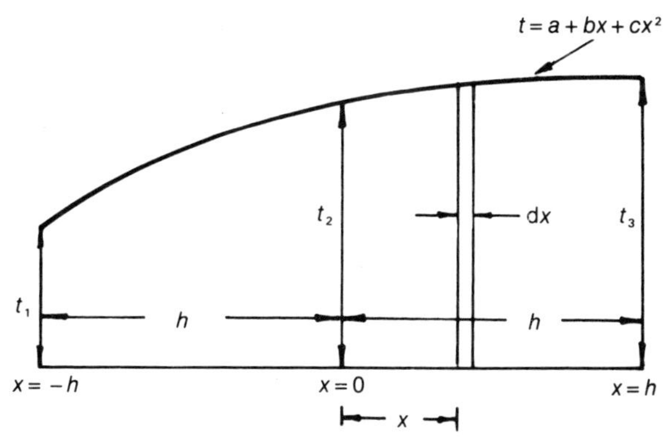

Fig. 5.14.

Now from (5.46), the maximum deflection of the membrane

$$= t^2/4$$

so that the volume of an elemental strip "dx"

$$= \tfrac{2}{3} t * \frac{t^2}{4} * \mathrm{d}x$$

Hence, the total volume under membrane

$$= \int \frac{t^3}{6} \mathrm{d}x$$

$$= \tfrac{1}{6} \int_{-h}^{h} (a^3 + 3a^2bx + 3a^2cx^2 + 3ab^2x^2$$
$$+ 6abc\,x^3 + 3ac^2x^4 + b^3x^3 + 3b^2cx^4$$
$$+ 3bc^2x^5 + c^3x^6)\mathrm{d}x \tag{5.52}$$

but as,

$$@\, x = -h, \quad t = t_1$$
$$@\, x = 0, \quad t = t_2$$

and

$$@\, x = h, \quad t = t_3$$

then,

$$a = t_2$$
$$b = \frac{(t_3 - t_1)}{2h} \tag{5.53}$$
$$c = \frac{t_1 + t_3 - 2t_2}{2h^2}$$

Substituting (5.53) into (5.52), and rearranging,

$$\text{Volume under membrane} = \frac{h}{420}\{(13t_1^3 + 64t_2^3 + 13t_3^3)$$
$$+ 16t_2(t_2t_1 + t_2t_3 - t_1t_3)$$
$$+ 20t_2(t_1^2 + t_3^2) - 3t_1(t_1t_3 + t_3^2)\} \tag{5.54}$$

and,

$$J = 2 * \text{volume under membrane}$$

To check the precision of equation (5.54), consider the triangle of Fig. 5.15.

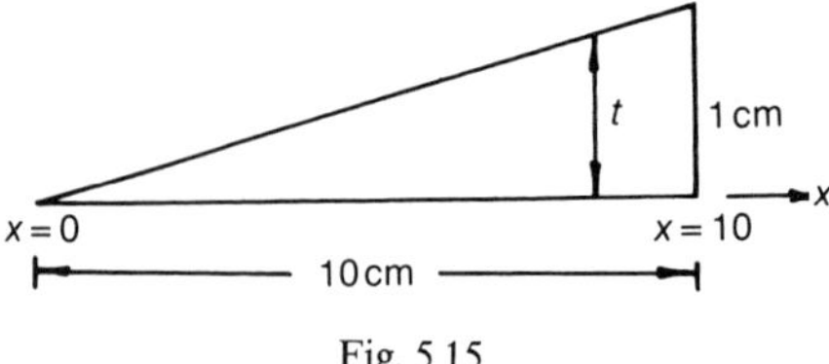

Fig. 5.15.

From Fig. 5.15, it can be seen that:

$$t = x/10$$

so that,

$$J = \tfrac{1}{3}\int_0^{10} (\tfrac{1}{10})^3 x^3 \mathrm{d}x$$
$$= 0.833\ \mathrm{cm}^4$$
$$\underline{J = 8.33\text{E-9}\ \mathrm{m}^4}$$

Substituting the following into equation (5.54):

$$t_1 = 0; \quad t_2 = 0.5\text{E-2}\ \mathrm{m}; \quad t_3 = 1\text{E-2}\ \mathrm{m}$$
$$\underline{J = 8.33\text{E-9}\ \mathrm{m}^4 \text{ (as required)}}$$

5.11.1 EXAMPLE 5.5 *J* FOR SLIM AEROFOIL SECTIONS

Calculate J for the sections of Figs. 5.16(a) and (b), by equations (5.50) and (5.54), where all the dimensions are in cm.

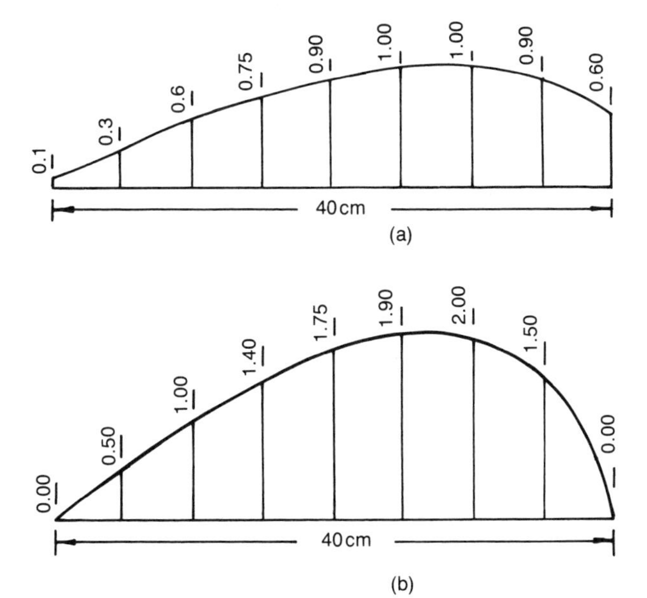

Fig. 5.16 Aerofoil sections.

5.11.2

The results are as follows:

Aerofoil Fig. 5.16(a)
J\{formula (5.50)\} = 7.233E-8 m^4
J\{formula (5.54)\} = 7.107E-8 m^4

Aerofoil Fig. 5.16(b)
J\{formula (5.50)\} = 4.577E-7 m^4
J\{formula (5.54)\} = 4.532E-7 m^4

From the above, it can be seen that the approximate formula (5.50) overestimates the value of J by 1.8% for aerofoil Fig. 5.16(a) and by 1% for aerofoil Fig. 5.16(b).

The maximum shear stress in the aerofoil section =

$$\hat{\tau} = T\hat{t}/J \tag{5.55}$$

5.12.1 THE TORSION OF A THIN-WALLED CLOSED TUBE

Consider a thin-walled closed tube, where the cross-section is covered by a thin weightless membrane, which is subjected to a uniform lateral pressure P.

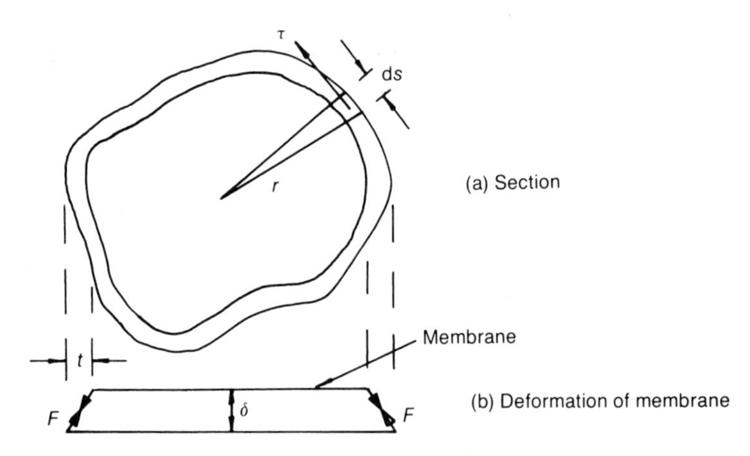

Fig. 5.17. Single cell.

Now the shear stress τ acts tangentially to the boundary, so that the torque

$$T = \int \tau t \cdot r ds$$

but

$$\tau t = q = \text{shear flow} = \text{a constant}$$

therefore

$$T = \tau t \int r ds$$

but

$$\int r ds = 2 * \text{enclosed area of section}$$

$$= 2A$$

where,

$$A = \text{enclosed area of section}$$

therefore

$$\tau = \frac{T}{2At} \tag{5.56}$$

and,

$$q = \frac{T}{2A} \tag{5.57}$$

5.12.2

From equation (5.57), it can be seen that for a thin-walled tube, the shear flow q, due to T, is constant.

To determine J, resolve the forces vertically in Fig. 5.17.

$$P * A = \oint F * \mathrm{d}s * \text{slope}$$

$$= \oint F\left(\frac{\delta}{t}\right)\mathrm{d}s$$

or,

$$\frac{PA}{F} = \delta \oint \mathrm{d}s/t$$

but,

$$\frac{P}{F} = 2$$

therefore

$$\delta = \frac{2A}{\oint \mathrm{d}s/t}$$

Now, volume under membrane,

$$V = A * \delta = 2A^2 \Big/ \oint \mathrm{d}s/t$$

but,

$$J = 2V$$

$$= 4A^2 \Big/ \oint \mathrm{d}s/t$$

$$J = \frac{4A^2 t}{L} \tag{5.58}$$

where,

L = peripheral length of the single cell of Fig. 5.17

Now,

$$T = 2G\theta * \text{volume}$$

$$T = 4G\theta A^2 t/L \tag{5.59}$$

5.13.1 BATHO–BREDT THEORY

This theory is suitable for the torsion of thin-walled closed tubes, and, in particular, for multi-cell tubes.

Consider an element of a thin-walled closed tube under the stress system of Fig. 5.18.

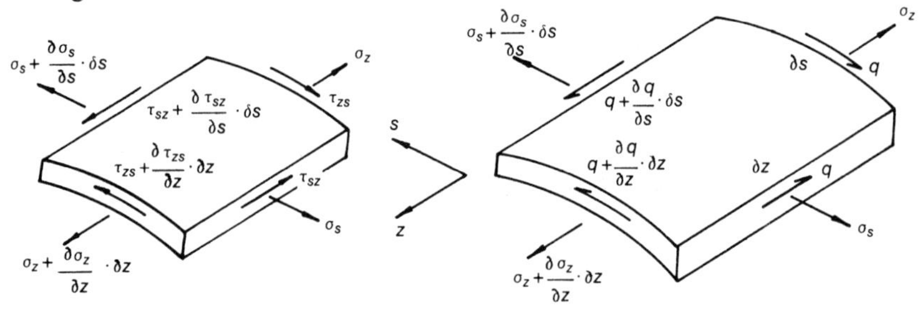

Fig. 5.18. Element of a thin-walled closed tube.

Considering equilibrium in the z direction,

$$\left(\sigma_z + \frac{\partial \sigma_z}{\partial z} \cdot \delta z\right) t\delta s - \sigma_z t\delta s + \left(q + \frac{\partial q}{\partial s}\delta s\right)\delta z - q\delta z = 0$$

which reduces to:

$$\frac{\partial q}{\partial s} + t\frac{\partial \sigma_z}{\partial z} = 0 \qquad (5.60)$$

Similarly for equilibrium in the s direction:

$$\frac{\partial q}{\partial z} + t\frac{\partial \sigma_s}{\partial s} = 0 \qquad (5.61)$$

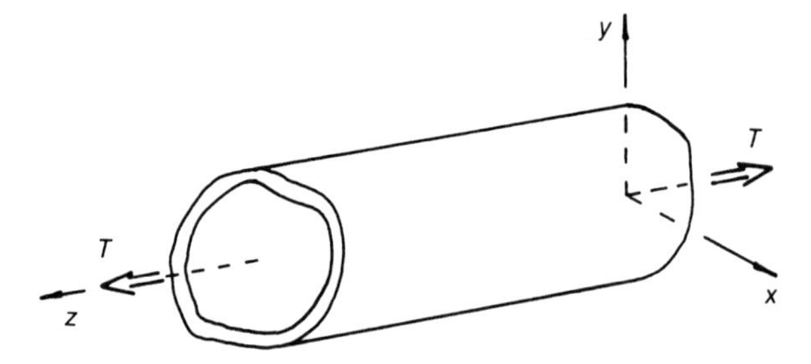

Fig. 5.19. Thin-walled closed tube.

If this closed tube is subjected to a pure torque, and there is no axial restraint, then there is no direct stress in the s–z system. Hence, (5.60) and (5.61) become:

$$\frac{\partial q}{\partial s} = 0$$

and,

$$\frac{\partial q}{\partial z} = 0$$

i.e.

$$\underline{q = \text{shear flow} = \text{constant, and } \tau = \text{shear stress} = q/t} \qquad (5.62)$$

5.13.2

Consider the closed tube shown in Fig. 5.20.

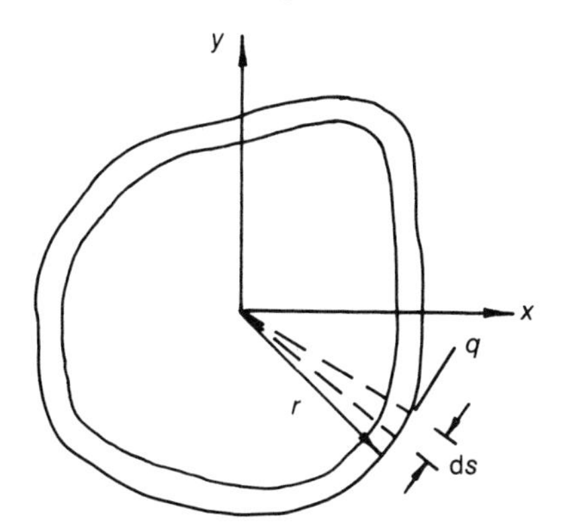

Fig. 5.20. Single-cell closed tube.

$$T = \oint rq\,\mathrm{d}s$$

but,

$$\oint r\,\mathrm{d}s = 2A$$

therefore

$$T = 2Aq \tag{5.63}$$

where,

$$A = \text{enclosed area}$$

The above theory is known as the Batho–Bredt theory, and (5.63) as the Batho–Bredt formula, which agrees with the membrane theory formula of (5.57).

New from reference [8], and Chapter 2,

$$\frac{\mathrm{d}\theta}{\mathrm{d}z} = \frac{q}{2A}\oint\frac{\mathrm{d}s}{Gt} \tag{5.64}$$

$$= \frac{T}{4A^2}\oint\frac{\mathrm{d}s}{Gt} \tag{5.65}$$

5.13.3

Equation (5.64) can be used for a *multi-cell tube*, where,

$$\frac{\mathrm{d}\theta}{\mathrm{d}z} = \frac{1}{2A_R G}\oint q_R\frac{\mathrm{d}s}{t}\ \text{for the } R\text{th cell}$$

i.e. for the Rth cell of the multi-cell tube of Fig. 5.21,

$$\frac{\mathrm{d}\theta}{\mathrm{d}z} = \frac{1}{2A_R G}[q_R(\delta_{12} + \delta_{34}) + (q_R - q_{R-1})\delta_{23} + (q_R - q_{R-1})\delta_{14}]$$

$$= \frac{1}{2A_R G}[-q_{R-1}\delta_{23} + q_R(\delta_{12} + \delta_{23} + \delta_{34} + \delta_{14}) - q_{R+1}\delta_{14}]$$

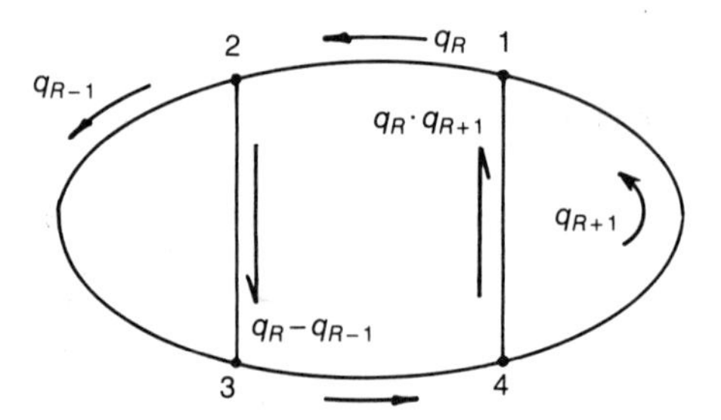

Fig. 5.21. Multi-cell tube.

or,

$$\frac{\mathrm{d}\theta}{\mathrm{d}z} = \frac{1}{2A_R G}[-q_{R-1}\delta_{R-1,R} + q_R\delta_R - q_{R+1}\delta_{R+1,R}] \tag{5.66}$$

where,

A_R = enclosed area of Rth cell
q_R = shear flow in Rth cell
$\delta_{R-1,R} = \int \mathrm{d}s/t$

for the wall common to the Rth and $(R-1)$th cells.

5.14.1 EXAMPLE 5.6 MULTI-CELL TUBE UNDER TORSION

Determine the shear stresses for the multi-cell tube under torsion, assuming that the tube has the properties given in Table 5.2, where,

ij^E = external wall defined by i and j
ij^I = internal wall defined by i and j
l = peripheral length
t = wall thickness
G = rigidity modulus
A_I, A_{II}, A_{III} = enclosed cell area of cells I, II and III (see Fig. 5.22)
T = applied torque = 12 000 N m

Determine, also, the angle of twist over a length of 3 m.

Table 5.2. Properties of cells.

Wall i–j	l (mm)	t (mm)	G (N/mm^2)	Cell area (mm^2)
12^E	1700	1.25	25,000	$A_I = 250\,000$
12^I	700	2.05	24,000	
13 and 24	800	1.3	26,000	$A_{II} = 350\,000$
34	600	1.75	25,500	
35 and 46	700	1.0	26,500	$A_{III} = 300\,000$
56	750	0.9	24,500	

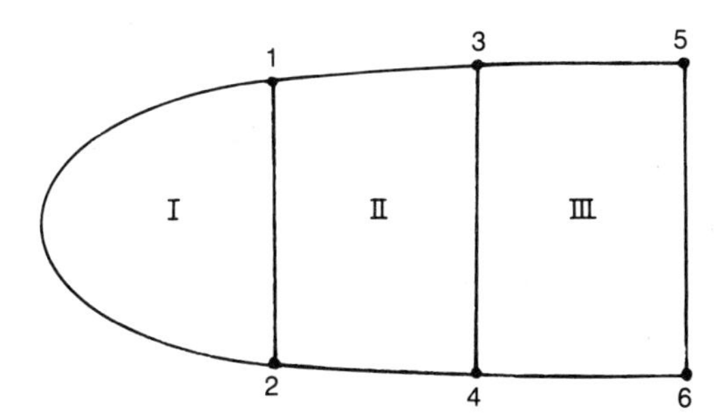

Fig. 5.22. Multi-cell structure.

5.14.2

Calculations for (δ/G) are given in Table 5.3, which are required to obtain the required number of simultaneous equations.

Table 5.3. Calculations for δ/G.

Wall	$\delta = l/t$	δ/G
12^E	1360.0	5.44E-2
12^I	341.5	1.42E-2
13 and 24	615.4	2.37E-2
34	342.9	1.34E-2
35 and 46	700.0	2.64E-2
56	833.3	3.40E-2

Now the shear flow for this tube is assumed to take the form of Fig. 5.23.

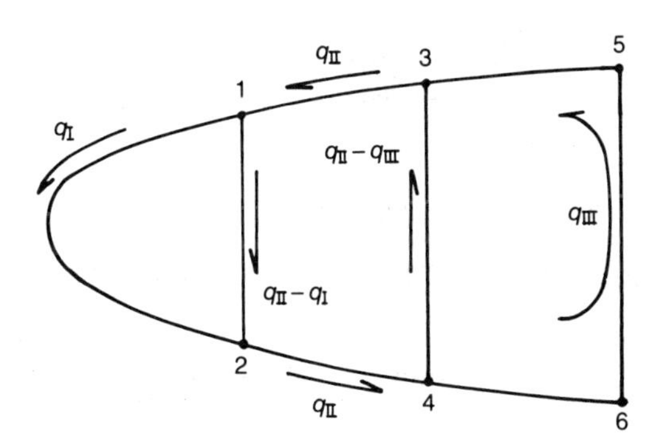

Fig. 5.23. Assumed shear flow.

Three of the four simultaneous equations can be obtained by applying (5.66) to each of the three cells, as follows, where the data are obtained from Tables 5.2 and 5.3:

Cell I

$$\frac{d\theta}{dz}=\frac{10^{-4}}{2\times 250\,000}\{q_I(544+142)-142q_{II}\} \tag{5.67}$$

Cell II

$$\frac{d\theta}{dz}=\frac{10^{-4}}{2\times 350\,000}\{-142q_I+q_{II}(142+237\times 2+134)-134q_{III}\} \tag{5.68}$$

Cell III

$$\frac{d\theta}{dz}=\frac{10^{-4}}{2\times 300\,000}\{-q_{II}\times 134+q_{III}\times(134+264\times 2+340)\} \tag{5.69}$$

The fourth simultaneous equation can be obtained from (5.63), i.e.

$$T=2\sum A_R q_R$$
$$T=2\times(250\,000q_I+350\,000q_{II}+300\,000q_{III}) \tag{5.70}$$

Equations (5.67) to (5.69) can be rewritten, as follows:

$$\frac{10^{-4}}{500\,000}(686q_I-142q_{II})=\frac{10^{-4}}{700\,000}(-142q_I+750q_{II}-134q_{III})$$

or,

$$157.5q_I-135.5q_{II}+19.14q_{III}=0 \tag{5.71}$$

and,

$$\frac{10^{-4}}{500\,000}(686q_I-142q_{II})=\frac{10^{-4}}{600\,000}(-134q_{II}+1002q_{III})$$
$$137.2q_I-6.07q_{II}-167q_{III}=0 \tag{5.72}$$

and,

$$\frac{10^{-4}}{700\,000}(-142q_I+750q_{II}-134q_{III})=\frac{10^{-4}}{600\,000}(-134q_{II}+1002q_{III})$$
$$-20.3q_I+129.5q_{II}-186q_{III}=0 \tag{5.73}$$

and from (5.70),

$$41.67q_I+58.33q_{II}+50q_{III}=1000 \tag{5.74}$$

If equations (5.71), (5.72) and (5.74) are used, the values of shear flow are as follows:

$$\underline{q_I=6.47\ \text{N/mm}\quad q_{II}=8.23\ \text{N/mm}\quad q_{III}=5.0\ \text{N/mm}} \tag{5.75}$$

To check the precision of these values of shear flow, use can be made of equation (5.73), as follows:

$$-20.3\times 6.47+129.5\times 8.23-186\times 5=$$
$$-131.3+1065.8-930=0\ \text{(OK)}$$

5.14.3

From (5.67),

$$\frac{d\theta}{dz} = 2E\text{-}10(686 \times 6.47 - 142 \times 8.23)$$

$$= 6.54E\text{-}10 \text{ rads/mm}$$

therefore

$$\text{Angle of twist} = 6.54E\text{-}10 \times 3000$$

$$= 1.962E\text{-}3 \text{ rads}$$

$$\underline{\text{angle of twist} = 0.112^\circ}$$

5.14.4

The shearing stress can be calculated from equation (5.75), as follows:

$$\tau_{12(E)} = \frac{6.47}{1.25} = 5.18 \text{ MPa} \tag{5.76}$$

$$\tau_{12(I)} = \frac{(8.23 - 6.47)}{2.05} = 0.86 \text{ MPa} \tag{5.77}$$

$$\tau_{13} \text{ and } \tau_{24} = \frac{8.23}{1.3} = 6.33 \text{ MPa} \tag{5.78}$$

$$\tau_{34} = \frac{(8.23 - 5)}{1.75} = 1.85 \text{ MPa} \tag{5.79}$$

$$\tau_{35} \text{ and } \tau_{46} = \frac{5}{T} = 5 \text{ MPa} \tag{5.80}$$

$$\tau_{56} = \frac{5}{0.9} = 5.56 \text{ MPa} \tag{5.81}$$

The distribution of these shearing stresses across the section is shown in Fig. 5.24.

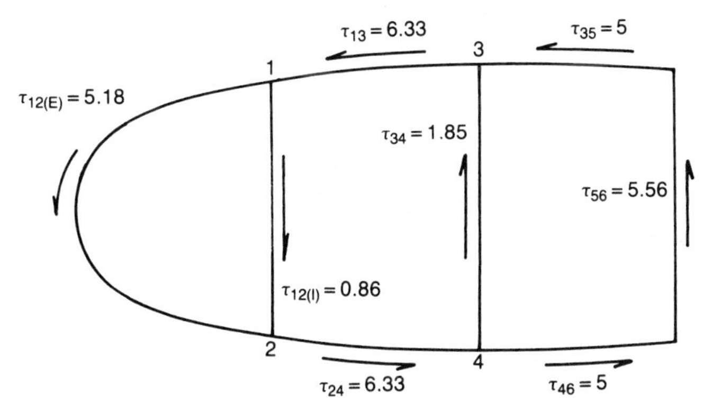

Fig. 5.24. Shear stresses due to torsion (MPa).

5.15.1 THE TORSION EQUATION FOR A SHAFT OF VARYING CIRCULAR CROSS-SECTION

Consider the varying circular section shaft of Fig. 5.25, and assume that,

$$u = w = 0$$

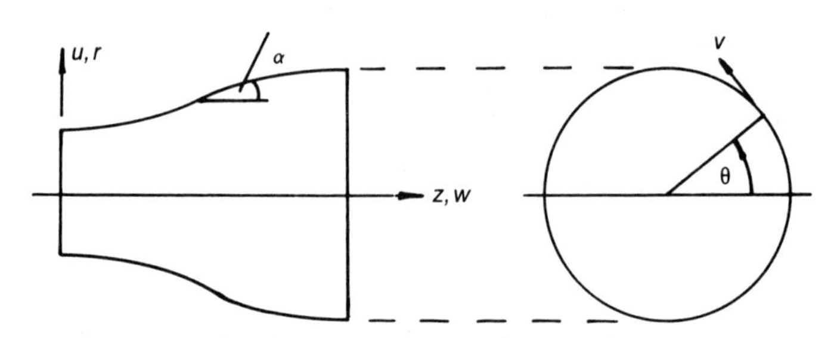

Fig. 5.25. Varying section shaft.

where,

u = radial deflection
v = circumferential deflection
w = axial deflection

As the section is circular, it will be convenient to use polar co-ordinates.
Let,

ε_r = radial strain = 0
ε_θ = hoop strain = 0
ε_z = axial strain = 0
γ_{rz} = shear strain in a longitudinal radial plane = 0
r = any radius on the cross-section

Thus, there are only two shear strains, $\gamma_{r\theta}$ and $\gamma_{\theta z}$, which are defined as follows:

$$\gamma_{r\theta} = \text{shear strain in the } r\text{–}\theta \text{ plane} = \frac{\partial v}{\partial r} - \frac{v}{r}$$

$$\gamma_{\theta z} = \text{shear strain in the } \theta\text{–}z \text{ plane} = \frac{\partial v}{\partial z}$$

But,

$$\tau_{r\theta} = G\gamma_{r\theta} = G\left(\frac{\partial v}{\partial r} - \frac{v}{r}\right) \tag{5.82}$$

and,

$$\tau_{\theta z} = G\gamma_{\theta z} = G\frac{\partial v}{\partial z} \tag{5.83}$$

From equilibrium considerations [3–7],

$$\frac{\partial \tau_{r\theta}}{\partial r} + \frac{\partial \tau_{\theta z}}{\partial z} + \frac{2\tau_{r\theta}}{r} = 0$$

which when rearranged becomes:

$$\frac{\partial}{\partial r}(r^2\tau_{r\theta}) + \frac{\partial}{\partial z}(r^2\tau_{\theta z}) = 0 \tag{5.84}$$

Let,

$$\kappa = \text{shear stress function}$$

where,

$$\frac{\partial \kappa}{\partial r} = r^2\tau_{\theta z} \tag{5.85}$$

and,

$$\frac{\partial \kappa}{\partial z} = -r^2\tau_{r\theta} \tag{5.86}$$

which satisfies (5.84).

From compatibility considerations [3–7],

$$\frac{\partial \gamma_{r\theta}}{\partial z} = \frac{\partial \gamma_{\theta z}}{\partial r} - \frac{\gamma_{\theta z}}{r}$$

or,

$$\frac{\partial \tau_{r\theta}}{\partial z} = \frac{\partial \tau_{\theta z}}{\partial r} - \frac{\tau_{\theta z}}{r} \tag{5.87}$$

From (5.86),

$$\frac{\partial \tau_{r\theta}}{\partial z} = -\frac{1}{r^2}\frac{\partial^2 \kappa}{\partial z^2} \tag{5.88}$$

From (5.85),

$$\frac{\partial \tau_{\theta z}}{\partial r} = \frac{1}{r^2}\frac{\partial^2 \kappa}{\partial r^2} - \frac{2}{r^3}\frac{\partial \kappa}{\partial r} \tag{5.89}$$

Substituting (5.88) and (5.89) into (5.87):

$$-\frac{1}{r^2}\frac{\partial^2 \kappa}{\partial z^2} = \frac{1}{r^2}\frac{\partial^2 \kappa}{\partial r^2} - \frac{2}{r^3}\frac{\partial \kappa}{\partial r} - \frac{1}{r^3}\frac{\partial \kappa}{\partial r}$$

or,

$$\frac{\partial^2 \kappa}{\partial r^2} - \frac{3}{r}\frac{\partial \kappa}{\partial r} + \frac{\partial^2 \kappa}{\partial z^2} = 0 \tag{5.90}$$

From considerations of equilibrium on the boundary,

$$\tau_{r\theta}\cos\alpha - \tau_{\theta z}\sin\alpha = 0 \tag{5.91}$$

where,

$$\cos\alpha = \frac{dz}{ds} \tag{5.92}$$

$$\sin\alpha = \frac{dr}{ds}$$

Substituting (5.85), (5.86) and (5.92) into (5.91),

$$-\frac{1}{r^2}\frac{\partial\kappa}{\partial z}\cdot\frac{dz}{ds}-\frac{1}{r^2}\frac{\partial\kappa}{\partial r}\cdot\frac{dr}{ds}$$

or,

$$\frac{2}{r^2}\frac{d\kappa}{ds}=0$$

i.e.

$$\kappa = \text{a constant on the boundary, as required.}$$

Equation (5.90) is the torsion equation for a tapered circular section, which is of similar form to equation (5.11).

5.16.1 PLASTIC TORSION

The assumption made in this section is that the material is ideally elastic–plastic, as described in Chapter 4, so that the shear stress is everywhere equal to τ_{yp}, the yield shear stress. As the shear stress is constant, the slope of the "membrane" must be constant, and for this reason, the "membrane" analogy is now referred to as a sand-hill analogy.

5.16.2

Consider a *circular section*, where the sand-hill is shown in Fig. 5.26.

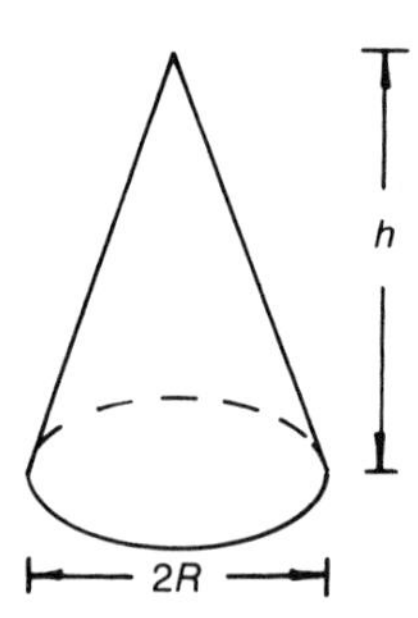

Fig. 5.26. Sand-hill for a circular section.

From Fig. 5.26, it can be seen that the volume of the sand-hill =

$$\text{Vol} = \tfrac{1}{3}\pi R^2 h$$

but,

$$\tau_{yp} = G\theta * \text{slope of sand-hill}$$

where

$$\theta = \text{twist/unit length}$$

or,

$$\tau_{yp} = G\theta\frac{h}{R}$$

therefore

$$h = R\tau_{yp}/G\theta$$

and,

$$\text{Vol} = \frac{\pi R^3 \tau_{yp}}{3G\theta}$$

Now,

$$J = 2 * \text{vol} = 2\pi R^3 \tau_{yp}/(3G\theta)$$

and,

$$T_p = G\theta J$$

therefore

$$T_p = 2\pi R^3 \tau_{yp}/3$$

where,

T_p = fully plastic torsional moment of resistance of the section, which agrees with the value obtained in Chapter 4.

5.16.3

Consider a *rectangular section*, where the sand-hill is shown in Fig. 5.27.

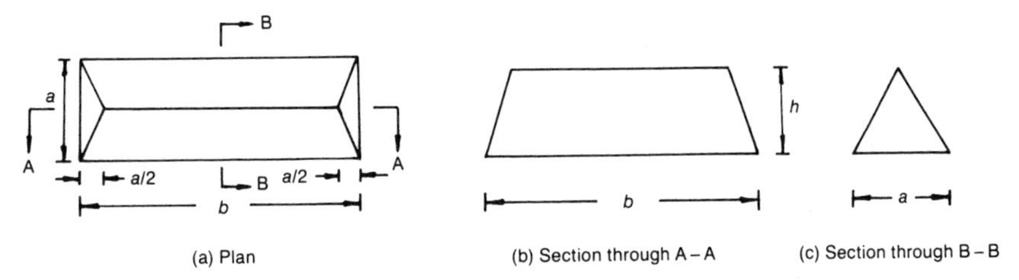

Fig. 5.27. Sand-hill for rectangular section.

Volume under sand-hill =

$$\text{Vol} = \tfrac{1}{2}abh - \tfrac{1}{3}\left(\tfrac{1}{2}a \times \frac{a}{2}\right) \times h \times 2$$

$$= \tfrac{1}{2}abh - \frac{a^2 h}{6}$$

$$= \frac{ah}{6}(3b - a)$$

and,

$$\tau_{yp} = G\theta * \text{slope of sand-hill} = G\theta * 2h/a$$

or,

$$h = \frac{a\tau_{yp}}{2G\theta}$$

therefore

$$\text{Vol} = \frac{a(3b - a)a\tau_{yp}}{12G\theta}$$

Now,

$$J = 2 * \text{vol} = a^2(3b - a)\tau_{yp}/(6G\theta)$$

and,

$$T_p = G\theta J$$

therefore

$$\underline{\underline{T_p = a^2(3b - a)\tau_{yp}/6}}$$

where,

T_p = fully plastic moment of resistance of the rectangular section

5.16.4

Consider an equilateral triangular section, where the sand-hill is shown in Fig. 5.28.

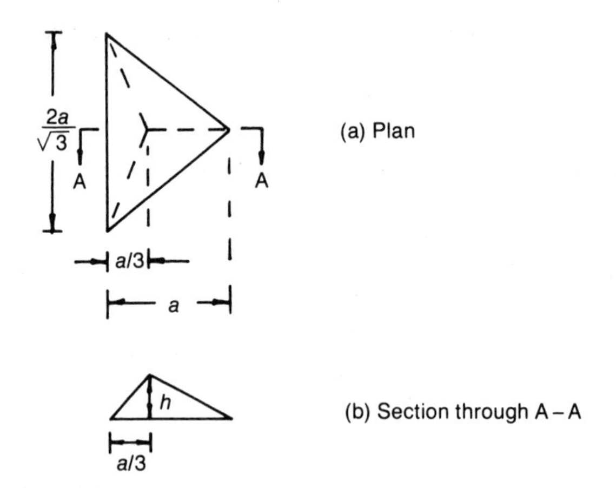

Fig. 5.28. Sand-hill for triangular section.

Now,

$$\tau_{yp} = G\theta * \text{slope of sand-hill}$$

or,

$$\tau_{yp} = G\theta * \frac{3h}{a}$$

and,

$$h = \frac{a\tau_{yp}}{3G\theta}$$

therefore

Volume of sand-hill =

$$\text{Vol} = \tfrac{1}{3}\left(\tfrac{1}{2} * \frac{2a}{\sqrt{3}} * a\right) * h$$

$$= \frac{a^2}{3\sqrt{3}} * \frac{a\tau_{yp}}{3G\theta}$$

or,

$$\text{Vol} = \frac{a^3 \tau_{yp}}{9\sqrt{3}G\theta}$$

and,

$$T_p = 2G\theta * \frac{a^3 \tau_{yp}}{9\sqrt{3}G\theta}$$

$$T_p = \frac{2a^3 \tau_{yp}}{9\sqrt{3}}$$

where,

T_p = fully plastic torsional resistance of the triangular section.

EXAMPLES FOR PRACTICE 5

1. Calculate the torsional constants for the thin-walled open sections of Figs. Q.5.1(a) and (b).

   Determine, also, the maximum shearing stresses for these sections and the angle of twist over a length of 1.25 m, if they are subjected to torques of magnitude 1 kN m. $G = 7.7E10\ N/m^2$.

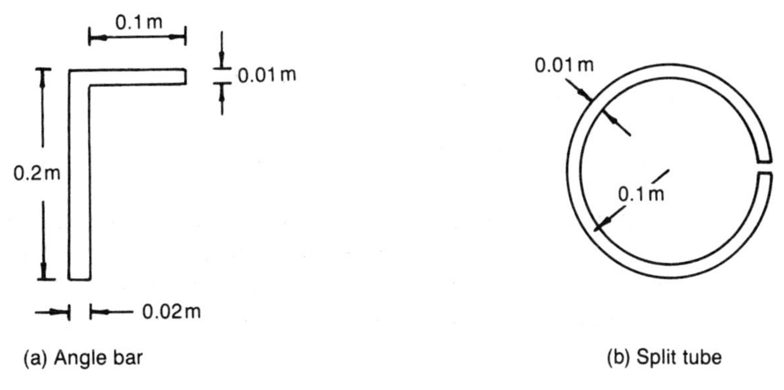

Fig. Q.5.1.

{(a) 5.667E-7 m^4; 35.29 MPa; 1.64°
(b) 2.094E-7 m^4; 47.76 MPa; 4.44°}

2. The single-cell rectangular tube of Fig. Q.5.2 is subjected to a torque of 2000 N m.

   Determine the torsional constant, the maximum shearing stress and the angle of twist over a length of 1.25 m. $G = 7.7E10\ N/m^2$.

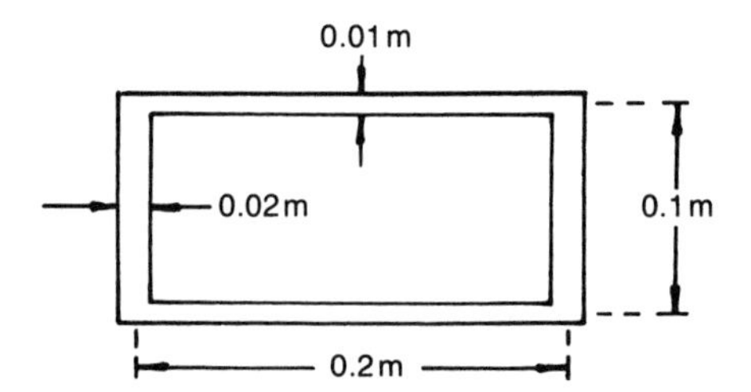

Fig. Q.5.2. Thin-walled closed tube.

{8E-3 m^4; 5 MPa; 2.325E-4°}

3. Determine the values of the shear stress for the two-cell thin-walled torque bar of Fig. Q.5.3 due to the application of a torque of 2000 N m. Hence, or otherwise, determine the angle of twist/length. $G = 7.7\text{E}10\ \text{N/m}^2$.

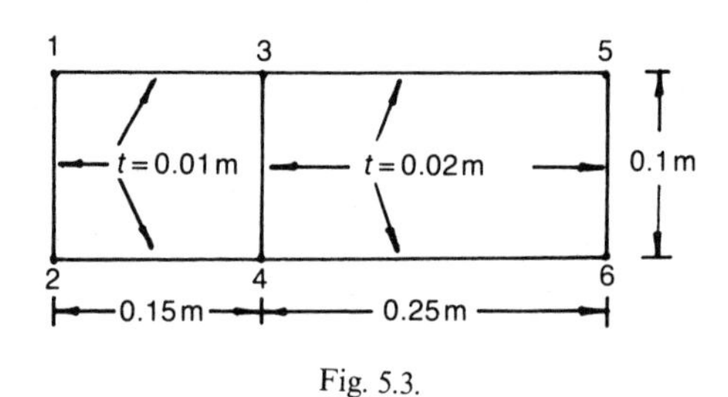

Fig. 5.3.

$$\{\tau_{(1-2,1-3\&2-4)} = 1.64\ \text{MPa}; \quad \tau_{(3-4)} = 0.69\ \text{MPa};$$

$$\tau_{(3-5,4-6\&5-6)} = 1.51\ \text{MPa}; \quad \frac{d\theta}{dz} = 0.0145°/\text{m}\}$$

4. Determine the values of the shear stress in the three-cell thin-walled closed tube of Fig. Q.5.4, due to the application of a torque of 4000 N m, assuming the geometrical properties of the section are as in Table Q.5.4. Hence, or otherwise, determine the twist/length. $G = 26\,000$ MPa.

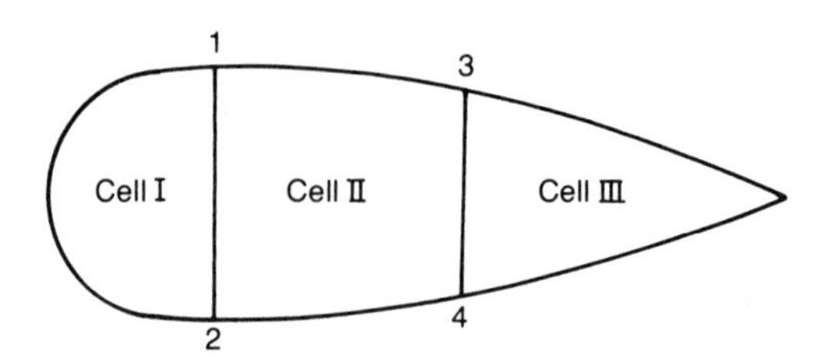

Fig. Q.5.4. Three-cell aerofoil.

Table Q.5.4. Geometrical properties of cell.

Wall	l (mm)	t (mm)	Cell area (mm^2)
$1\text{–}2^E$	1500	1.5	
$1\text{–}2^I$	500	1.75	Cell I = 260 000
1–3 and 2–4	600	1.5	
3–4	400	1.75	Cell II = 250 000
3–5 and 4–5	800	1.5	Cell III = 180 000

$$\{\tau_{1\text{-}2(E)} = 2.041\,\text{MPa};\quad \tau_{1\,2(I)} = 0.106\,\text{MPa};\quad \tau_{1\text{-}3} \text{ and } \tau_{2\text{-}4} = 2.16\,\text{MPa};$$

$$\tau_{3\text{-}4} = 0.609\,\text{MPa};\quad \tau_{3\text{-}5} \text{ and } \tau_{4\text{-}5} = 1.454\,\text{MPa};\quad \frac{d\theta}{dz} = 0.0127°/\text{m}\}$$

5. Determine the fully plastic moment of resistance of the sections of Fig. Q.5.1(a) and (b), given that $\tau_{yp} = 173$ MPa.
   Determine, also, τ_{yp} for both sections.

$\{T_p = 7785$ N m and 5435 N m; $T_{yp} = 4902$ N m and 3623 N m$\}$

6

The Buckling of Struts

6.1.1 STRUTS AND TIES

A *strut*, in its most usual form, can be described as a column under axial compression, as shown in Fig. 6.1.

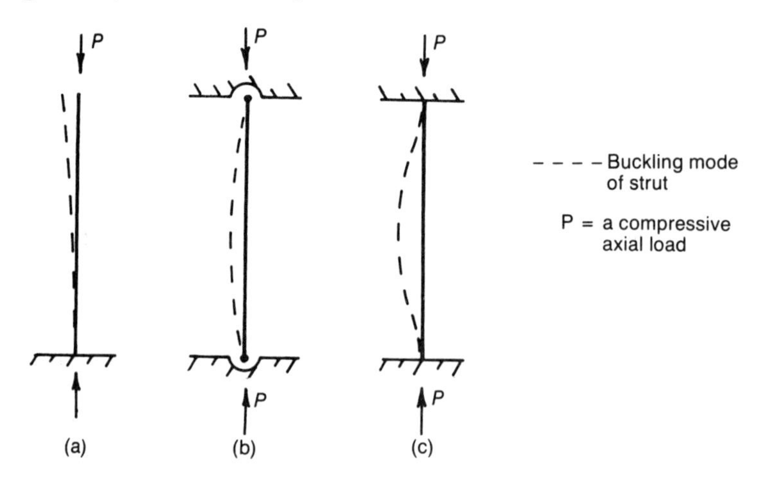

Fig. 6.1. Some axially loaded struts.

From Fig. 6.1, it can be seen that despite the fact that struts are loaded axially, they fail owing to bending moments caused by lateral movement of the struts. The reason for this is that under increasing axial compression, the bending resistance of the strut decreases, until a point is reached where the bending stiffness of the strut is so small that the slightest offset of load, or geometrical imperfection of the strut, causes catastrophic failure (or *instability*).

6.1.2

Similarly for a beam (or length of rubber) under increasing axial tension, its bending stiffness increases, until a point is reached where failure occurs. Beams or rods under tension are called *beam-ties* or *ties*, respectively, and beams under compression are often called *beam-columns*; in this text, considerations will only be made of the latter.

6.1.3

Struts appear in many and various forms, from the pillars in the hold of a ship to the forks of a bicycle, and from pillars supporting the roofs of ancient monuments to those supporting the roofs of modern football stadiums.

For some struts, the design against failure is one of guarding against instability, whilst for others, it is one of determining the stresses due to the combined action of axial and lateral loading.

6.2.1 AXIALLY LOADED STRUTS

Initially "straight" struts, which are subjected to axial compression, but without additional lateral loading, as shown in Fig. 6.1, are classified under the following three broad headings:

6.2.2

Very short struts, as shown in Fig. 6.2, which fail owing to excessive stress values. For such cases, instability is not of any importance.

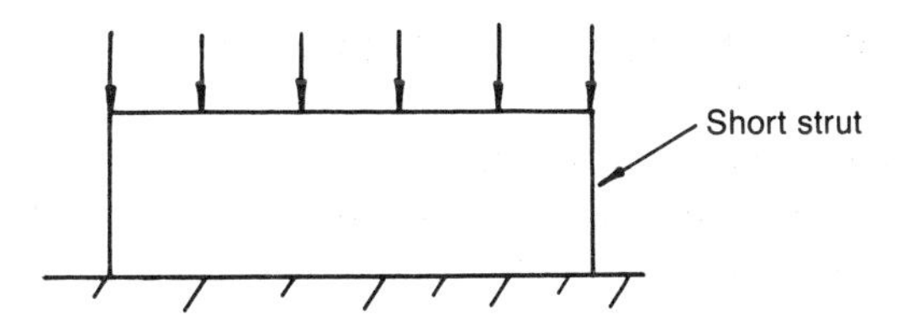

Fig. 6.2. Very short strut.

6.2.3

Very long and slender struts, which fail owing to elastic instability. For such cases, the calculation of stresses is not important, and the only material properties of interest are elastic modulus and in some cases Poisson's ratio.

6.2.4

Intermediate struts, whose slenderness is somewhere between the previous two extremes. Most struts tend to fall into this category, where both the elastic properties of the material and its failure stress are important. Another feature of much importance for this class of strut is its initial geometrical imperfections.

The initial geometrical imperfections of a strut can cause it to fail at a buckling load which may be a small fraction of the elastic buckling load, and the difference between these two buckling loads is sometimes called *elastic knockdown.*

The buckling behaviour of intermediate struts is sometimes termed *inelastic instability.*

6.3.1 ELASTIC INSTABILITY OF VERY LONG SLENDER STRUTS

This theory is due to Euler [22], and it breaks down the following classes of strut.

(a) Very short struts.
(b) Intermediate struts.
(c) Eccentrically loaded struts.
(d) Struts with initial curvature.
(e) Laterally loaded struts.

Despite the fact that the theory presented in this section is only applicable to very long slender struts, which are made from homogeneous, isotropic and elastic materials, the theory can be extended to cater for axially loaded intermediate struts. In addition to this, the differential equation describing the behaviour of this class of strut can be extended to deal with eccentrically loaded, initially curved and laterally loaded struts.

The Euler theory for the elastic instability of a number of initially straight struts will now be considered.

6.4.1 EXAMPLE 6.1 STRUT PINNED AT ITS ENDS

Determine the Euler buckling load for an axially loaded strut, pinned at its ends, as shown in Fig. 6.3. It may be assumed that the ends of the strut are free to move axially towards each other.

6.4.2

In the case of this strut, lateral movement of the struts is prevented at its ends, but the strut is free to rotate at these points (i.e. it is "position fixed" at both ends).

Experiments have shown that such struts tend to have a buckling mode, as shown by the dashed line of Fig. 6.3.

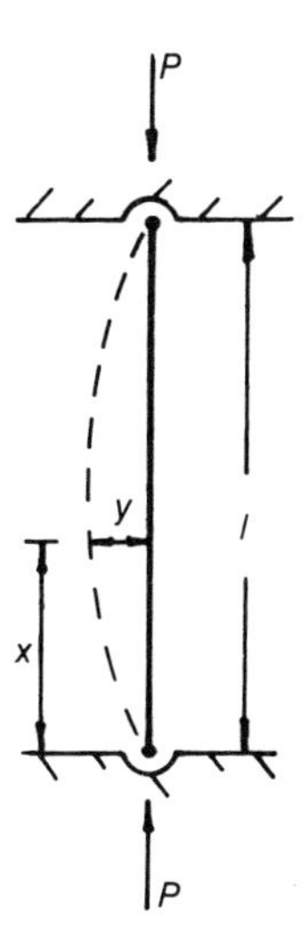

Fig. 6.3. Axially loaded trust, pinned at its ends.

Just prior to instability, let the lateral deflection of the strut be y, at a distance x from its base.

Now,

$$EI\frac{d^2y}{dx^2} = M = -Py \tag{6.1}$$

For struts, it is necessary for the sign convention for bending moment to be such that the *bending moment due to the product Py is always negative* and vice versa for a tie.

The mathematical reasons why the product Py must have a negative sign before it for a strut and why it must have a positive sign before it for a tie are dealt with in a number of texts [23] and will not be dealt with here, as this book is concerned with applying mathematics to some problems in engineering.

The student must, however, remember to define the sign convention for bending moment, based on the assumption that the product Py causes a negative bending moment for all struts.

Equation (6.1) can now be rewritten in the form:

$$\frac{d^2y}{dx^2} + \frac{Py}{EI} = 0 \tag{6.2}$$

Let $\alpha^2 = P/EI$, so that (6.2) becomes:

$$\frac{d^2y}{dx^2} + \alpha^2 y = 0 \tag{6.3}$$

Let

$$y = Ae^{\lambda x} \tag{6.4}$$

so that

$$\frac{d^2y}{dx^2} = \lambda^2 Ae^{\lambda x} = \lambda^2 y \tag{6.5}$$

Substituting (6.4) and (6.5) into (6.3),

$$\lambda^2 y + \alpha^2 y = 0$$

or,

$$\lambda^2 + \alpha^2 = 0$$

therefore

$$\lambda = \pm j\alpha$$

where,

$$j = \sqrt{-1}$$

i.e.

$$\underline{y = A\cos\alpha x + B\sin\alpha x} \tag{6.6}$$

As there are two constants, namely A and B, it will be necessary to apply two boundary conditions to (6.6).

By inspection, it can be seen that

$$\left.\begin{aligned} &\text{at } x = 0, \quad y = 0 \\ \text{and} \\ &\text{at } x = l, \quad y = 0 \end{aligned}\right. \tag{6.7}$$

From (6.7a),

$$A = 0,$$

and from (6.7b),

$$B\sin\alpha l = 0 \tag{6.8}$$

Now the condition $B = 0$ in (6.8) is not of practical interest, as the strut will not suffer lateral deflections if this were so; therefore, the only possibility for equation (6.8) to apply is that

$$\sin\alpha l = 0$$

or

$$\alpha l = 0, \pi, 2\pi, 3\pi, \text{ etc.}$$

The value of αl that is of interest for the buckling of struts will be the lowest positive value, as nature has shown that the strut of Fig. 6.3 will buckle by this mode, i.e.

$$\alpha l = \pi$$

or

$$\sqrt{\frac{P}{EI}}\,l = \pi$$

or

$$P = \frac{\pi^2 EI}{l^2} \tag{6.9}$$

Equation (6.9) is often written in the form of (6.10), where,

P_e = Euler buckling load of a strut, pinned at its ends.

$$P_e = \frac{\pi^2 EI}{l^2} \tag{6.10}$$

6.5.1 EXAMPLE 6.2 STRUT CLAMPED AT ITS ENDS

Determine the Euler buckling load for an axially loaded strut, clamped at its ends, as shown in Fig. 6.4. It may be assumed that the ends of the strut are free to move axially towards each other.

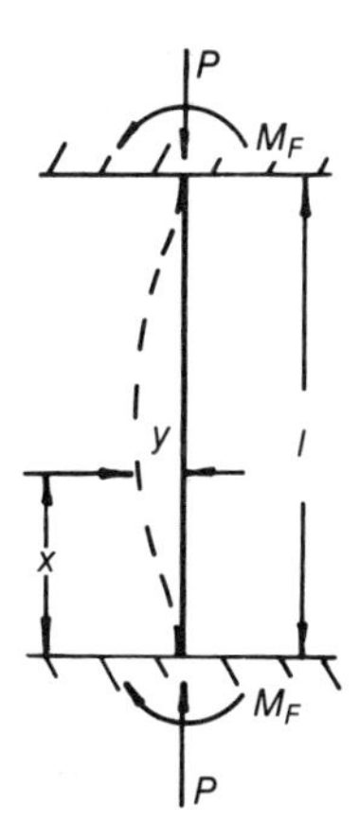

Fig. 6.4. Axially loaded strut, clamped at its ends.

6.5.2

At both its ends, the strut is prevented from lateral and rotational movement, where the latter is achieved through the reaction moments M_F (i.e. it is "direction fixed" at both ends).

Experiments have shown that the strut will have the buckled form shown by the dashed line.

Just prior to buckling, let the deflection at a distance x from the base of the strut be given by y, so that,

$$EI\frac{d^2y}{dx^2} = -Py + M_F \tag{6.11}$$

N.B. As the product Py is a negative bending moment, M_F will be a positive one.

Equation (6.11) can be rewritten in the form

$$\frac{d^2y}{dx^2} + \alpha^2 y = \frac{M_F}{EI} \tag{6.12}$$

From Example 6.1, it can be seen that the complementary function is

$$y = A\cos\alpha x + B\sin\alpha x \tag{6.13}$$

To obtain the particular integral, let

$$\mathrm{D} = \frac{\mathrm{d}}{\mathrm{d}x} = \text{operator "D"},$$

so that (6.12) becomes,

$$(\mathrm{D}^2 + \alpha^2)y = \frac{M_\mathrm{F}}{EI}$$

or

$$y = \frac{M_\mathrm{F}}{(\mathrm{D}^2 + \alpha^2)EI}$$

$$= \frac{\left(1 + \dfrac{\mathrm{D}^2}{\alpha^2}\right)^{-1}}{\alpha^2} \frac{M_\mathrm{F}}{EI}$$

i.e.

$$y = \frac{M_\mathrm{F}}{EI\alpha^2} \tag{6.14}$$

From (6.13) and (6.14), the complete solution is

$$y = A\cos\alpha x + B\sin\alpha x + \frac{M_\mathrm{F}}{EI\alpha^2} \tag{6.15}$$

and

$$\frac{\mathrm{d}y}{\mathrm{d}x} = -\alpha A\sin\alpha x + \alpha B\cos\alpha x$$

Now there are three unknowns, namely A, B and M_F; hence, it will be necessary to use three boundary conditions, as follows:

when

$$\underline{x = 0, \quad \frac{\mathrm{d}y}{\mathrm{d}x} = 0} \quad \text{therefore} \quad \underline{B = 0}$$

and when,

$$\underline{x = 0, \quad y = 0}$$

therefore

$$0 = A + \frac{M_\mathrm{F}}{EI\alpha^2}$$

or,

$$A = \frac{-M_\mathrm{F}}{EI\alpha^2}$$

The third boundary condition can be when

$$\underline{x = l, \quad y = 0}$$

or

$$0 = A\cos\alpha l + \frac{M_{\mathrm{F}}}{EI\alpha^2}$$

or

$$\frac{M_{\mathrm{F}}}{EI\alpha^2}(1 - \cos\alpha l) = 0$$

i.e.

$$\cos\alpha l = 1, \quad \text{or} \quad \alpha l = 0, 2\pi, 4\pi, 6\pi, \text{etc.}$$

Nature has shown that the value of interest is $\alpha l = 2\pi$ (the lowest positive root) or

$$\underline{P_{\mathrm{e}} = 4\pi^2 EI/l^2} \tag{6.16}$$

where,

P_{e} = Euler buckling load for clamped ends.

6.6.1 STRUT FIXED AT ONE END AND FREE AT THE OTHER

By a similar process, it can be seen that the Euler buckling load for a strut fixed at one end and free at the other, as shown in Fig. 6.5, is given by

$$P_{\mathrm{e}} = \frac{\pi^2 EI}{4l^2} \tag{6.17}$$

Fig. 6.5. Axially loaded strut, fixed at one end and free at the other.

Comparing equations (6.10), (6.16) and (6.17), it can be seen that the assumed end conditions play a significant role. Indeed, the “clamped ends” strut has an Euler buckling load four times greater than the “pinned ends” strut and sixteen times greater than the strut of Fig. 6.5.

In practice, however, it is very difficult to obtain the completely clamped condition of Fig. 6.4, as there is usually some rotation due to elasticity etc.

at the ends and the effect of this is to drastically reduce the predicted value for P_e from equation (6.16).

6.7.1 LIMIT OF APPLICATION OF EULER THEORY

The Euler formulae are obviously inapplicable where they predict elastic instability loads greater than the crushing strength of the strut material.

For example, assuming that a "pinned ends" strut is made from mild steel, with a crushing stress of 300 MN/m^2, the lower limit of the Euler theory is obtained, as follows:

$$P_e = \frac{\pi^2 EI}{l^2} = \frac{\pi^2 EAk^2}{l^2} \tag{6.18}$$

where,

A = cross-sectional area

k = least radius of gyration

$$\text{Yield load} = 300 \frac{\text{MN}}{\text{m}^2} \times A \tag{6.19}$$

Equating (6.18) and (6.19),

$$\frac{\pi^2 EA}{(l/k)^2} = 300\,A \tag{6.20}$$

Let $E = 2 \times 10^{11}$ N/m^2 = 2×10^5 MN/m^2, and substitute this into (6.20) to give:

$$(l/k)^2 = \frac{\pi^2 \times 2 \times 10^5}{300} = 6579.7$$

or

$$\underline{(l/k) = 81 = \text{slenderness ratio}} \tag{6.21}$$

From (6.21), it can be seen that, for mild steel, the Euler theory is obviously inapplicable where the *slenderness ratio* is less than 80, and in practice, the theory needs correction for some values of slenderness ratios greater than this.

6.8.1 RANKINE–GORDON FORMULA

This formula is applicable to intermediate struts, and extends the Euler formula to take account of the crushing stress of the strut material by a semi-empirical approach, as follows.

For a very short strut, the crushing load

$$P_c = \sigma_c A \tag{6.22}$$

where,

A = cross-section

σ_c = crushing stress

Let,

$$P_R = \text{Rankine–Gordon load}$$

where,

$$\frac{1}{P_R} \propto \frac{1}{P_e} + \frac{1}{P_c}$$

By experiment, it was found that

$$P_R = \frac{\sigma_c A}{\left[1 + a\left(\frac{l}{k}\right)^2\right]} \tag{6.23}$$

where,

a = denominator constant in the Rankine–Gordon formula, which is dependent on boundary conditions and material property.

For mild steel,

$a = \frac{1}{7500}$ for pin-jointed ends

$a = \frac{1}{4} \times \frac{1}{7500}$ for both ends clamped

$a = 4 \times \frac{1}{7500}$ for one end fixed and the other free

A comparison of the Rankine–Gordon formula with Euler, for geometrically perfect struts, is given in Fig. 6.6.

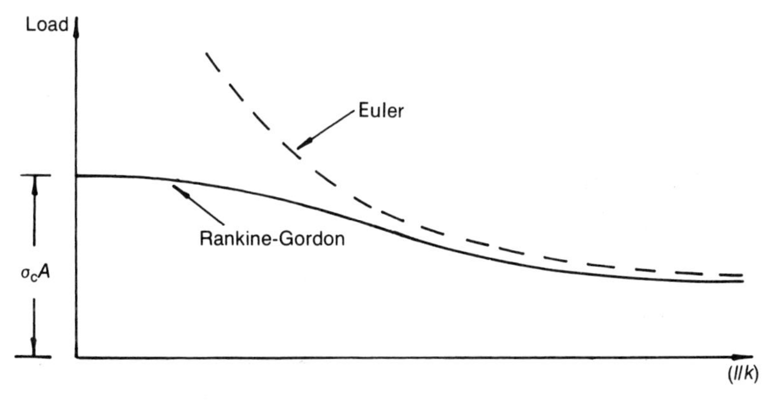

Fig. 6.6. Comparison of Euler with Rankine–Gordon.

6.9.1 EFFECTS OF GEOMETRICAL IMPERFECTIONS

For intermediate struts with geometrical imperfections, the buckling load is further decreased, as shown in Fig. 6.7.

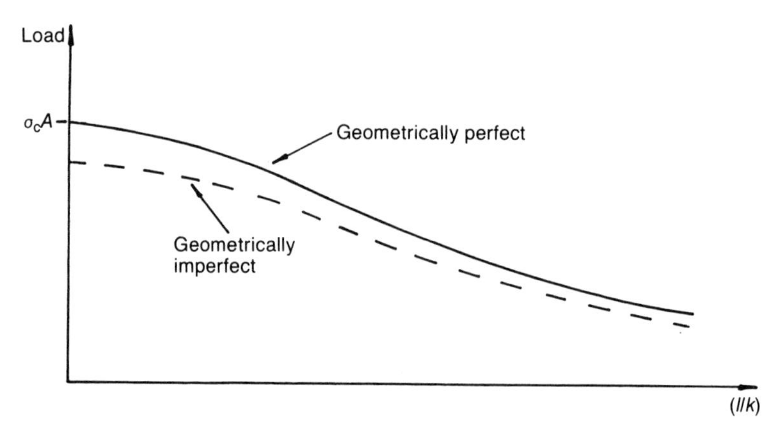

Fig. 6.7. Rankine–Gordon loads for perfect and imperfect struts.

6.10.1 Johnson's parabolic formula

This is a simplified version of the Rankine–Gordon expression, and it proved to be popular in design offices, prior to the invention of the hand-held calculator.

$$\text{Johnson buckling stress} = \sigma_c - b\,(l/k)^2$$

where,

b = a constant depending on end conditions, material properties, etc.

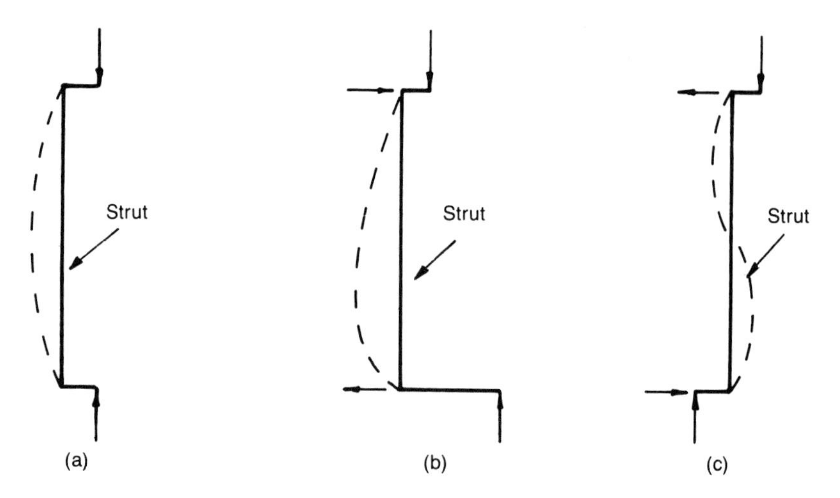

(a) Eccentricity equal and on the same side of the strut

(b) Eccentricity unequal and on the same side of the strut

(c) Eccentricity on opposite sides of the strut

- - - - Deflected form of strut

Fig. 6.8. Eccentrically loaded struts.

6.11.1 ECCENTRICALLY LOADED STRUTS

Struts in this category, examples of which are shown in Fig. 6.8, frequently occur in practice. For such problems, the main object in stress analysis is to determine the stresses due to the combined effects of bending and axial load, unlike the struts of Sections 6.3 to 6.6, where the main object was to obtain a crippling load.

6.12.1 EXAMPLE 6.3 ECCENTRICALLY LOADED STRUT

The pillar in the hold of ship is in the form of a tube of external diameter 25 cm, internal diameter 22.5 cm and length 10 m. If the pillar is subjected to an eccentric load of 20 tonnes, as shown in Fig. 6.9, calculate the maximum permissible eccentricity, if the maximum permissible stress is 75 MN/m^2. It may be assumed that $E = 2 \times 10^{11}$ N/m^2.

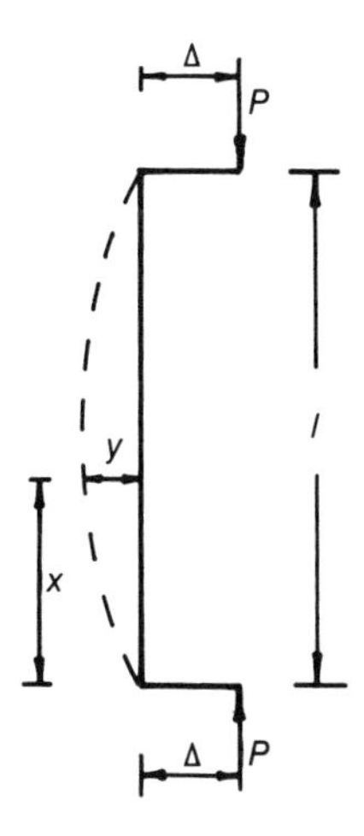

Fig. 6.9. Eccentrically loaded strut.

6.12.2

At any distance x from the base,

$$EI\frac{d^2y}{dx^2} = M = -P(y + \Delta)$$

or

$$\frac{d^2y}{dx} + \frac{P}{EI}y = -\frac{P}{EI}\Delta \tag{6.24}$$

Let,

$$\alpha^2 = \frac{P}{EI}$$

so that (6.24) becomes

$$\frac{d^2y}{dx^2} + \alpha^2 y = -\alpha^2 \Delta \tag{6.25}$$

The complementary function of (6.25) is

$$y = A\cos\alpha x + B\sin\alpha x \tag{6.26}$$

and the particular integral of (6.25) is

$$y = \frac{-\alpha^2\Delta}{(D^2 + \alpha^2)}$$

giving,

$$y = -\Delta \tag{6.27}$$

Combining (6.26) and (6.27), the complete solution is

$$y = A\cos\alpha x + B\sin\alpha x - \Delta \tag{6.28}$$

Now there are two unknowns in (6.28); therefore, two boundary conditions will be required, as follows:

@ $\quad x = 0, \quad y = 0$

and (6.29)

@ $\quad x = l, \quad y = 0$

From (6.29a):

$$\underline{A = \Delta} \tag{6.30}$$

From (6.29b)

$$0 = \Delta\cos\alpha l + B\sin\alpha l - \Delta$$

or

$$B = \frac{\Delta(1 - \cos\alpha l)}{\sin\alpha l} = \frac{\Delta 2\sin^2(\alpha l/2)}{2\sin(\alpha l/2)\cos(\alpha l/2)}$$

$$\underline{B = \Delta\tan(\alpha l/2)} \tag{6.31}$$

Substituting (6.30) and (6.31) into (6.28), the deflected form of the strut, at any distance x, is given by:

$$\underline{\underline{y = \Delta[\cos\alpha x + \tan(\alpha l/2)\sin(\alpha x) - 1]}} \tag{6.32}$$

The maximum deflection δ occurs at $x = l/2$, i.e.

$$\begin{aligned}\delta &= \Delta[\cos(\alpha l/2) + \tan(\alpha l/2)\sin(\alpha l/2) - 1] \\ &= \Delta\cos(\alpha l/2)[1 + \tan^2(\alpha l/2) - 1/\cos(\alpha l/2)] \\ &= \Delta\cos(\alpha l/2)[\sec^2(\alpha l/2) - 1/\cos(\alpha l/2)]\end{aligned}$$

$$\underline{\delta = \Delta[\sec(\alpha l/2) - 1]} \tag{6.33}$$

By inspection, the maximum bending moment $\hat{M}$ is given by:

$$\hat{M} = P(\delta + \Delta)$$

$$\underline{\hat{M} = P\Delta \sec(\alpha l/2)} \tag{6.34}$$

$$I = \frac{\pi}{64}[(25 \times 10^{-2})^4 - (22.5 \times 10^{-2})^4]$$

$$\underline{I = 6.594 \times 10^{-5}\,\text{m}^4}$$

$$A = \frac{\pi}{4}[(25 \times 10^{-2})^2 - (22.5 \times 10^{-2})^2]$$

$$\underline{A = 9.327 \times 10^{-3}\,\text{m}^2}$$

$$\frac{\alpha l}{2} = \sqrt{\frac{P}{EI}}\frac{l}{2} = \sqrt{\frac{20 \times 10^3 \times 9.81}{2 \times 10^{11} \times 6.594 \times 10^{-5}}} \times 5$$

$$\underline{\frac{\alpha l}{2} = 0.6099\,\text{rads} = 34.94^\circ}$$

From (6.34):

$$\hat{M} = 20\,000 \times \Delta \times 1/[\cos(34.94)] \times 9.81$$

$$\underline{\hat{M} = 239\,341\Delta}$$

Now,

$$\hat{\sigma} = \sigma(\text{direct}) \pm \sigma(\text{bending})$$

or

$$-75 \times 10^6 = \frac{-20 \times 10^3 \times 9.81}{9.327 \times 10^{-3}} - \frac{239\,341\Delta}{6.594 \times 10^{-5}} \times 12.5 \times 10^{-2}$$

$$= -2.104 \times 10^7 - 4.537 \times 10^8\Delta$$

or

$$\Delta = 0.1189\,\text{m} = \underline{\underline{11.9\,\text{cm}}}$$

6.13.1 EXAMPLE 6.4 ECCENTRIC LOADING AT BOTH ENDS

If the pillar of Example 6.3 were subjected to the eccentric loading of Fig. 6.10, determine the maximum permissible value of Δ.

6.13.2

The horizontal reactions R are required to achieve equilibrium, and the relationship between them and P can be obtained by taking moments, as follows:

$$Rl = P \times 2\Delta$$

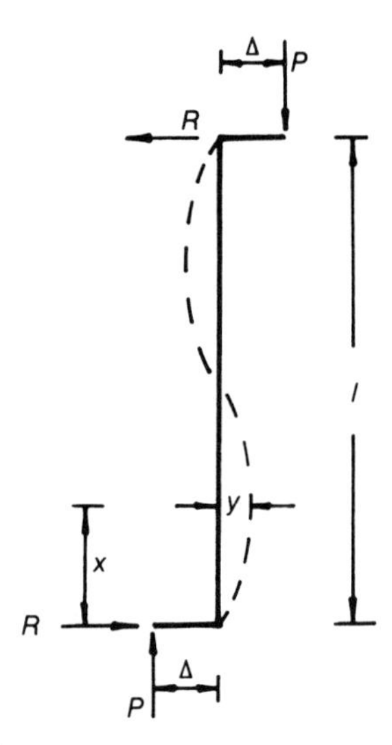

Fig. 6.10. Eccentrically loaded strut.

or

$$R = 2P\Delta/l \tag{6.35}$$

At any distance x from the base,

$$EI\frac{d^2y}{dx^2} = -P(y+\Delta) + Rx$$

or

$$\frac{d^2y}{dx^2} + \alpha^2 y = \frac{R\alpha^2 x}{P} - \alpha^2\Delta \tag{6.36}$$

The complete solution of (6.36) is

$$\underline{y = A\cos\alpha x + B\sin\alpha x + \frac{Rx}{P} - \Delta} \tag{6.37}$$

There are two unknowns; therefore two boundary conditions are required, as follows:

@ $\quad x = 0, \quad y = 0$ therefore $\underline{A = \Delta}$ (6.38)

@ $\quad x = l/2, \quad y = 0$

therefore

$$0 = \Delta\cos(\alpha l/2) + B\sin(\alpha l/2) + \frac{2P\Delta l}{2Pl} - \Delta$$

or

$$\underline{B = -\Delta\cot(\alpha l/2)} \tag{6.39}$$

Substituting (6.38) and (6.39) into (6.37):

$$y = \Delta[\cos\alpha x - \cot(\alpha l/2)\sin\alpha x] + \frac{2P\Delta x}{Pl} - \Delta$$

$$\underline{y = \Delta[\cos\alpha x - \cot(\alpha l/2)\sin\alpha x + (2x/l) - 1]} \quad (6.40)$$

Now,

$$M = EI\frac{d^2y}{dx^2}$$

or

$$M = EI\alpha^2\Delta[-\cos\alpha x + \cot(\alpha l/2)\sin\alpha x]$$
$$= -P\Delta[\cos\alpha x - \cot(\alpha l/2)\sin\alpha x]$$
$$= \frac{-P\Delta[\sin(\alpha l/2)\cos\alpha x - \cos(\alpha l/2)\sin\alpha x]}{\sin(\alpha l/2)}$$

or

$$M = \frac{-P\Delta\sin(\alpha l/2 - \alpha x)}{\sin(\alpha l/2)}$$

The maximum bending moment $\hat{M}$ occurs when $\alpha l/2 - \alpha x = \pm\pi/2$, i.e.

$$\underline{\hat{M} = \pm P\Delta/[\sin(\alpha l/2)] = \pm P\Delta\,\mathrm{cosec}(\alpha l)}$$

Now,

$$\hat{\sigma} = \sigma(\text{direct}) \pm \sigma(\text{bending})$$

i.e.

$$-75\times10^6 = -21.04\times10^6 - \frac{20\,000\times9.81\,\mathrm{cosec}(34.94)\Delta\times12.5\times10^{-2}}{6.594\times10^{-5}}$$

or,

$$53.96\times10^6 = 649.4\times10^6\Delta$$
$$\underline{\Delta = 0.0831\text{ m} = 8.31\text{ cm}}$$

6.14.1 STRUTS WITH INITIAL CURVATURE

Struts in this category are usually assumed to have an initial sinusoidal or parabolic shape of the form shown in Fig. 6.11. As in Section 6.6, stresses are of more importance for this class of strut than are crippling loads.

Let,

$$y_0 = f(x) = \text{initial equation of strut}$$

and,

$$R_0 = \text{initial radius of curvature of strut of } x$$
$$= \frac{d^2y_0}{dx^2}$$

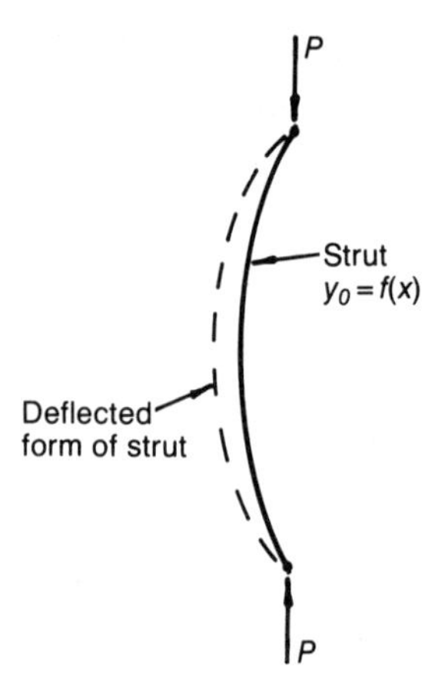

Fig. 6.11. Initially curved strut.

From the relationship,

$$EI\left(\frac{1}{R}-\frac{1}{R_0}\right)=M$$

$$EI\left(\frac{d^2y}{dx^2}-\frac{d^2y_0}{dx^2}\right)=M \tag{6.41}$$

Normally, solution of (6.41) is achieved by assuming the initial shape of the strut y_0 to be either sinusoidal or parabolic or circular.

6.15.1 EXAMPLE 6.5 STRUT WITH INITIAL SINUSOIDAL CURVATURE

Determine the maximum deflection and bending moment for a strut with pinned ends, which has an initial sinusoidal curvature of the form:

$$y_0 = \Delta \sin\left(\frac{\pi x}{l}\right) \tag{6.42}$$

where,

Δ = a small initial central deflection

6.15.2

Substituting (6.42) and its second derivative into (6.41), the following is obtained:

$$EI\left[\frac{d^2y}{dx^2}+\frac{\pi^2}{l^2}\Delta \sin\left(\frac{\pi x}{l}\right)\right] = -Py$$

or,

$$\frac{d^2y}{dx^2}+\alpha^2 y = -\frac{\pi^2}{l^2}\Delta \sin\left(\frac{\pi x}{l}\right) \tag{6.43}$$

The complementary function of (6.43) is

$$y = A\cos\alpha x + B\sin\alpha x \tag{6.44}$$

and the particular integral is

$$y = \frac{-\frac{\pi^2}{l^2}\Delta\sin\left(\frac{\pi x}{l}\right)}{(\mathrm{D}^2 + \alpha^2)}$$

$$= \frac{-\frac{\pi^2}{l^2}\Delta\sin\left(\frac{\pi x}{l}\right)}{[(-\pi^2/l^2) + \alpha^2]} \tag{6.45}$$

From (6.44) and (6.45), the complete solution is:

$$\underline{y = A\cos\alpha x + B\sin\alpha x - \frac{(\pi^2/l^2)\Delta\sin(\pi x/l)}{[(-\pi^2/l^2) + \alpha^2]}} \tag{6.46}$$

Two suitable boundary conditions are:

@ $\quad x = 0, \quad y = 0$ therefore $\underline{A = 0}$

@ $\quad x = l/2, \quad \frac{\mathrm{d}y}{\mathrm{d}x} = 0$ therefore $\underline{B = 0}$

i.e.

$$y = \frac{(\pi^2/l^2)\Delta\sin(\pi x/l)}{[(\pi^2/l^2) - \alpha^2]} = \frac{\Delta P_e\sin(\pi x/l)}{(P_e - P)} \tag{6.47}$$

The maximum deflection δ occurs at $x = l/2$:

$$\delta = \Delta P_e/(P_e - P) \tag{6.48}$$

and

$$\hat{M} = \Delta P P_e/(P_e - P) \tag{6.49}$$

where,

$$\underline{P_e = \pi^2 EI/l^2}$$

6.16.1 EXAMPLE 6.6 SECOND STRUT WITH INITIAL SINUSOIDAL CURVATURE

Determine the maximum permissible value of Δ for an initially curved strut of sinusoidal curvature, given the following:

(a) Sectional and material properties are as in Example 6.4.
(b) Length of strut = 10 m.
(c) The strut is pinned at its ends.

6.16.2

Now,

$$\hat{\sigma} = \sigma(\text{direct}) \pm \sigma(\text{bending})$$

$$P_e = \pi^2 EI/l^2 = 1.302\,\text{MN}$$

$$P = 20\,000 \times 9.81 = 0.196\,\text{MN}$$

therefore

$$\hat{\sigma} = -75 \times 10^6 = -21.04 \times 10^6 - \Delta \times \frac{1.302 \times 0.196 \times 10^6 \times 12.5 \times 10^{-2}}{1.106 \times 6.594 \times 10^{-5}}$$

or,

$$53.96 \times 10^6 = 437.49\Delta \times 10^6$$

or,

$$\underline{\Delta = 0.123\,\text{m} = 12.3\,\text{cm}}$$

6.17.1 PERRY–ROBERTSON FORMULA

From Section 6.15.1, it can be seen that the maximum stress for a strut with initial sinusoidal curvature is given by:

$$\sigma_c = \frac{P}{A} + \frac{\Delta P P_e \bar{y}}{(P_e - P)I} \tag{6.50}$$

Putting

$$\left.\begin{aligned} \gamma &= \Delta\bar{y}/k^2 \\ \text{and} \quad \sigma_e &= P_e/A \end{aligned}\right\} \tag{6.51}$$

where,

k = least radius of gyration of the strut's cross-section.

Substituting (6.51) into (6.50):

$$\sigma_c = \frac{P}{A} + \frac{\gamma(P/A)\sigma_e}{(\sigma_e - P/A)}$$

therefore.

$$\sigma_c(\sigma_e - P/A) = \frac{P}{A}(\sigma_e - P/A) + \gamma\left(\frac{P}{A}\right)\sigma_e$$

or,

$$\left(\frac{P}{A}\right)^2 - [\sigma_c + (\gamma + 1)\sigma_e]\frac{P}{A} + \sigma_c\sigma_e = 0$$

i.e.

$$\underline{\left(\frac{P}{A}\right) = \tfrac{1}{2}[\sigma_c + (\gamma + 1)\sigma_e] - \tfrac{1}{2}\sqrt{[\sigma_c + (\gamma + 1)\sigma_e]^2 - 4\sigma_c\sigma_e}} \tag{6.52}$$

From B.S.S. Code of Practice 113,

$$\underline{\gamma = 0.003(l/k)}$$

6.18.1 EXAMPLE 6.7 APPLICATION OF PERRY–ROBERTSON FORMULA

Calculate the maximum permissible value for P for the strut of Example 6.6, using γ from B.S.S. 113.

$$k = \sqrt{\frac{I}{A}} = 0.084\,\text{m}$$

$$\underline{\gamma = 0.357}$$

6.18.2

From (6.52):

$$\frac{P}{A} = \tfrac{1}{2}\left[75 \times 10^6 + 1.357 \times \frac{1.302 \times 10^6}{9.327 \times 10^{-3}}\right]$$

$$- \tfrac{1}{2}\sqrt{[75 \times 10^6 + 1.357 \times 139.59 \times 10^6]^2}$$

$$- 4 \times 75 \times 10^6 \times 139.59 \times 10^6$$

$$= 132.7 \times 10^6 - 83.73 \times 10^6$$

$$= 48.97 \times 10^6 \text{N}/m^2$$

i.e.

$$P = 0.456\,\text{MN} = \underline{46.56\,\text{tonnes}}$$

6.19.1 LATERALLY LOADED STRUTS

Beam-columns, subjected to a combination of lateral and axial loads, are a very common form of laterally loaded strut. In such cases, the effects of the lateral load are so large that the problem is one of determining safe stresses rather than the crippling loads.

6.20.1 EXAMPLE 6.8 LATERALLY LOADED STRUT WITH A UNIFORMLY DISTRIBUTED LOAD

A longitudinal structural component on the bottom of a ship is subjected to a uniformly distributed lateral load of 1 kN/m, together with a compressive axial load of 0.1 MN, as shown in Fig. 6.12. Determine the maximum stress in the member, given the following:

$$l = 10\,\text{m} \qquad A = 9 \times 10^{-3}\,\text{m}^2 \qquad I = 5.13 \times 10^{-5}\,\text{m}^4$$

$$Z = \text{sectional modulus} = 2 \times 10^{-4}\,\text{m}^3$$

$$E = 2 \times 10^{11}\,\text{N/m}^2$$

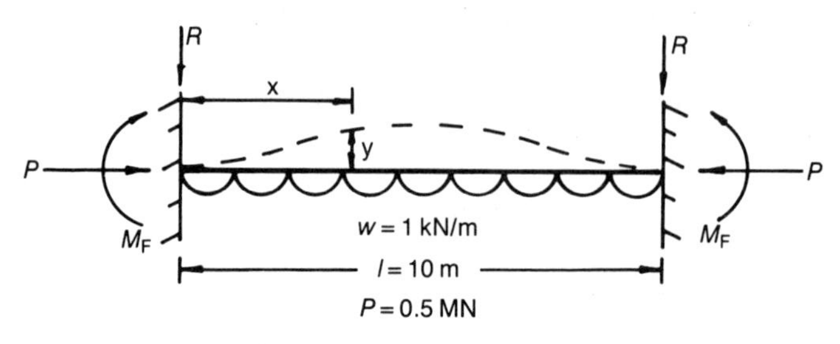

Fig. 6.12. Laterally loaded strut.

6.20.2

Taking moments about the left end,

$$M_F + Rl = M_F + \frac{wl^2}{2}$$

or

$$R = \frac{wl}{2} \tag{6.53}$$

Now,

$$EI\frac{d^2y}{dx^2} = M$$

$$= -Py - Rx + M_F + \frac{wx^2}{2}$$

or,

$$\frac{d^2y}{dx^2} + \frac{P}{EI}y = -\frac{R}{EI}x + \frac{M_F}{EI} + \frac{wx^2}{2EI}$$

Putting $\alpha^2 = P/EI$, the complete solution is given by:

$$y = A\cos\alpha x + B\sin\alpha x + \frac{[-Rx + M_F + wx^2/2]}{EI(D^2 + \alpha^2)}$$

$$= A\cos\alpha x + B\sin\alpha x + \frac{[-Rx + M_F + wx^2/2]}{EI\alpha^2\left[1 + \left(\frac{D}{\alpha}\right)^2\right]}$$

$$= A\cos\alpha x + B\sin\alpha x + \frac{\left[1 + \left(\frac{D}{\alpha}\right)^2\right]^{-1}[-Rx + M_F + wx^2/2]}{EI\alpha^2}$$

$$y = A\cos\alpha x + B\sin\alpha x + \frac{[-Rx + M_F + wx^2/2 - w/\alpha^2]}{EI\alpha^2}$$

or,

$$y = A\cos\alpha x + B\sin\alpha x + \frac{[-wlx/2 + M_F + wx^2/2 - w/\alpha^2]}{EI\alpha^2} \quad (6.54)$$

$$\frac{dy}{dx} = -\alpha A\sin\alpha x + \alpha B\cos\alpha x + \frac{[-wl/2 + wx]}{EI\alpha^2} \quad (6.55)$$

Boundary conditions

@ $x = 0, \quad y = 0$

therefore

$$0 = A + \frac{[M_F - w/\alpha^2]}{EI\alpha^2}$$

or,

$$\underline{A = -[M_F - w/\alpha^2]/[EI\alpha^2]} \quad (6.65)$$

@ $x = 0, \quad \dfrac{dy}{dx} = 0$

therefore

$$0 = \alpha B - \frac{wl}{2EI\alpha^2}$$

$$\underline{B = wl/(2EI\alpha^3)} \quad (6.57)$$

@ $x = l, \quad y = 0$

therefore

$$0 = A\cos\alpha l + B\sin\alpha l + \frac{[-wl^2/2 + M_F + wl^2/2 - w/\alpha^2]}{EI\alpha^2}$$

$$0 = -\frac{[M_F - w/\alpha^2]}{EI\alpha^2}\cos\alpha l + \frac{wl}{(2EI\alpha^3)}\sin\alpha l + \frac{[M_F - w/\alpha^2]}{EI\alpha^2}$$

$$0 = \frac{M_F}{EI\alpha^2}(1 - \cos\alpha l) + \frac{w/\alpha^2}{EI\alpha^2}(\cos\alpha l - 1) + \frac{wl}{2EI\alpha^3}\sin\alpha l$$

therefore

$$\underline{M_F = \frac{w}{\alpha^2(1 - \cos\alpha l)}(1 - \cos\alpha l - \alpha l\sin(\alpha l)/2)} \quad (6.58)$$

Now,

$$\frac{d^2y}{dx^2} = -\alpha^2 A\cos\alpha x - \alpha^2 B\sin\alpha x + w/(EI\alpha^2)$$

Hence, bending moment at the centre

$$\underline{M_{(x=l/2)} = -EI\left[\alpha^2\left(A\cos\left(\frac{\alpha l}{2}\right) + B\sin\left(\frac{\alpha l}{2}\right)\right) - w/(EI\alpha^2)\right]} \tag{6.59}$$

The maximum bending moment will either be the value given in (6.58) or that given in (6.59).

$$\alpha^2 = \frac{P}{EI} = \frac{100\,000}{2 \times 10^{11} \times 5.13 \times 10^{-5}} = 9.747 \times 10^{-3}$$

$$\underline{\alpha = 0.09872}$$

$$\underline{\alpha l = 0.9872 \text{ rads} = 56.57^\circ}$$

$$\underline{\frac{\alpha l}{2} = 28.28^\circ}$$

$$M_F = \frac{1000}{9.746 \times 10^{-3}(1 - 0.5509)}(1 - 0.5509 - 0.4119)$$

$$= 228\,497 \times 0.0372$$

$$\underline{M_F = 8499 \text{ N m}}$$

To determine $M_{(l/2)}$

$$A = -[8499 - 1000/9.7456 \times 10^{-3}]/[1.026 \times 10^7 \times 9.7456 \times 10^{-3}]$$

$$= +94\,111/99\,990$$

$$\underline{A = 0.941}$$

$$\underline{B = 0.507}$$

$$M_{(l/2)} = -1.026 \times 10^7[9.746 \times 10^{-3}(0.829 + 0.240) - 0.01]$$

$$\underline{M_{(l/2)} = -4293 \text{ N m}}$$

i.e.

$$\underline{\underline{\hat{M} = 8499 \text{ N m}}}$$

and,

$$\hat{\sigma} = \frac{-0.1}{9 \times 10^{-3}} - \frac{8499 \times 10^{-6}}{2 \times 10^{-4}}$$

$$= -11.11 - 42.5$$

$$\underline{\underline{\hat{\sigma} = -53.61 \text{ MN/m}^2}}$$

If the *beam-column* effect where neglected,

$$\hat{M} = \frac{wl^2}{12} = 8333\,\text{N m}$$

i.e. the *beam-column* effect caused an additional bending moment of 166 N m for the case.

6.21.1 EXAMPLE 6.9 LATERALLY LOADED STRUT WITH A HYDROSTATIC LOAD

A vertical structural component, which may be assumed to be pinned at its ends, is subjected to the loading shown in Fig. 6.13. Assuming the member details are as in Example 6.8, determine the position and value of the maximum stress in the member.

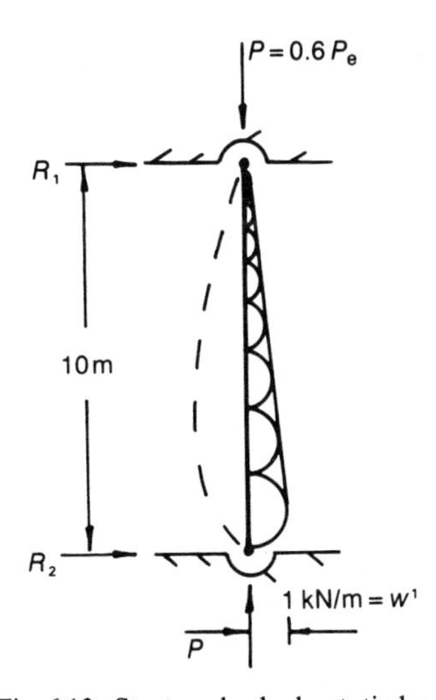

Fig. 6.13. Strut under hydrostatic load.

6.21.2

An alternative approach will now be used for the solution of this problem, as the bending moments are known at its ends.

Let,

M_w = bending moment due to the transverse loading, i.e.

$$M = -Py + M_w$$

or,

$$\frac{d^2M}{dx^2} = -P\frac{d^2y}{dx^2} + \frac{d^2M_w}{dx^2} \tag{6.60}$$

but,

$$EI\frac{d^2y}{dx^2} = M \quad \text{and} \quad \frac{d^2M_w}{dx^2} = w$$

where,

$$w = \text{load/unit length at } x$$

Hence, equation (6.60) becomes:

$$\frac{d^2M}{dx^2} + \alpha^2 M = w \tag{6.61}$$

where,

$$\alpha^2 = P/EI$$

For the present problem,

$$\frac{d^2M}{dx^2} + \alpha^2 M = 1000\left(\frac{x}{l}\right) \tag{6.62}$$

the complete solution of which is

$$M = A\cos\alpha x + B\sin\alpha x + \frac{1000}{\alpha^2}\left(\frac{x}{l}\right) \tag{6.63}$$

@ $\quad x = 0, \quad M = 0 \quad$ therefore $\quad \underline{A = 0}$ (6.64)

@ $\quad x = l, \quad M = 0 \quad$ therefore $\quad \underline{B = -w'/(\alpha^2 \sin\alpha l)}$ (6.65)

i.e.

$$M = -\frac{w'\sin\alpha x}{\alpha^2\sin\alpha l} + \frac{w'x}{\alpha^2 l}$$

where $w' = 1000\,\text{N/m}$. For $\hat{M}$,

$$\frac{dM}{dx} = 0$$

or,

$$0 = -\frac{w'\cos\alpha x}{\alpha\sin\alpha l} + \frac{w'}{\alpha^2 l}$$

or,

$$\underline{\cos\alpha x = \frac{\sin\alpha l}{\alpha l}}$$

$$\alpha^2 = \frac{0.6\pi^2 EI}{l^2 EI}$$

$$\underline{\alpha = 0.243} \quad \underline{\alpha^2 = 0.0592}$$

$$\underline{\alpha l = 2.433\,\text{rads} = 139.427^\circ}$$

therefore

$$\cos\alpha x = 0.267$$

$$\alpha x = 74.49^\circ = 1.3\,\text{rads}$$

$$\underline{x = 5.344\,\text{m}} \tag{6.66}$$

Substituting (6.64) to (6.66) into (6.63):

$$\hat{M} = -\frac{1000 \times 0.964}{0.0385} + \frac{1000 \times 5.344}{0.592}$$

$$= -25\,039 + 9027$$

$$\underline{\hat{M} = -16\,012\,\mathrm{N\,m}}$$

$$\hat{\sigma} = -\frac{0.6P_e}{9 \times 10^{-3}} - \frac{16\,012}{2 \times 10^{-4}}$$

but,

$$P_e = \frac{\pi^2 EI}{l^2} = 1.013 \times 10^6\,\mathrm{N}$$

therefore

$$\hat{\sigma} = -67.51\frac{\mathrm{MN}}{\mathrm{m}^2} - 80.06\frac{\mathrm{MN}}{\mathrm{m}^2}$$

$$\underline{\underline{\hat{\sigma} = -147.6\,\mathrm{MN/m}^2}}$$

If beam-column effect were neglected, then,

$$\hat{M} = 0.0642wl^2 = \underline{6415\,\mathrm{N\,m}}$$

i.e. the beam-column effect has increased the value of $\hat{M}$ by a factor of 2.496.

N.B. This method is also suitable if the shearing forces are known at points on the strut, as the following example illustrates.

6.22.1 EXAMPLE 6.10 ALTERNATIVE SOLUTION FOR AN ECCENTRICALLY LOADED STRUT

Determine $\hat{M}$ for the eccentrically loaded strut of Example 6.3.

6.22.2

$$\frac{d^2M}{dx^2} + \alpha^2 M = 0$$

or,

$$M = A\cos\alpha x + B\sin\alpha x$$

$$\frac{dM}{dx} = \text{shearing force} = -\alpha A\sin\alpha x + \alpha B\cos\alpha x$$

@ $x = 0, \quad M = P\Delta$

therefore

$$\underline{A = P\Delta}$$

@ $x = l/2, \quad \dfrac{dM}{dx} = 0$

$$0 = -\not{\alpha} P\Delta \sin\left(\frac{\alpha l}{2}\right) + \not{\alpha} B \cos\left(\frac{\alpha l}{2}\right)$$

or,

$$\underline{B = P\Delta \tan\left(\frac{\alpha l}{2}\right)}$$

therefore

$$\underline{M = P\Delta\left[\cos \alpha x + \tan\left(\frac{\alpha l}{2}\right)\sin \alpha x\right]}$$

$\hat{M}$ occurs @ $x = l/2$

$$\hat{M} = P\Delta\left[\cos\frac{\alpha l}{2} + \sin^2\left(\frac{\alpha l}{2}\right)\Big/\cos\left(\frac{\alpha l}{2}\right)\right]$$

$$= P\Delta\left\{\cos\left(\frac{\alpha l}{2}\right) + \left[1 - \cos^2\left(\frac{\alpha l}{2}\right)\right]\Big/\cos\left(\frac{\alpha l}{2}\right)\right\}$$

$$\underline{\hat{M} = P\Delta \sec\left(\frac{\alpha l}{2}\right)} \text{ (as required)}$$

6.23.1 LAPLACE TRANSFORM METHOD FOR LATERALLY LOADED STRUTS

This method [2] is particularly suitable for initially straight struts, loaded with complex combinations of lateral loads, as the solution of the resulting differential equations is relatively simple.

First, however, it will be necessary to define the Laplace transforms of the derivatives of y.

$$\mathscr{L}\left(\frac{dy}{dx}\right) = s\bar{Y} - y_0$$

$$\mathscr{L}\left(\frac{d^2y}{dx^2}\right) = s^2\bar{Y} - sy_0 - y_1$$

$$\mathscr{L}\left(\frac{d^3y}{dx^3}\right) = s^3\bar{Y} - s^2y_0 - sy_1 - y_2 \tag{6.67}$$

$$\mathscr{L}\left(\frac{d^4y}{dx^4}\right) = s^4\bar{Y} - s^3y_0 - s^2y_1 - sy_2 - y_3$$

$$\mathscr{L}\left(\frac{d^ny}{dx^n}\right) = s^n\bar{Y} - s^{n-1}y_0 - s^{n-2}y_1 - s^{n-3}y_2 - \cdots - y_{n-1}$$

where,

$\bar{Y}$ = L.T. of y.

y_0 = value of y @ $x = 0$

$y_1, y_2, y_3, \ldots, y_{n-1}$ = successive values of derivatives @ $x = 0$

L.T. ≡ Laplace Transform

Table 6.1 gives a list of some useful Laplace transforms.

Table 6.1. Some useful Laplace transforms.

$F(x)$	$F(s)$
1	$\frac{1}{s}$
x	$\frac{1}{s^2}$
$x^{n-1}/n-1$	$\frac{1}{s^n}$
$e^{\omega x}$	$1/(s-\omega)$
$\frac{\sin(\omega x)}{\omega}$	$1/(s^2+\omega^2)$
$\cos(\omega x)$	$s/(s^2+\omega^2)$
$\frac{1-\cos(\omega x)}{\omega^2}$	$1/s(s^2+\omega^2)$

6.23.2

Prior to the application of Laplace transforms to struts, it will be necessary to derive various functions, which are required to represent loads.

6.23.3 Step Functions

The *unit step function*, $H(t)$, is defined as

$$H(t) = 0 \quad \text{when } t < 0$$
$$ 1 \quad \text{when } t > 0$$

i.e. it has a discontinuity at $t = 0$ (see Fig. 6.14).

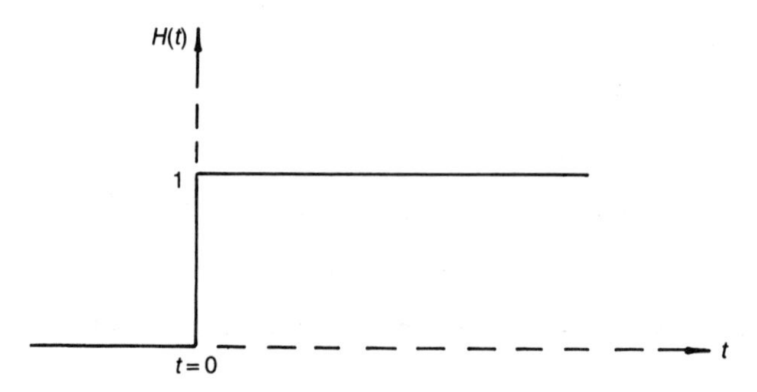

Fig. 6.14. Unit step function.

If,

$$c = \text{constant},$$

then $cH(t)$ has a step discontinuity of magnitude c at $t = 0$.

6.23.4

Similarly, the step function $H(t - a)$ (see Fig. 6.15) is zero when $t < a$ and unity when $t > a$.

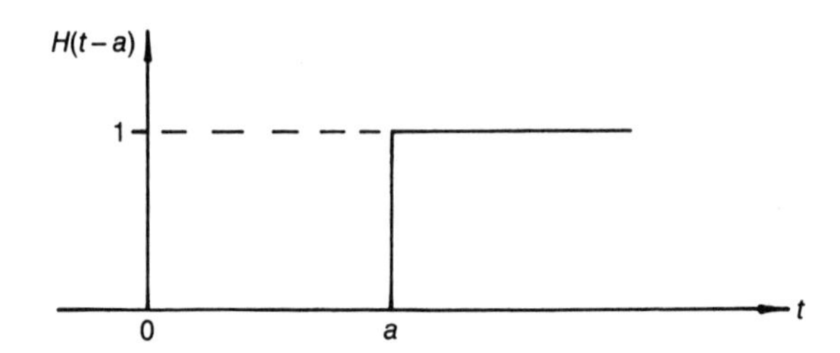

Fig. 6.15. Unit step function, commencing at $t = a$.

Since the integral defining a Laplace transform involves only positive values of t:

$$\mathscr{L}\{H(t)\} = \frac{1}{s}$$

and

$$\mathscr{L}^{-1}\left\{\frac{1}{s}\right\} = H(t)$$

Similarly,

$$\mathscr{L}\{H(t-a)\} = \int_0^\infty e^{-st} H(t-a)\,dt$$

$$= \int_a^\infty e^{-st}\,dt \quad (\text{if } a > 0)$$

since $H(t-a)=0$ when $t<a$; therefore

$$\mathscr{L}\{H(t-a)\}=\left[-\frac{e^{-st}}{s}\right]_a^\infty=\frac{e^{-as}}{s} \tag{6.68}$$

6.23.5

Equation (6.68) can be extended to represent the finite step function shown in Fig. 6.16, where $f(t)$ can be represented as follows:

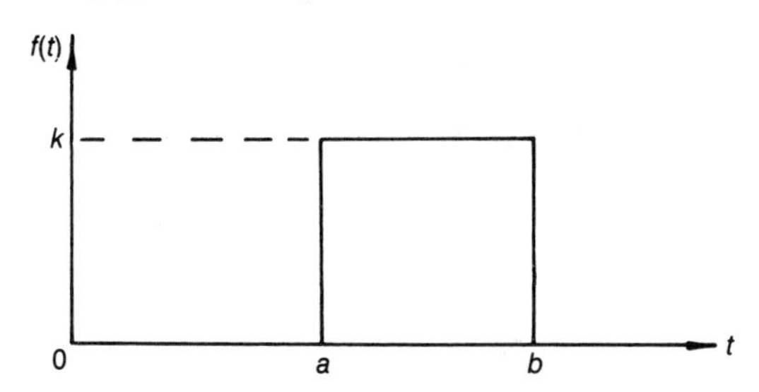

Fig. 6.16. Unit step function.

$$f(t)=\begin{cases}0 & \text{when } t<a\\ k & \text{when } a<t<b\\ 0 & \text{when } t>b\end{cases}$$

or,

$$\mathscr{L}\{k[H(t-a)-H(t-b)]\}=k\left(\frac{e^{-as}-e^{-bs}}{s}\right)$$

N.B. It should be noted that this function can be used for partial or continuous uniformly distributed loads acting on beams and struts.

6.23.6 Impulse Functions

The previous theory can be extended to represent impulsive functions of the type shown in Fig. 6.17.

Let,

$$\begin{aligned}\delta(t)&=\text{unit impulse function}\\ &=\lim_{\alpha\to 0}\left[\frac{1}{\alpha}\{H(t)-H(t-\alpha)\}\right]\end{aligned}$$

$$\mathscr{L}\{\delta(t)\}=\frac{1}{\alpha}\left\{\frac{1}{s}-\frac{e^{-\alpha s}}{s}\right\}=\frac{1-e^{-\alpha s}}{\alpha s}$$

$$=\lim_{\alpha\to 0}\frac{\alpha s-\dfrac{\alpha^2 s^2}{2!}+\cdots}{\alpha s}$$

$$=1$$

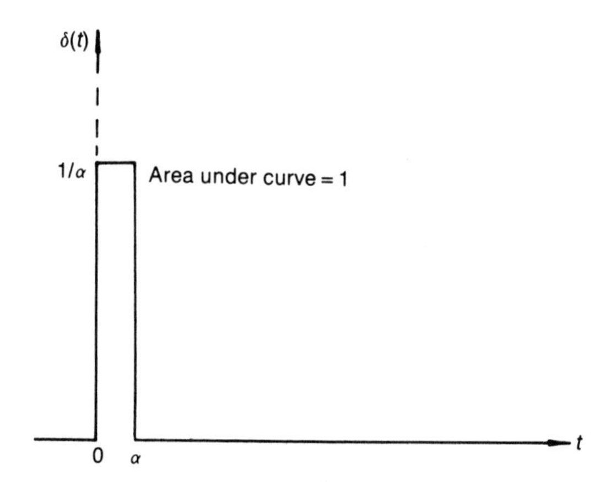

Fig. 6.17. Impulsive function.

thus,

$$\mathscr{L}(\delta t) = 1$$

and,

$$\mathscr{L}^{-1}(1) = \delta(t)$$

Similarly, at $t = \text{a}$

$$\delta(t-a) = \lim_{\alpha \to 0}\left[\frac{1}{\alpha}\{H(t-a) - H(t-a-\alpha)\}\right]$$

and

$$\{\delta(t-a)\} = \lim_{\alpha \to 0}\left\{\frac{\mathrm{e}^{-as} - \mathrm{e}^{-(a+\alpha)s}}{\alpha s}\right\}$$

$$= \mathrm{e}^{-as}\lim_{\alpha \to 0}\left(\frac{1-\mathrm{e}^{-\alpha s}}{\alpha s}\right)$$

$$\underline{\{\delta(t-a)\} = \mathrm{e}^{-as}} \tag{6.69}$$

Equation (6.69) can be used for concentrated lateral loads acting on beams and struts.

From (6.68) and (6.69), it can be seen that,

$$\mathscr{L}\{H(t)\} = \frac{1}{s}\mathscr{L}\{\delta(t)\}$$

6.23.7

The function $\delta'(t)$ is used for representing couples, as shown in Fig. 6.18, where it can be seen that this function is an extension of Section 6.23.6.

$$\delta'(t) = c(t) = \lim_{\alpha \to 0}\frac{\delta(t) - \delta(t-\alpha)}{\alpha}$$

This is a double pulse, of strengths $\pm 1/\alpha$ at an interval of time α apart, i.e. a doublet of *unit moment* @ $t = 0$.

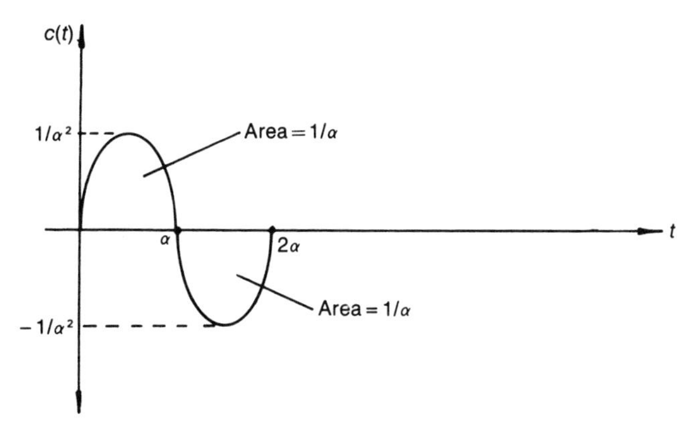

Fig. 6.18. The function $\delta(t)$

Similarly @ $t = a$

$$\delta'(t-a) = \lim_{\alpha \to 0} \left\{ \frac{\delta(t-a) - \delta(t-a-\alpha)}{\alpha} \right\}$$

From the formula for the L.T of a derivative,

$$\mathscr{L}\{\delta'(t)\} = s \quad \text{and} \quad \mathscr{L}\{\delta''(t)\} = s^2, \text{etc.}$$

$$\underline{\mathscr{L}\{\delta'(t-a)\} = s e^{-as}} \tag{6.70}$$

The above expression can be used to represent couples acting on beams and struts.

6.23.8 Beams Under Transverse Loads

Equations (6.68) to (6.70) can be used to represent lateral loads on the beam of Fig. 6.19.

Equation (6.70) represents the couple T.

Equation (6.69) represents the concentrated load W.

Equation (6.68) represents the uniformly distributed load w.

That is,

$$EI \frac{d^4 y}{dx^4} = T \cdot c(x-a) + W \cdot \delta(x-b) + w \cdot H(x-c)$$

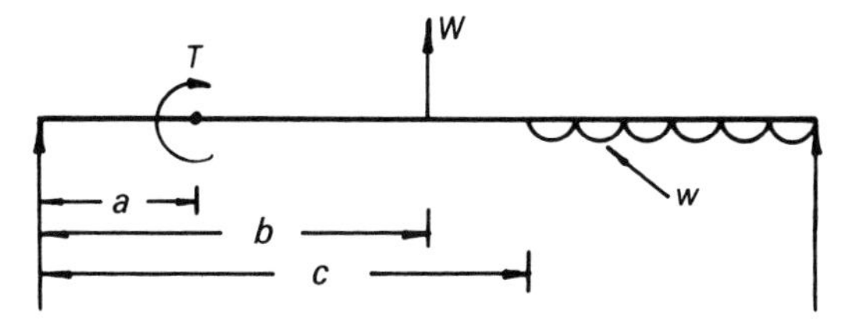

Fig. 6.19. Laterally loaded beam.

or,

$$EI(s^4\bar{Y} - s^3y_0 - s^2y_1 - sy_2 - y_3)$$

$$= T\cdot s\cdot \mathrm{e}^{-as} + W\cdot 1\cdot \mathrm{e}^{-bs} + w\cdot\frac{1}{s}\cdot\mathrm{e}^{-cs}$$

$$= \mathscr{W}(s)$$

$$EI\bar{Y} = \frac{\mathscr{W}(s)}{s^4} + EI\left(y_0 + y_1x + y_2\frac{x^2}{2!} + y_3\frac{x^3}{3!}\right)$$

therefore

$$EIy = \mathscr{L}^{-1}\left\{\frac{\mathscr{W}(s)}{s^4}\right\} + A + Bx + Cx^2 + Dx^3$$

Now,

$$\mathscr{L}^{-1}\left\{\frac{W\mathrm{e}^{-bs}}{s^4}\right\} = W\mathscr{L}^{-1}\frac{\mathrm{e}^{-(b-s)}}{s^4}$$

$$= \underline{\frac{W(x-b)^3}{3!}H(x-b)}$$

Also,

$$\mathscr{L}^{-1}\left\{\frac{w\mathrm{e}^{-cs}}{s^5}\right\} = w\mathscr{L}^{-1}\left\{\frac{\mathrm{e}^{-cs}}{s^5}\right\}$$

$$= \frac{w(x-c)^4}{4!}H(x-c)$$

Similarly,

$$\mathscr{L}^{-1}\left\{\frac{T\cdot s\cdot \mathrm{e}^{-as}}{s^4}\right\} = T\mathscr{L}^{-1}\left\{\frac{\mathrm{e}^{-as}}{s^3}\right\}$$

$$= T\frac{(x-a)^2}{2!}H(x-a)$$

therefore

$$\mathscr{L}^{-1}\left\{\frac{\mathscr{W}(s)}{s^4}\right\} = \frac{T(x-a)^2}{2!}H(x-a) + \frac{W(x-b)^3}{3!}H(x-b)$$

$$+ \frac{w(x-c)^4}{4!}H(x-c)$$

i.e.

$$EIy = \frac{T(x-a)^2}{2!}H(x-a)$$

$$+ \frac{W(x-b)^3}{3!}H(x-b) + \frac{w(x-c)^4}{4!}H(x-c)$$

$$+ A + Bx + Cx^2 + Dx^3 \tag{6.71}$$

Equation (6.61) can also be expressed in the form of Laplace transforms, as is shown in the next section.

6.23.9 Laterally Loaded Struts (Fig. 6.20)

From (6.61),

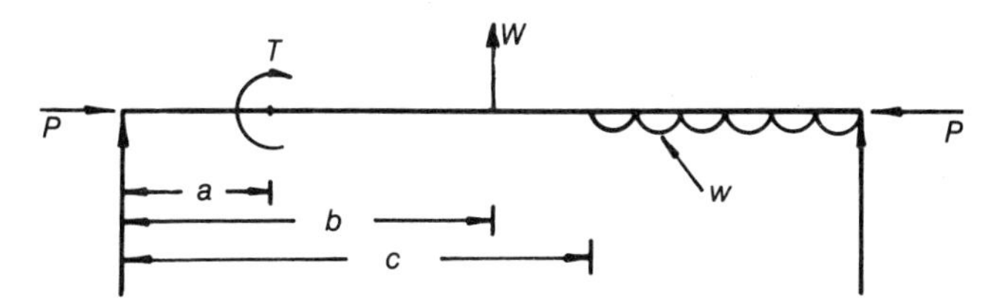

Fig. 6.20. Laterally loaded strut.

$$\frac{d^2M}{dx^2} + \alpha^2 M = T \cdot c(x-a) + W \cdot \delta(x-b) + w \cdot H(x-c)$$

or,

$$s^2\bar{M} - sM_0 - M_1 + \alpha^2\bar{M} = T \cdot s \cdot e^{-as} + W \cdot 1 \cdot e^{-bs} + w \cdot \frac{1}{s} e^{-cs}$$

$$\bar{M} = T\frac{s}{s^2+\alpha^2}e^{-as} + W\frac{1}{s^2+\alpha^2}e^{-bs} + w\frac{1}{s(s^2+\alpha^2)}e^{-cs}$$

$$+ M_0\frac{s}{s^2+\alpha^2} + M_1\frac{1}{s^2+\alpha^2}$$

therefore

$$M = T\cos\alpha(x-a)H(x-a) + \frac{W}{\alpha}\sin\alpha(x-b)H(x-b)$$

$$+ \frac{w}{\alpha^2}[1-\cos\alpha(x-c)]H(x-c) + M_0\cos\alpha x + \frac{M_1}{\alpha}\sin\alpha x \tag{6.72}$$

where,

T = couple (+ve clockwise)
W = concentrated load (+ve upwards)
w = load/unit length (+ve upwards)

Equation (6.72) is very useful for solving struts of various types, as is shown by its application to a number of different problems.

6.24.1 EXAMPLE 6.11 USE OF LAPLACE TRANSFORMS FOR DETERMINING EULER LOADS

Determine the Euler buckling loads for the axially loaded struts of Example 6.1 and 6.2.

6.24.2 (a) Pinned-end strut

@ $x=0, \quad M = \underline{M_0 = 0}$

and

@ $x = l, \quad M = 0$

or

$$0 = \frac{M_1}{\alpha}\sin(\alpha l)$$

but

$$M_1 \neq 0$$

therefore

$$\sin(\alpha l) = 0$$

or

$$\alpha l = \pi$$

therefore

$$\underline{P_e = \pi^2 EI/l^2} \text{ (as required)}$$

6.24.3 (b) Clamped-end strut

From (6.72),

$$M = M_0\cos(\alpha x) + \frac{M_1}{\alpha}\sin(\alpha x)$$

and

$$\frac{dM}{dx} = -P\frac{dy}{dx} = \alpha_0 M_0 \sin(\alpha x) + M_1\cos(\alpha x)$$

@ $x = 0, \quad \dfrac{dM}{dx} = -P\dfrac{dy}{dx} = 0$ therefore $\underline{M_1 = 0}$

@ $x = l/2, \quad -P\dfrac{dy}{dx} = 0$

$$0 = -\alpha M_0 \sin\left(\frac{\alpha l}{2}\right)$$

i.e.

$$\frac{\alpha l}{2} = \pi$$

or,

$$\underline{P_e = \frac{4\pi^2 EI}{l^2}} \text{ (as required)}$$

6.25.1 EXAMPLE 6.12 USE OF LAPLACE TRANSFORMS FOR ECCENTRICALLY LOADED STRUTS

Determine the maximum bending moment for the eccentrically loaded strut of Example 6.3.

6.25.2

$$\frac{d^2M}{dx^2} + \alpha^2 M = 0$$

Hence from (6.72):

$$M = M_0 \cos(\alpha x) + \frac{M_1}{\alpha} \sin(\alpha x)$$

@ $x = 0, \quad M = P\Delta \quad \text{therefore} \quad \underline{M_0 = P\Delta}$

@ $x = l, \quad M = P\Delta$

therefore

$$P\Delta = P\Delta \cos(\alpha l) + \frac{M_1}{\alpha} \sin(\alpha l)$$

$$M_1 = \frac{P\Delta(1 - \cos \alpha l)\alpha}{\sin \alpha l} = F(@\ x = 0)$$

therefore

$$M = P\Delta \cos \alpha x + \frac{P\Delta(1 - \cos \alpha l)}{\sin \alpha l} \sin \alpha x$$

@ $x = l/2$

$$\hat{M} = P\Delta \cos\frac{\alpha l}{2} + \frac{P\Delta}{\sin \alpha l}(1 - \cos \alpha l) \sin\frac{\alpha l}{2}$$

$$= P\Delta \cos\frac{\alpha l}{2} + P\Delta \frac{2\sin^2\frac{\alpha l}{2}}{2\sin\frac{\alpha l}{2}\cos\frac{\alpha l}{2}} \sin\frac{\alpha l}{2}$$

$$= P\Delta \cos\frac{\alpha l}{2} + P\Delta \frac{\sin^2\frac{\alpha l}{2}}{\cos\frac{\alpha l}{2}} = \frac{P\Delta}{\cos\frac{\alpha l}{2}}\left(\cos^2\frac{\alpha l}{2} + \sin^2\frac{\alpha l}{2}\right)$$

$$\underline{\hat{M} = P\Delta \sec\frac{\alpha l}{2}} \text{ (as required)}$$

6.26.1 EXAMPLE 6.13 COMBINED AXIAL AND LATERAL LOADING

Determine the bending moment at the point "C" for the laterally loaded strut of Fig. 6.21.

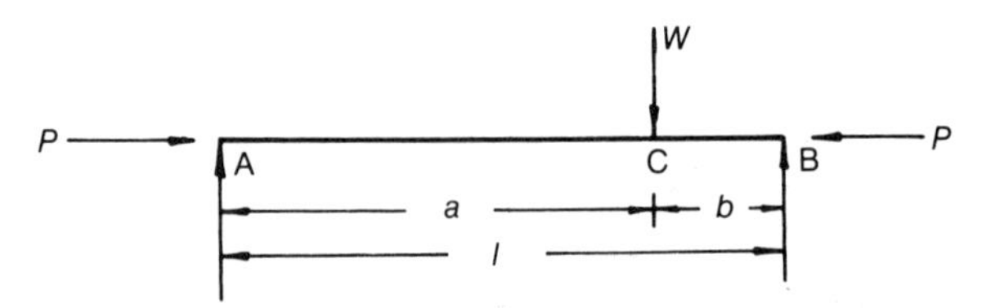

Fig. 6.21. Laterally loaded strut.

6.26.2

$$M = -\frac{W}{2}\sin\alpha(x-a)H(x-a) + M_0\cos\alpha x + \frac{M_1}{\alpha}\sin\alpha x$$

@ $x = 0$, $M = 0$ therefore $\underline{M_0 = 0}$

@ $x = l$, $M = 0$

therefore

$$0 = -\frac{W}{\alpha}\sin\alpha b + \frac{M_1}{\alpha}\sin\alpha l$$

or,

$$M_1 = W\frac{\sin\alpha b}{\sin\alpha l}$$

$$\underline{M = -\frac{W}{\alpha}\sin\alpha(x-a)H(x-a) + \frac{W}{\alpha}\frac{\sin\alpha b}{\sin\alpha l}\sin\alpha x}$$

@ $x = b$, $M = M_c$

therefore

$$\underline{M_c = \frac{W}{\alpha}\frac{\sin\alpha b\sin\alpha a}{\sin\alpha l}}$$

6.27.1 EXAMPLE 6.14 COMBINED DISTRIBUTED AND AXIAL LOADING

Determine an expression for the bending moment for the laterally loaded strut of Fig. 6.22.

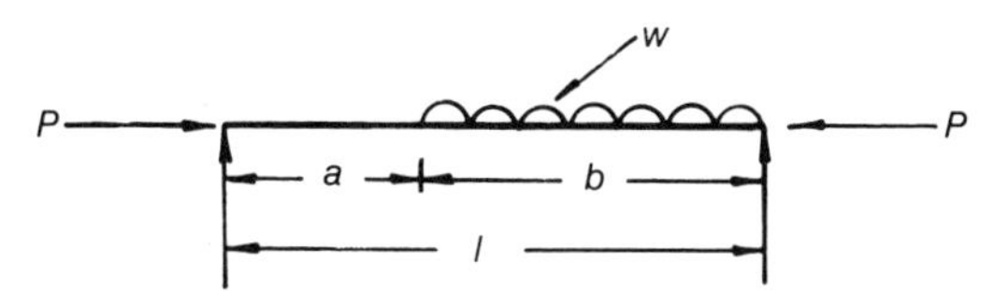

Fig. 6.22. Laterally loaded strut.

6.27.2

$$M = -\frac{w}{\alpha^2}[1 - \cos \alpha(x-a)]H(x-a) + M_0 \cos \alpha x + \frac{M_1}{\alpha} \sin \alpha x$$

@ $x = 0, \quad M = M_0 = 0$

@ $x = l, \quad M = 0$

$$0 = -\frac{w}{\alpha^2}(1 - \cos \alpha b) + \frac{M_1}{\alpha} \sin \alpha l$$

$$M_1 = \frac{w}{\alpha} \frac{(1 - \cos \alpha b)}{\sin \alpha l}$$

and,

$$M = -\frac{w}{\alpha^2}[1 - \cos \alpha(x-\alpha)]H(x-a) + \frac{w}{\alpha^2} \frac{(1 - \cos \alpha b)}{\sin \alpha l} \sin \alpha x$$

6.28.1 BUCKLING OF STRUTS, USING ENERGY METHODS

Another useful method for determining the elastic instability loads of axially loaded struts is through considerations of energy. For such methods, it is normal to assume a satisfactory buckling form that meets the boundary conditions etc. of the strut. Buckling forms involving sinusoidal ones are particularly popular.

6.28.2

Prior to applying the method, it will be necessary to obtain P_e in terms of the first and second derivatives of y. Let,

x = distance measured along the undeflected line of strut, as shown in Fig. 6.23

s = distance measured along the axis of the deflected strut

Δ = axial movement

Based on small deflection theory,

$$\Delta = \int_0^l (\mathrm{d}s - \mathrm{d}x)$$

$$= \int_0^l \mathrm{d}s - \int_0^l \mathrm{d}x$$

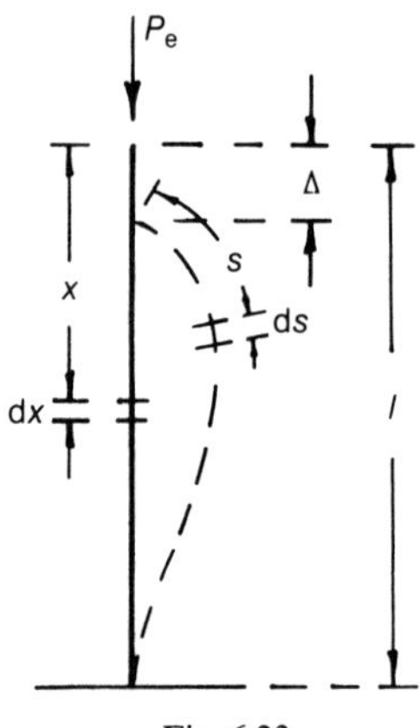

Fig. 6.23.

but,

$$ds^2 = dx^2 + dy^2$$

$$= dx^2\left(1 + \left(\frac{dy}{dx}\right)^2\right)$$

i.e.

$$ds = \sqrt{\left[1 + \left(\frac{dy}{dx}\right)^2\right]}\,dx$$

therefore

$$\Delta = \int_0^l \left[1 + \left(\frac{dy}{dx}\right)^2\right]^{\frac{1}{2}} dx - \int_0^l dx$$

$$= \int_0^l \left[1 + \tfrac{1}{2}\left(\frac{dy}{dx}\right)^2 + \cdots\right] dx - \int_0^l dx$$

$$\underline{\Delta = \tfrac{1}{2}\int_0^l \left(\frac{dy}{dx}\right)^2 dx}$$

Work done (by P_e) $= P_e\Delta$

$$= \frac{P_e}{2}\int_0^l \left(\frac{dy}{dx}\right)^2 dx \tag{6.73}$$

Bending strain energy = BSE, where,

$$\text{BSE} = \int \frac{M^2}{2EI} dx \tag{6.74}$$

Eating (6.73) and (6.74):

$$\frac{P_e}{2} * \int_0^l \left(\frac{dy}{dx}\right)^2 dx = \tfrac{1}{2}\int_0^l \frac{M^2}{EI} dx$$

therefore

$$P_e = \frac{\int_0^l \frac{M^2}{EI} \mathrm{d}x}{\int_0^l \left(\frac{\mathrm{d}y}{\mathrm{d}x}\right)^2 \mathrm{d}x}$$

but,

$$M = EI\frac{\mathrm{d}^2y}{\mathrm{d}x^2},$$

so that

$$P_e = \frac{\int_0^l EI\left(\frac{\mathrm{d}^2y}{\mathrm{d}x^2}\right)^2 \mathrm{d}x}{\int_0^l \left(\frac{\mathrm{d}y}{\mathrm{d}x}\right)^2 \mathrm{d}x} \tag{6.75}$$

6.29.1 EXAMPLE 6.15 PINNED-ENDS STRUT

Determine the Euler buckling load for the pinned-end strut of Example 6.1, using equation (6.75).

6.29.2

A suitable buckling form for this strut will be:

$$y = A\sin\frac{\pi x}{l} \tag{6.76}$$

which will have the following derivatives:

$$\frac{\mathrm{d}y}{\mathrm{d}x} = \frac{\pi}{l}A\cos\frac{\pi x}{l} \tag{6.77}$$

and

$$\frac{\mathrm{d}^2y}{\mathrm{d}x^2} = -\frac{\pi^2}{l^2}A\sin\frac{\pi x}{l} \tag{6.78}$$

From (6.76), it can be seen that $y=0$ at $x=0$ and $x=l$, and that y is a maximum at $x=l/2$, as required.

Furthermore, from (6.77) and (6.78), it can be seen that $\mathrm{d}y/\mathrm{d}x \neq 0$ at $x=0$ and $x=l$, and that $\mathrm{d}^2y/\mathrm{d}x^2=0$ at $x=0$, and that at $x=l/2$, $\mathrm{d}^2y/\mathrm{d}x^2$ has a maximum value, as required.

Substituting (6.77) and (6.78) into (6.75):

$$P_e = \frac{\pi^4 EI}{l^4}\frac{l^2}{\pi^2}\frac{\int_0^l \sin^2\left(\frac{\pi x}{l}\right)\mathrm{d}x}{\int \cos^2\left(\frac{\pi x}{l}\right)\mathrm{d}x}$$

or,

$$P_e = \pi^2 EI/l^2 \text{ (as required)}$$

6.30.1 DYNAMIC INSTABILITY

The theory covered in this chapter has been based on static stability, but it must be pointed out that for struts subjected to compressive periodic axial forces, there is a possibility that dynamic instability can occur when the lateral critical frequency of the beam-column is reached.

The study of dynamic stability is beyond the scope of this book, but is dealt with in much detail in reference [32].

EXAMPLES FOR PRACTICE 6

1. Determine the Euler buckling load for the axially loaded strut of Fig. 6.5 by each of the following approaches:

   (a) Method of Section 6.3.
   (b) Use of Laplace transforms.
   (c) Equation (6.75).

   $\{P_e = \pi^2 EI/(4l^2)\}$

2. Determine the Euler buckling load for an initially straight axially loaded strut, which is pinned at one end and fixed at the other.

   $\{P_e = 20.2EI/l^2\}$

3. Using equation (6.75), determine the Euler buckling load for a strut, clamped at its ends.

   $\{P_e = 4\pi^2 EI/l^2\}$

4. Find the Euler crushing load for a hollow cylindrical cast-iron column, 0.15 m external diameter and 20 mm thick, if it is 6 m long and hinged at both ends. $E = 75 \times 10^9$ N/m^2. Compare this load with that given by the Rankine formula using constants of 540 MN/m^2 and 1/1600. For what length of column would these two formulae give the same crushing load?

   {363.2 kN, 386.7 kN, 4.55 m}

5. A short steel tube of 0.1 m outside diameter, when tested in compression, was found to fail under an axial load of 800 kN. A 15 m length of the same tube when tested as a pin-jointed strut failed under a load of 30 kN. Assuming that the Euler and Rankine–Gordon formulae apply to the strut, calculate

   (i) the tube inner diameter,

(ii) the denominator constant in the Rankine–Gordon formula.

$E = 196.5\,\text{GN/m}^2$.

$\{0.0734\,\text{m}, \frac{1}{9114}\}$

6. A steel pipe, 36 mm inner diameter, 6 mm thick and 1 m long, is supported so that the ends are hinged, but all expansion is prevented. The pipe is unstressed at 0 °C. Calculate the temperature at which buckling will occur.

$$\sigma_c = 325\,\text{MN/m}^2, \quad a = \frac{1}{7500}, \quad E = 200\,\text{GN/m}^2, \quad \alpha = 11.1 \times 10^{-6}/°\text{C}$$

{68.4 C}

7. The table below shows the results of a series of buckling tests carried out on a steel tube of external diameter 35 mm and internal diameter 25 mm.

   Assuming the Rankine–Gordon formula to apply, determine the numerator and denominator constants for this tube.

l(mm)	600	1000	1400	1800
$P_R(kN)$	150	125	110	88

$\{350\,\text{MN/m}^2, \frac{1}{32\,000}\}$

8. The results of two tests on steel struts with pinned ends were found to be:

Test number	1	2
Slenderness ratio	50	80
Average stress at failure (MN/m^2)	266.7	194.4

$A = 1\,\text{m}^2$

(a) Assuming that the Rankine–Gordon formula applies to both struts, determine the numerator and denominator constants of the Rankine–Gordon formula.

(b) If a steel bar of rectangular section 0.06 m × 0.019 m and of length 0.4 m is used as a strut with both ends clamped, determine the safe load, using the constants derived in (a) and employing a safety factor of 4.

$\{350\,\text{MN/m}^2, \frac{1}{8000}, 342\,\text{kN}\}$

9. A long slender strut of length L is encastré at one end and pin-jointed at the other. At its pinned end, it carries an axial load P, together with a couple M. Show that the magnitude of the couple at the clamped end is given by the expression:

$$M\left(\frac{\alpha L - \sin \alpha L}{\alpha L \cos \alpha L - \sin \alpha L}\right)$$

Determine the value of this couple if P is one-quarter of the Euler buckling load for this class of strut.

$\{M(\alpha L - \sin \alpha L)/(\alpha L \cos \alpha L - \sin \alpha L)\}$

10. A long strut, initially straight, securely fixed at one end and free at the other, is loaded at the free end with an eccentric load whose line of action is parallel to the original axis. Deduce an expression for the deviation of the free end from its original position.

$\{\Delta(\sec \alpha L - 1)\}$ and $\alpha = \sqrt{P/EI}$

$\Delta =$ eccentricity

11. A tubular steel strut of 70 mm external diameter and 50 mm internal diameter is 3.25 m long. The line of action of the compressive forces is parallel to, but eccentric from, the axis of the tube, as shown in Fig. Q.6.11.

Find the maximum allowable eccentricity of these forces, if the maximum permissible deflection (total) is not greater than 15 mm. $E = 2 \times 10^{11}$ N/m^2.

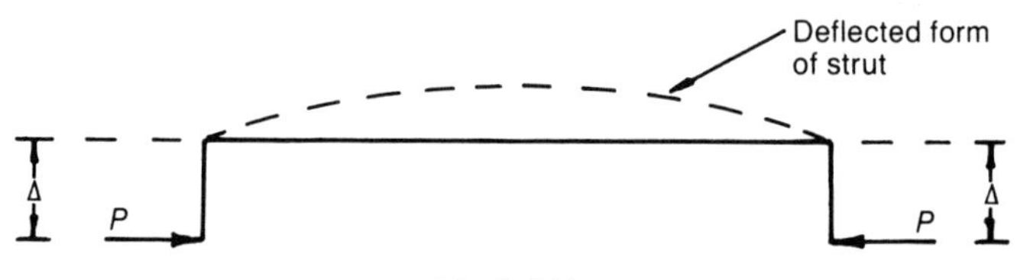

Fig. Q.6.11.

$\{\Delta = 5\,\text{mm}\}$

12. The eccentrically loaded strut of Fig. Q.6.12 is subjected to a compressive load P. If $EI = 20\,000$ N m^2, determine the position and value of the maximum deflection assuming the following data apply:

$P = 5000$ N

$l = 3$ m

$\Delta = 0.01$ m

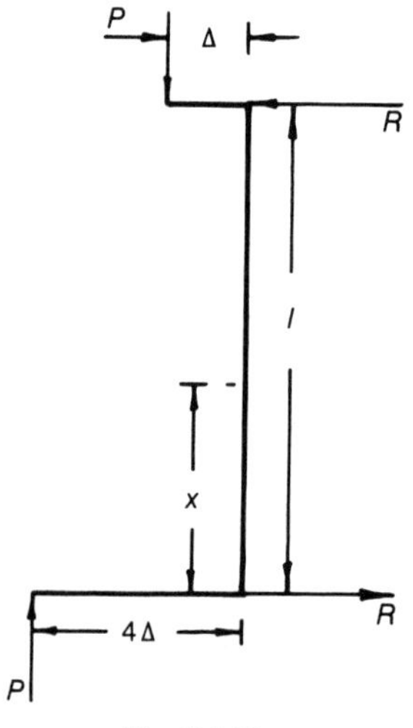

Fig. Q.6.12.

$\{x = 1.501\,\text{m},\ \delta = 0.0608\,\text{m}\}$

13. Show that for the eccentrically loaded strut of Fig. Q.6.13, the bending moment at any distance x is given by:

$$M = P\Delta\left[-2\cos\alpha x + \frac{(1 + 2\cos\alpha l)}{\sin\alpha l}\sin\alpha x\right]$$

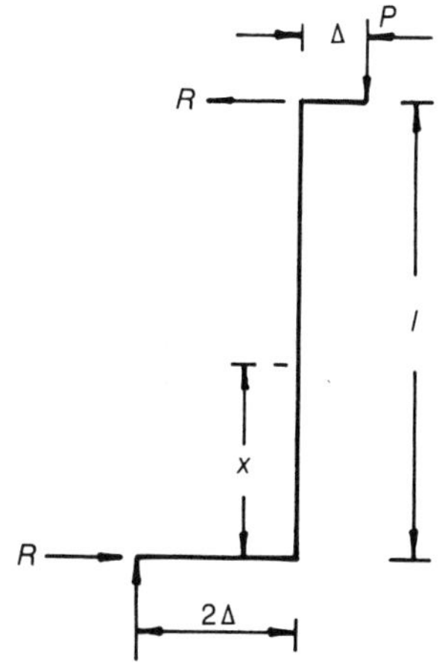

Fig. Q.6.13.

14. An initially curved strut, whose initial deflection forms is small and parabolic, is symmetrical about its mid-point. If the strut is subjected to a compressive axial load P at its pinned ends, show that the maximum compressive stress is given by:

$$\frac{P}{A}\left\{1 + \frac{\Delta\bar{y}}{k^2}\frac{8EI}{Pl^2}\left(\sec\frac{\alpha l}{2} - 1\right)\right\}$$

where,

Δ = initial central deflection

k = least radius of gyration

Determine Δ for such a strut, assuming the geometrical and material properties of Example 6.6 apply.

{12.3 cm}

15. An initially curved strut, whose initial deflected form is small and circular, is subjected to a compressive axial load P at its pinned ends. Show that the total deflection y at any distance x is given by:

$$y = -\frac{8\Delta}{\alpha^2 l^2}\{\cos\alpha x + \tan(\alpha l/2)\sin\alpha x - 1\}$$

where,

Δ = initial central deflection

Determine Δ for such a strut, assuming the geometrical and material properties of Example 6.6 apply.

{12.3 cm}

16. The eccentrically loaded strut of Fig. Q.6.16 is prevented from buckling by a restraining force F, applied to its mid-span, as shown.

Prove that

$$F = \frac{2P\Delta\{1 - \sec(\alpha l/2)\}}{\{(l/2) - (1/\alpha)\tan(\alpha l/2)\}}$$

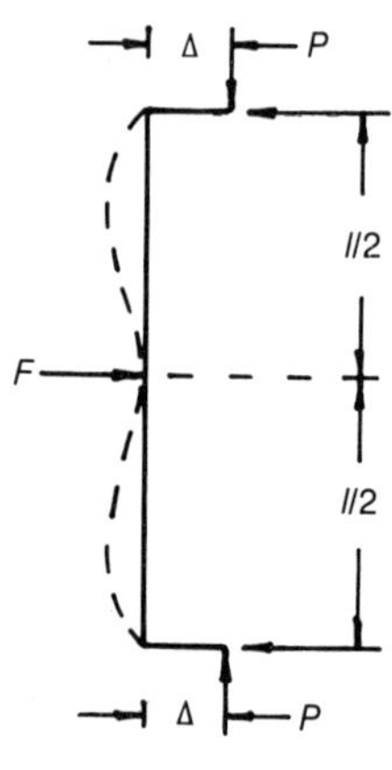

Fig. Q.6.16.

17. A vertical stiffner is subjected to a compressive axial load of 200 kN, together with a lateral hydrostatic load, which gradually increases from zero at the top to an intensity of 30 kN/m at the bottom.

If the ends of the stiffner are pin-jointed, determine the position and value of the maximum bending moment, assuming the following to apply:

$$\text{Height of stiffner} = 3\,\text{m}$$

$$I = 1 \times 10^{-4}\,\text{m}^4, \quad E = 2 \times 10^{11}\,\text{N/m}^2$$

$\{1.73\,\text{m}, \ -14.16\,\text{kN m}\}$

18. A laterally loaded strut with end couples is subjected to the loading shown in Fig. Q.6.18. Determine the bending moments at the points B, C and D, given that $E = 2 \times 10^{11}\,\text{N/m}^2$ and $I = 1 \times 10^{-6}\,\text{m}^4$.

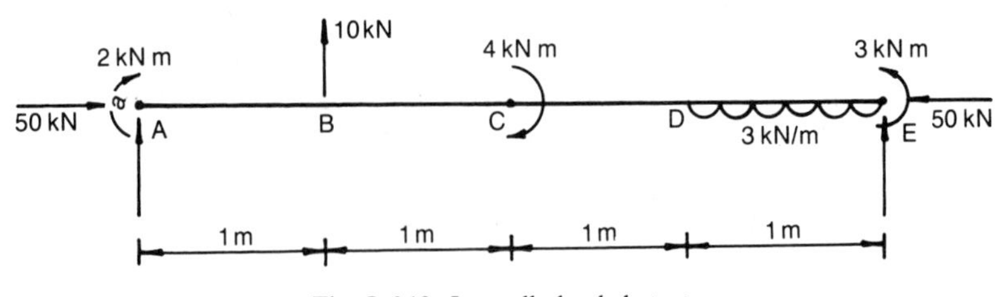

Fig. Q.6.18. Laterally loaded strut.

$\{-8656\,\text{N m}, \ -7603\,\text{N m}, \ -1179\,\text{N m}\}$

7

Thick Curved Beams

7.1.1 THICK AND THIN CURVED BEAMS

The theories for straight beams and thin curved beams cannot be applied to thick curved beams, because for thick curved beams, the neutral axis in bending is no longer at the centroid of the section, and the strain no longer varies linearly across the section. The reason for this is that the radius of curvature for a thick curved beam varies substantially across the depth of the beam.

7.2.1 WINKLER'S THEORY FOR THICK CURVED BEAMS

Consider the thick curved beam of Fig. 7.1(a), which is subjected to a positive bending moment M that causes its curvature to increase, as shown in Fig. 7.1(b), where,

$$\text{Curvature} = 1/R$$

$$R = \text{radius of curvature}$$

$$R_0 = \text{initial radius of curvature of the neutral layer “NL”}$$

$$R_1 = \text{final radius of curvature of the neutral layer “N}'\text{L}'\text{”}$$

Consider the layer CD, which increases its length to C′D′, so that the strain at a distance y from the neutral layer =

$$\varepsilon = \frac{\mathrm{C'D'} - \mathrm{CD}}{\mathrm{CD}} = \frac{(R_1 + y)(\theta + \delta\theta) - (R_0 + y)\theta}{(R_0 + y)\theta}$$

$$= \frac{R_1(\theta + \delta\theta) - R_0\theta + y\delta\theta}{(R_0 + y)\theta}$$

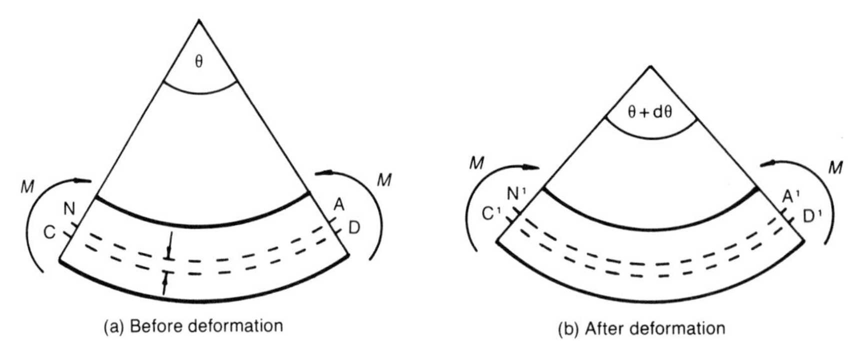

Fig. 7.1. Thick curved beam.

but as NL is the neutral layer,

the length NL = the length N′L′

or,

$$R_1(\theta + \delta\theta) = R_0\theta$$

therefore

$$\varepsilon = \frac{y\delta\theta}{(R_0 + y)\theta}$$

Now,

$$\sigma = \text{bending stress at any distance } y \text{ from NL}$$

$$= \frac{Ey\delta\theta}{(R_0 + y)\theta} \tag{7.1}$$

but from equilibrium considerations

$$\int \sigma * \mathrm{d}A = 0$$

or,

$$\frac{E * \delta\theta}{\theta} \int \frac{y\,\mathrm{d}A}{(R_0 + y)} = 0$$

or,

$$\int \frac{y\,\mathrm{d}A}{(R_0 + y)} = 0 \tag{7.2}$$

Now the moment of resistance =

$$M = \int \sigma y\,\mathrm{d}A$$

$$= \frac{E * \delta\theta}{\theta} \int \frac{y^2\,\mathrm{d}A}{(R_0 + y)}$$

$$= \frac{E * \delta\theta}{\theta} \int \frac{[y(y + R_0) - R_0 y]\mathrm{d}A}{(R_0 + y)}$$

$$= \frac{E * \delta\theta}{\theta} \left\{ \int y\,\mathrm{d}A - R_0 \int \frac{y}{(R_0 + y)}\mathrm{d}A \right\}$$

but from (7.2)

$$\int \frac{y}{(R_0 + y)} \mathrm{d}A = 0$$

therefore

$$M = \frac{E * \delta\theta}{\theta} \int y \, \mathrm{d}A \tag{7.3}$$

where,

$$\int \mathrm{d}A = A = \text{cross-sectional area}$$

Let,

e = distance between the neutral axis and the centroidal axis, so that,

$$\int y \, \mathrm{d}A = A * e$$

but from (7.1),

$$\sigma = \frac{E * y * \delta\theta}{(R_0 + y)\theta}$$

or,

$$\frac{E * \delta\theta}{\theta} = \frac{(R_0 + y)\sigma}{y} \tag{7.4}$$

Substituting (7.4) into (7.3), and rearranging,

$$\sigma = \frac{M * y}{\{Ae(R_0 + y)\}} \tag{7.5}$$

7.3.1 EXAMPLE 7.1 BENDING OF A THICK CURVED BEAM OF RECTANGULAR SECTION

Determine e for a thick curved beam of rectangular section. Hence, or otherwise, determine the stresses on the inner and outer surfaces of this beam, when it is subjected to a bending moment of 400 N m, assuming that this bending action tends to cause the beam to straighten. Plot the stress distribution across the section.

7.3.2 To Determine e

Let,

R_m = radius of curvature of the centroid of the section
R_0 = radius of curvature of the neutral axis of the section (NA)
y = distance of any fibre on the section from NA
z = distance of any fibre on the section from the centroidal axis

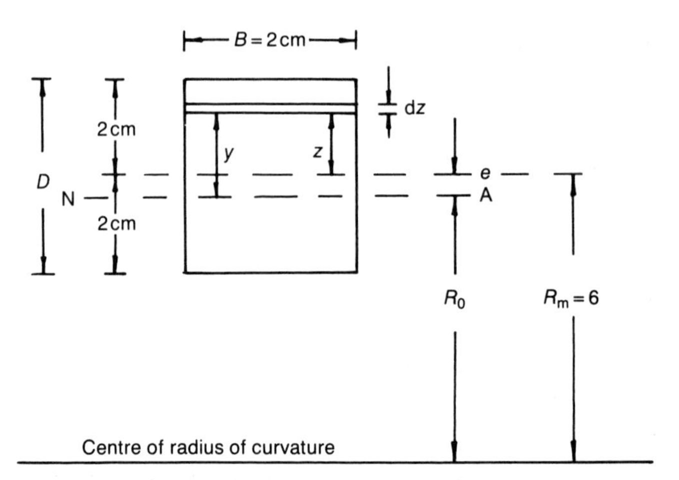

Fig. 7.2. Rectangular section.

Now from (7.2),

$$\int \frac{y \cdot \mathrm{d}A}{(R_0 + y)} = 0$$

or,

$$\int \frac{(z+e)}{(R_m + z)} \mathrm{d}A = 0$$

or,

$$\int \frac{(z+e)}{(R_m + z)} B \cdot \mathrm{d}z = 0$$

or,

$$\int \frac{[(R_m + z) - (R_m - e)]}{(R_m + z)} \mathrm{d}z = 0$$

or,

$$\int_{-D/2}^{D/2} \mathrm{d}z - (R_m - e) \int_{-d/2}^{d/2} \frac{\mathrm{d}z}{(R_m + z)} = 0$$

or,

$$[z - (R_m - e) \ln (R_m + z)]_{-D/2}^{D/2} = 0$$

or,

$$D - (R_m - e) \ln \left(\frac{R_m + D/2}{R_m - D/2} \right) = 0$$

or,

$$R_m - e = D / \ln \left(\frac{R_m + D/2}{R_m - D/2} \right)$$

therefore

$$\underline{e = R_m - D / \ln \left(\frac{R_m + D/2}{R_m - D/2} \right)} \tag{7.6}$$

For the present problem,

$$R_m = 60\,\text{mm}$$
$$D = 40\,\text{mm}$$
$$B = 20\,\text{mm}$$

therefore

$$e = 60 - 40/\ln\left(\frac{60+20}{60-20}\right)$$
$$= 60 - 40/0.693$$
$$\underline{e = 2.292\,\text{mm}}$$

Now,

$$R_0 = R_m - e$$
$$= 60 - 2.292 = \underline{57.708\,\text{mm}}$$

As the bending moment tends to straighten the beam (i.e. it decreases curvature),

$$M = -400\,\text{N m}$$

and from (7.5),

$$\sigma = \frac{-400\,y}{Ae(R_0 + y)} \tag{7.7}$$

where,

$$A = 40\,\text{mm} \times 20\,\text{mm}$$
$$\underline{A = 800\,\text{mm}^2}$$

Now on the internal face,

$$y = -(D/2 - e) = -(20 - 2.292)$$
$$\underline{y = -17.708\,\text{mm}}$$

and,

$$\sigma_I = \text{stress on the internal face}$$
$$= \frac{-400\,\text{E3} \times (-17.708)}{800 \times 2.292 \times (57.708 - 17.708)}$$
$$= 7083.2\,\text{E3}/73\,344$$
$$\underline{\sigma_I = 96.58\,\text{MPa (tensile)}}$$

On the external face,

$$y = D/2 + e = 20 + 2.292 = 22.292$$

and,

$$\sigma_E = \text{bending stress on the outer face of the curved beam}$$

$$= \frac{-400\,\text{E3} \times 22.292}{800 \times 2.292 \times (57.708 + 22.292)}$$

$$\underline{\sigma_E = -60.78\ \text{MPa}}$$

The stress distribution across the section is shown in Fig. 7.3.

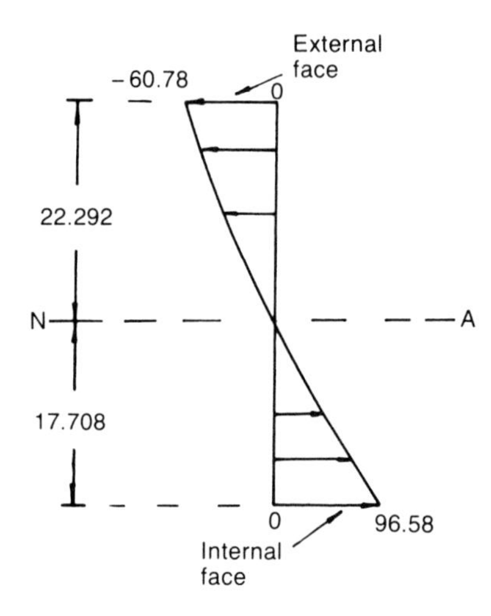

Fig. 7.3. Stress distribution across section.

N.B. For the above rectangular section, it can be seen that the stress due to bending is considerably greater on the internal face of the curved beam than on its external face. One method of making these stresses more equal is to let the section on the internal face be wider than that on the external face, as can be achieved by the use of a trapezoidal section. Similar sections are used for crane hooks.

7.4.1 EXAMPLE 7.2 CRANE HOOKS

Determine the maximum stresses on the internal and external faces of the crane hook of Fig. 7.4, and plot the stress distribution across this section.

7.4.2

To find D_1 and D_2, take moments about XX, i.e.

$$\frac{(B_1 + B_2)}{2} * D * D_1 = B_2 * D * \frac{D}{2} + \frac{(B_1 - B_2)}{2} * D * \frac{D}{3}$$

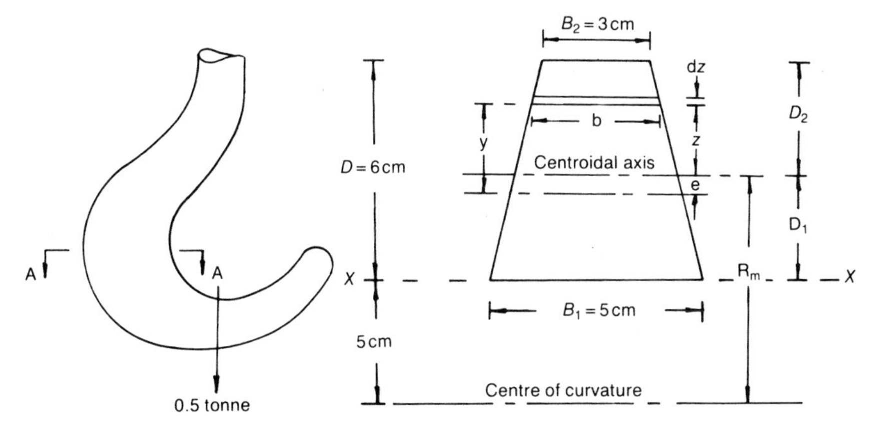

Fig. 7.4. Crane hook.

or,

$$D_1 = \frac{\dfrac{B_2 D}{2} + \dfrac{B_1 D}{6} - \dfrac{B_2 D}{6}}{[(B_1 + B_2)/2]}$$

$$= \frac{\left(\dfrac{B_2 D}{3} + \dfrac{B_1 D}{6}\right) \times 2}{(B_1 + B_2)}$$

therefore

$$\underline{D_1 = (2B_2 + B_1) * D/[3(B_1 + B_2)]} \tag{7.8}$$

Now,

$$D = D_1 + D_2 \tag{7.9}$$

or,

$$D_2 = D - D_1 = \frac{[3(B_1 + B_2)]D}{[3(B_1 + B_2)]} - \frac{(2B_2 + B_1)D}{[3(B_1 + B_2)]}$$

i.e.

$$\underline{D_2 = \frac{(2B_1 + B_2)D}{[3(B_1 + B_2)]}} \tag{7.10}$$

From (7.2),

$$\int \frac{y\,\mathrm{d}A}{(R_0 + y)} = \int \frac{(z + e)}{(R_m + z)}\,\mathrm{d}A = 0$$

or,

$$\int \frac{[(R_m + z) + (e - R_m)]}{(R_m + z)}\,\mathrm{d}A = 0$$

or,

$$\int \mathrm{d}A - (R_m - e)\int \frac{\mathrm{d}A}{(R_m + z)} = 0$$

or,

$$A - (R_m - e)\int \frac{\mathrm{d}A}{(R_m + z)} = 0$$

or,

$$\frac{A}{\int \frac{\mathrm{d}A}{(R_m + z)}} = R_m - e$$

therefore

$$e = R_m - \frac{A}{\int \frac{\mathrm{d}A}{(R_m + z)}} \tag{7.11}$$

In this case,

$$\int \frac{\mathrm{d}A}{(R_m + z)} = \int \frac{b \cdot \mathrm{d}z}{(R_m + z)}$$

but,

$$b = B_2 + (B_1 - B_2)(D_2 - z)/D$$

therefore

$$\int \frac{\mathrm{d}A}{(R_m + z)} = \int_{-D_1}^{D_2} \frac{\{B_2 + [(B_1 - B_2)D_2 - (B_1 - B_2)z]/D\}\mathrm{d}z}{(R_m + z)}$$

$$= \int_{D_1}^{D_2} \frac{[B_2 + (B_1 - B_2)D_2/D + (B_1 - B_2)R_m/D - (B_1 - B_2)(R_m + z)/D]\mathrm{d}z}{(R_m + z)}$$

$$= [B_2 + (B_1 - B_2)D_2/D + (B_1 - B_2)R_m/D]\ln\left(\frac{R_m + D/2}{R_m - D/2}\right)$$

$$- (B_1 - B_2)(D_2 + D_1)/D$$

$$= [B_2 + (B_1 - B_2)(D_2 + R_m)/D]\ln\left(\frac{R_m + D_2}{R_m - D_1}\right) - (B_1 - B_2) \tag{7.12}$$

From (7.8),

$$D_1 = (2 \times 30 + 50) \times 60/[3 \times (50 + 30)]$$

$$D_1 = 110 \times 60/240 = \underline{27.5\,\mathrm{mm}}$$

From (7.9),

$$\underline{D_2 = 32.5\,\mathrm{mm}}$$

From Fig. 7.4,

$$\underline{R_m = 50 + D_1 = 77.5\,\mathrm{mm}}$$

Now,

$$\int \frac{dA}{(R_m + z)} = [30 + (50 - 30)(32.5 + 77.5)/60] \ln\left(\frac{77.5 + 32.5}{77.5 - 27.5}\right) - (50 - 30)$$

$$= 66.67 \times \ln\left(\frac{110}{50}\right) - 20$$

$$\underline{\int \frac{dA}{(R_m + z)} = 32.563}$$

and,

$$A = (30 + 50) \times 60/2 = 2400$$

Hence, from (7.11),

$$e = 77.5 - 2400/32.56$$

$$= 77.5 - 73.71$$

$$\underline{e = 3.79\,\text{mm}}$$

From (7.5), the stress due to bending =

$$\sigma = \frac{My}{Ae(R_0 + y)}$$

Now,

$$R_0 = R_m - e = 77.5 - 3.79 = 73.71\,\text{mm}$$

On the *internal face*,

$$y = 3.79 - 27.5 = -23.71\,\text{mm}$$

Now,

$$M = -500 \times 9.81 \times 73.71 = \underline{-3.615\text{E}5\,\text{N m}}$$

therefore

$$\sigma_{IB} = \text{maximum bending stress on the internal face}$$

$$= \frac{-3.615\text{E}5 \times (-23.71)}{2400 \times 3.79 \times (73.71 - 23.71)}$$

$$\underline{\sigma_{IB} = 18.85\,\text{MPa}}$$

therefore

$$\sigma_I = \text{maximum stress on the inner face}$$

$$= 18.85 + \frac{500 \times 9.81}{2400} = 18.85 + 2.04$$

$$\underline{\sigma_I = 20.89\,\text{MPa}}$$

On the *outer face*,

$$y = 60 - 23.71 = 36.29\,\text{mm}$$

and

$$\sigma_E = \text{maximum stress on the outer face}$$
$$= \frac{-3.615E5 \times (36.29)}{2400 \times 3.79 \times (73.71 + 36.29)} + 2.04$$
$$\underline{\sigma_E = -11.07\,\text{MPa}}$$

The stress distribution is shown in Fig. 7.5.

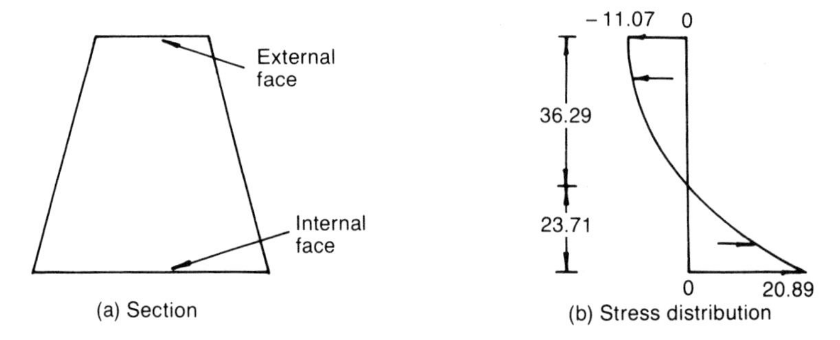

Fig. 7.5. Stress distribution across the section (MPa).

EXAMPLES FOR PRACTICE 7

1. Determine e for the "I" section of the thick curved beam of Fig. Q.7.1.

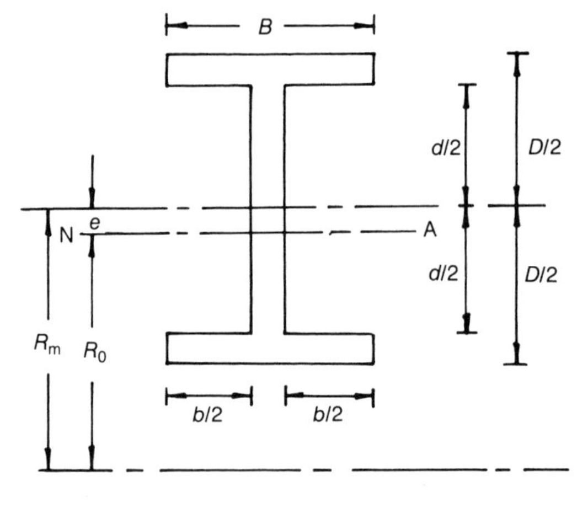

Fig. Q.7.1.

$$\left\{ e = R_m - (BD - bd) \Big/ \left[B \ln\left(\frac{R_m + D/2}{R_m - D/2}\right) - b \ln\left(\frac{R_m + d/2}{R_m - d/2}\right)\right]\right\}$$

2. The thick curved semi-circular arc of Fig. Q.7.2 is of circular cross-section and it is subjected to the forces shown.

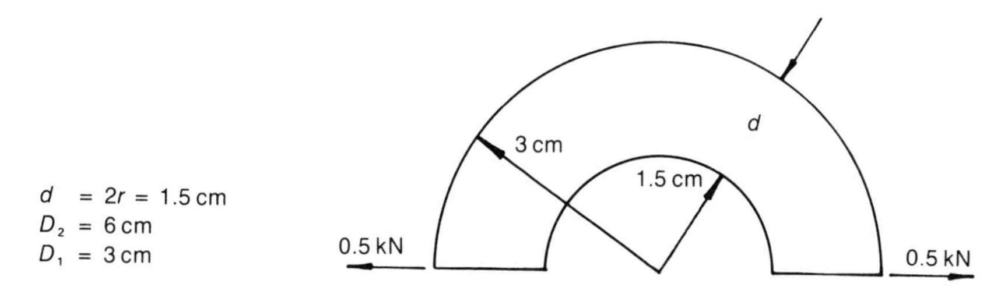

Fig. Q.7.2. Circular ring, under a diametral load.

Determine the maximum stresses that occur on its internal and external faces.

$\{e = R_m - r^2/[2(R_m - (R_m^2 - r^2)^{\frac{1}{2}})]; \sigma_I = 79.73\ \text{MPa}; \sigma_I = -56.91\ \text{MPa}\}$

8

Circular Plates

8.1.1 LARGE AND SMALL DEFLECTIONS OF PLATES

In this chapter, considerations will be made of three classes of plate problem, namely

(1) small deflections of plates, where the maximum deflection does not exceed half the plate thickness, and the deflections are mainly due to the effects of flexure;
(2) large deflections of plates, where the maximum deflection exceeds half the plate thickness, and membrane effects become significant; and
(3) very thick plates, where shear deflections are significant.

8.1.2

Plates take many and various forms [24], from circular plates to rectangular ones, and from plates on ships' decks to ones of arbitrary shape with cut-outs etc., but in this chapter, considerations will be made mostly of the small deflections of circular plates.

8.2.1 PLATE DIFFERENTIAL EQUATION, BASED ON SMALL DEFLECTION ELASTIC THEORY

Let,

w = out-of-plane deflection at any radius r, so that,

$$\frac{\mathrm{d}w}{\mathrm{d}r} = \theta$$

and

$$\frac{d^2w}{dr^2} = \frac{d\theta}{dr}$$

R_t = tangential or circumferential radius of curvature at $r = AC$ (see Fig. 8.1).

R_r = radial or meridional radius of curvature at $r = BC$.

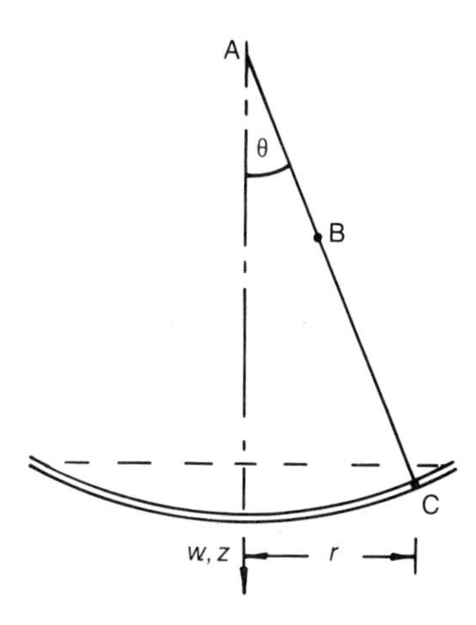

Fig. 8.1. Deflected form of a circular plate.

From standard small deflection theory of beams [1], it is evident that,

$$R_r = 1\bigg/\frac{d^2w}{dr^2} = 1\bigg/\frac{d\theta}{dr} \tag{8.1}$$

or,

$$\frac{1}{R_r} = \frac{d\theta}{dr} \tag{8.2}$$

From Fig. 8.1, it can be seen that,

$$R_t = AC = r/\theta \tag{8.3}$$

or,

$$\frac{1}{R_t} = \frac{1}{r}\frac{dw}{dr} = \frac{\theta}{r} \tag{8.4}$$

Let,

z = the distance of any fibre on the plate from its neutral axis,

so that,

$$\varepsilon_r = \text{radial strain} = \frac{z}{R_r} = \frac{1}{E}(\sigma_r - \nu\sigma_t) \tag{8.5}$$

and,

$$\varepsilon_t = \text{circumferential strain} = \frac{z}{R_t} = \frac{1}{E}(\sigma_t - \nu\sigma_r) \tag{8.6}$$

From equations (8.1) to (8.6), it can be shown that,

$$\sigma_r = \frac{Ez}{(1-\nu^2)}\left(\frac{1}{R_r} + \frac{\nu}{R_t}\right) = \frac{Ez}{(1-\nu^2)}\left(\frac{d\theta}{dr} + \frac{\nu\theta}{r}\right) \quad (8.7)$$

$$\sigma_t = \frac{Ez}{(1-\nu^2)}\left(\frac{1}{R_t} + \frac{\nu}{R_r}\right) = \frac{Ez}{(1-\nu^2)}\left(\frac{\theta}{r} + \frac{d\theta}{dr}\right) \quad (8.8)$$

where,

σ_r = radial stress due to bending
σ_t = circumferential stress due to bending

8.2.2

The tangential or circumferential bending moment per unit radial length =

$$M_t = \int_{-t/2}^{+t/2} \sigma_t z \cdot dz$$

$$= \int_{-t/2}^{+t/2} \frac{E}{(1-\nu^2)}\left(\frac{\theta}{r} + \frac{d\theta}{dr}\right) z^2\, dz$$

$$= \frac{E}{(1-\nu^2)}\left(\frac{\theta}{r} + \nu\frac{d\theta}{dr}\right)\left[\frac{z^3}{3}\right]_{-t/2}^{+t/2}$$

$$= \frac{Et^3}{12(1-\nu^2)}\left(\frac{\theta}{r} + \nu\frac{d\theta}{dr}\right)$$

therefore

$$\underline{\underline{M_t = D\left(\frac{\theta}{r} + \nu\frac{d\theta}{dr}\right) = D\left(\frac{1}{r}\frac{dw}{dr} + \nu\frac{d^2w}{dr^2}\right)}} \quad (8.9)$$

where,

t = plate thickness

and

$$D = \frac{Et^3}{12(1-\nu^2)} = \underline{\text{flexural rigidity}}$$

Similarly, the radial bending moment per unit circumferential length,

$$\underline{\underline{M_r = D\left(\frac{d\theta}{dr} + \frac{\nu\theta}{r}\right) = D\left(\frac{d^2w}{dr^2} + \frac{\nu}{r}\frac{dw}{dr}\right)}} \quad (8.10)$$

Substituting (8.9) and (8.10) into (8.7) and (8.8), the bending stresses could be put in the following form:

$$\left.\begin{aligned} \sigma_t &= 12M_t * z/t^3 \\ \text{and} \quad \sigma_r &= 12M_r * z/t^3 \end{aligned}\right\} \quad (8.11)$$

and the maximum stresses $\hat{\sigma}_t$ and $\hat{\sigma}_r$ will occur at the outer surfaces of the plate (i.e. @ $z = \pm t/2$). Therefore

$$\underline{\hat{\sigma}_t = 6M_t/t^2} \tag{8.12}$$

and

$$\underline{\hat{\sigma}_r = 6M_r/t^2} \tag{8.13}$$

8.2.3

The plate differential equation can now be obtained by considering the equilibrium of the plate element of Fig. 8.2.

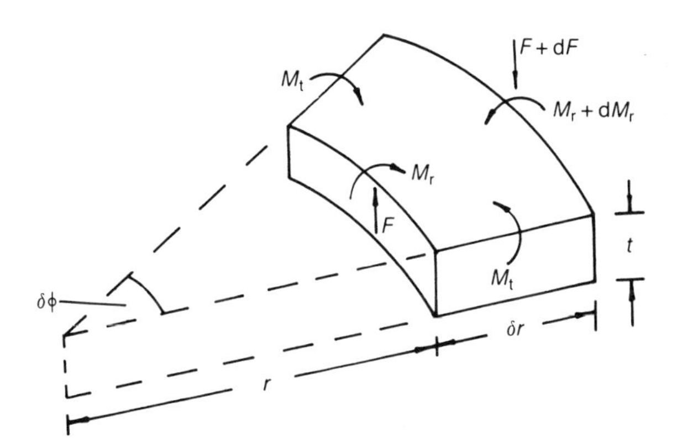

Fig. 8.2. Element of a circular plate.

Taking moments about the outer circumference of the element,

$$(M_r + \delta M_r)(r + \delta r)\delta\phi - M_r\ r\delta\phi - 2M_t \cdot \delta r \sin\frac{\delta\phi}{2} - F \cdot r \cdot \delta\phi\delta r = 0$$

In the limit, this becomes

$$M_r + r \cdot \frac{\mathrm{d}M_r}{\mathrm{d}r} - M_t - Fr = 0 \tag{8.14}$$

Substituting (8.9) and (8.10) into (8.14),

$$\left(\frac{\mathrm{d}\theta}{\mathrm{d}r} + \frac{\nu\theta}{r}\right) + \left(r \cdot \frac{\mathrm{d}^2\theta}{\mathrm{d}r^2} + r \cdot \frac{\nu\,\mathrm{d}\theta}{r\,\mathrm{d}r} - r \cdot \frac{\nu\theta}{r^2}\right) - \frac{\theta}{r} - \nu\frac{\mathrm{d}\theta}{\mathrm{d}r} = \frac{Fr}{D}$$

or,

$$\frac{\mathrm{d}^2\theta}{\mathrm{d}r^2} + \left(\frac{1}{r}\right)\frac{\mathrm{d}\theta}{\mathrm{d}r} - \frac{\theta}{r^2} = \frac{F}{D}$$

which can be rewritten in the form:

$$\frac{\mathrm{d}}{\mathrm{d}r}\left[(1/r) \cdot \frac{\mathrm{d}(r\theta)}{\mathrm{d}r}\right] = \frac{F}{D} \tag{8.15}$$

where,

$$F = \text{shearing force/unit circumferential length}$$

Equation (8.15) is known as the plate differential equation for circular plates.

8.2.4

For a plate subjected to a lateral pressure p per unit area and a concentrated load W at the centre, F can be obtained from equilibrium considerations. Resolving "vertically",

$$2\pi rF = \pi r^2 \cdot p + W$$

therefore

$$F = \frac{pr}{2} + \frac{W}{2\pi r} \text{(except at } r = 0) \tag{8.16}$$

Substituting (8.16) into (8.15),

$$\frac{d}{dr}\left[(1/r)\frac{d(r\theta)}{dr}\right] = \frac{1}{D}\left[\frac{pr}{2} + \frac{W}{2\pi r}\right]$$

therefore

$$\frac{1}{r}\frac{d(r\theta)}{dr} = \frac{1}{D}\left[\frac{pr^2}{4} + \frac{W}{2\pi}\ln r\right] + C_1$$

$$\frac{d(r\theta)}{dr} = \frac{1}{D}\left[\frac{pr^3}{4} + \frac{Wr}{2\pi}\ln r\right] + C_1 r$$

$$r\theta = \frac{1}{D}\left[\frac{pr^4}{16} + \frac{Wr^2}{4\pi}\ln r - \frac{Wr^2}{8\pi}\right] + \frac{C_1 r^2}{2} + C_2$$

$$\theta = \frac{1}{D}\left[\frac{pr^3}{16} + \frac{Wr}{4\pi}\ln r - \frac{Wr}{8\pi}\right] + \frac{C_1 r}{2} + \frac{C_2}{r} \tag{8.17}$$

since,

$$\frac{dw}{dr} = \theta$$

$$w = \int \theta \, dr + C_3$$

hence,

$$w = \frac{pr^4}{64D} + \frac{Wr^2}{8\pi D}(\ln r - 1) + \frac{C_1 r^2}{4} + C_2 \ln r + C_3 \tag{8.18}$$

N.B.

$$\int r \ln r \, dr = \int \frac{\ln r}{2} d(r^2)$$

$$= \frac{r^2}{2}\ln r - \int \frac{r^2}{2} d(\ln r) = \frac{r^2}{2}\ln r - \int \frac{r}{2} dr$$

$$= \underline{\frac{r^2}{2}\ln r - \frac{r^2}{4} + \text{a constant}} \tag{8.19}$$

8.3.1 EXAMPLE 8.1 CIRCULAR PLATE, CLAMPED AROUND ITS CIRCUMFERENCE, UNDER A CONCENTRATED LOAD

Determine the maximum deflection and stress in a circular plate, clamped around its circumference, when it is subjected to a centrally placed concentrated load W.

8.3.2

Putting $p = 0$ in (8.18):

$$w = \frac{Wr^2}{8\pi D}(\ln r - 1) + \frac{C_1 r^2}{4} + C_2 \ln r + C_3$$

$$\frac{dw}{dr} = \frac{Wr}{4\pi D}(\ln r - 1) + \frac{Wr}{8\pi D} + \frac{C_1 r}{2} + \frac{C_2}{r}$$

as dw/dr cannot equal ∞ @ $r = 0$, $C_2 = 0$

@ $r = R, \quad \frac{dw}{dr} = w = 0$

therefore

$$0 = \frac{WR^2}{8\pi D}\ln R - \frac{WR^2}{8\pi D} + \frac{C_1 R^2}{4} + C_3$$

and,

$$0 = \frac{WR}{4\pi D}\ln R - \frac{WR}{4\pi D} + \frac{WR}{8\pi D} + \frac{C_1 R}{2}$$

Hence,

$$C_1 = \frac{W}{4\pi D}(1 - 2\ln R)$$

$$C_3 = -\frac{WR^2}{8\pi D}\ln R + \frac{WR^2}{8\pi D} - \frac{WR^2}{16\pi D} + \frac{WR^2}{8\pi D}\ln R = \frac{WR^2}{16\pi D}$$

$$w = \frac{WR^2}{8\pi D}\ln r - \frac{Wr^2}{8\pi D} + \frac{Wr^2}{16\pi D} - \frac{Wr^2}{8\pi D}\ln R + \frac{WR^2}{16\pi D}$$

or,

$$\underline{w = \frac{WR^2}{16\pi D}\left[1 - \frac{r^2}{R^2} + \frac{2r^2}{R^2}\ln\left(\frac{r}{R}\right)\right]}$$

The maximum deflection ($\hat{w}$) occurs @ $r = 0$

$$\hat{w} = \frac{WR^2}{16\pi D}$$

Substituting the derivatives of w into (8.9) and (8.10),

$$M_r = \frac{W}{4\pi}\left[1 + \ln\left(\frac{r}{R}\right)\cdot(1 + \nu)\right]$$

$$M_t = \frac{W}{4\pi}\left[\nu + (1 + \nu)\cdot\ln\left(\frac{r}{R}\right)\right]$$

8.4.1 EXAMPLE 8.2 CLAMPED CIRCULAR PLATE UNDER A UNIFORM PRESSURE

Determine the maximum deflection and stress that occur when a circular plate clamped around its external circumference is subjected to a uniform lateral pressure p.

8.4.2

From (8.18),

$$w = \frac{pr^4}{64D} + \frac{C_1 r^2}{4} + C_2 \ln r + C_3$$

$$\frac{dw}{dr} = \frac{pr^3}{16D} + \frac{C_1 r}{2} + \frac{C_2}{r}$$

and,

$$\frac{d^2 w}{dr^2} = \frac{3pr^2}{16D} + \frac{C_1}{2} - \frac{C_2}{r^2}$$

@ $\quad r = 0, \quad \frac{dw}{dr} \neq \infty$ therefore $C_2 = 0$

@ $\quad r = R, \quad w = \frac{dw}{dr} = 0$

therefore

$$0 = \frac{pr^4}{64D} + \frac{C_1 R^2}{4} + C_3$$

$$0 = \frac{pR^3}{16D} + \frac{C_1 R}{2}$$

therefore

$$C_1 = -\frac{pR^2}{8D}$$

$$C_3 = \frac{pR^4}{64D}$$

therefore

$$w = \frac{pR^4}{64D}\left(1 - \frac{r^2}{R^2}\right)^2 \tag{8.20}$$

Substituting the appropriate derivatives of w into (8.9) and (8.10),

$$M_r = \frac{pR^2}{16}\left[-(1+v) + (3+v)\frac{r^2}{R^2}\right] \tag{8.21}$$

$$M_t = \frac{pR^2}{16}\left[-(1+v) + (1+3v)\frac{r^2}{R^2}\right] \tag{8.22}$$

Maximum deflection ($\hat{w}$) occurs at $r = 0$

$$\hat{w} = \frac{pR^4}{64D} \tag{8.23}$$

By inspection it can be seen that the maximum bending moment is obtained from (8.21), when $r = R$, i.e.

$$\hat{M}_r = pR^2/8$$

and

$$\hat{\sigma} = 6\hat{M}_r/t^2$$
$$= \underline{0.75\, pR^2/t^2}$$

8.5.1 EXAMPLE 8.3 ANNULAR DISC

Determine the expressions for M_r and M_t in an annular disc, simply-supported around its outer circumference, when it is subjected to a concentrated load W, distributed around its inner circumference, as shown in Fig. 8.3.

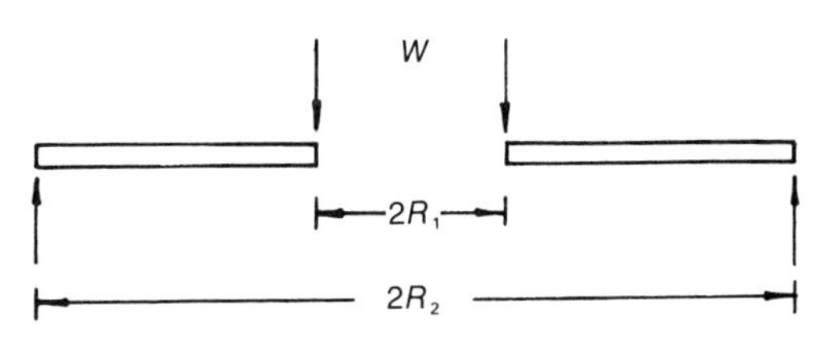

Fig. 8.3. Annular disc.

W = total load around the inner circumference.

8.5.2

From equation (8.18),

$$w = \frac{Wr^2}{8\pi D}(\ln r - 1) + \frac{C_1 r^2}{4} + C_2 \ln r + C_3$$

@ $r = R_2, \quad w = 0$

or,

$$0 = \frac{WR_2^2}{8\pi D}(\ln R_2 - 1) + \frac{C_1}{4} R_2^2 + C_2 \ln R_2 + C_3 \tag{8.24}$$

Now,

$$\frac{dw}{dr} = \frac{Wr}{4\pi D}(\ln r - 1) + \frac{Wr}{8\pi D} + \frac{C_1 r}{2} + \frac{C_2}{r} \tag{8.25}$$

and,

$$\frac{d^2w}{dr^2} = \frac{W}{4\pi D}(\ln r - 1) + \frac{W}{4\pi D} + \frac{W}{8\pi D} + \frac{C_1}{2} - \frac{C_2}{r^2} \tag{8.26}$$

A suitable boundary condition is that,

$$M_r = 0 \ @ \ r = R_1 \text{ and } @ \ r = R_2$$

but,

$$M_r = D\left(\frac{d^2w}{dr^2} + \frac{v}{r}\frac{dw}{dr}\right)$$

therefore

$$\frac{W}{4\pi D}(\ln R_1 - 1) + \frac{3W}{8\pi D} + \frac{C_1}{2} - \frac{C_2}{R_1^2} + \frac{v}{R_1}\left\{\frac{WR_1}{4\pi D}(\ln R_1 - 1) + \frac{WR_1}{8\pi D} + \frac{C_1R_1}{2} + \frac{C_2}{R_1}\right\} = 0 \tag{8.27}$$

and,

$$\frac{W}{4\pi D}(\ln R_2 - 1) + \frac{3W}{8\pi D} + \frac{C_1}{2} - \frac{C_2}{R_2^2} + \frac{v}{R_2}\left\{\frac{WR_2}{4\pi D}(\ln R_2 - 1) + \frac{WR_2}{8\pi D} + \frac{C_1R_2}{2} + \frac{C_2}{R_2}\right\} = 0 \tag{8.28}$$

Solving (8.27) and (8.28) for C_1 and C_2,

$$C_1 = \frac{-W}{4\pi D}\left\{\frac{2(R_2^2 \ln R_2 - R_1^2 \ln R_1)}{(R_2^2 - R_1^2)} + \frac{(1-v)}{(1+v)}\right\} \tag{8.29}$$

and,

$$C_2 = \frac{-W}{4\pi D} * \frac{(1+v)}{(1-v)} * \frac{(R_2^2 R_1^2)}{(R_2^2 - R_1^2)} \ln\left(\frac{R_2}{R_1}\right) \tag{8.30}$$

C_3 is not required to determine expressions for M_r and M_t. Hence,

$$M_r = D(W/8\pi D)\{(1+v)2\ln r + (1-v)\} + (C_1/2)(1+v) - (C_2/r^2)(1+v) \tag{8.31}$$

and,

$$M_t = D(W/8\pi D)\{(1+v)2\ln r - (1-v)\} + (C_1/2)(1+v) + (C_2/r^2)(1-v) \tag{8.32}$$

8.6.1 EXAMPLE 8.4 CIRCULAR PLATE WITH PARTIAL PRESSURE LOADING

A flat circular plate of radius R_2 is simply-supported concentrically by a tube of radius R_1, as shown in Fig. 8.4. If the "internal" portion of the plate is

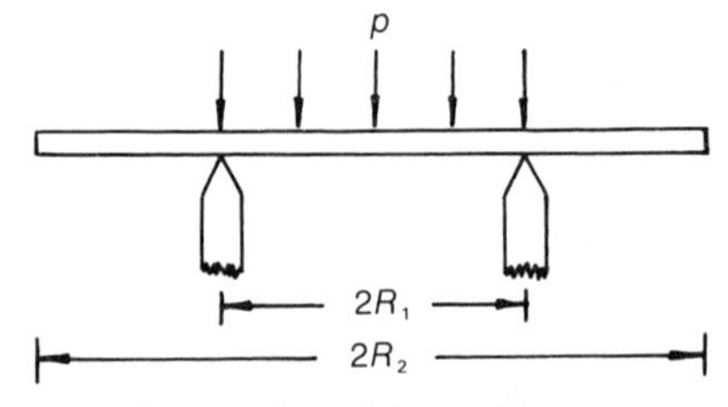

Fig. 8.4. Circular plate with a partial pressure load.

subjected to a uniform pressure p, show that the central deflection δ of the plate is given by:

$$\delta = \frac{pR_1^4}{64D}\left\{3 + 2\left(\frac{R_1}{R_2}\right)^2\left(\frac{1-\nu}{1+\nu}\right)\right\}$$

8.6.2

Now the shearing force/unit length F for $r > R_1$ is zero, and for $r < R_1$,

$$F = pr/2$$

so that the plate differential equation becomes:

$$r < R_1 \qquad\qquad r > R_1$$

$$\frac{d}{dr}\left\{\frac{1}{r}\frac{d}{dr}\left(r\frac{dw}{dr}\right)\right\} = \frac{pr}{2D} \qquad = 0$$

or

$$\frac{1}{r}\frac{d}{dr}\left(r\frac{dw}{dr}\right) = \frac{pr^2}{4D} + A \qquad = B \tag{8.33}$$

For continuity at $r = R_1$, the two expressions on the right of equation (8.33) must be equal, i.e.

$$\frac{pR_1^2}{4D} + A = B$$

or

$$\underline{B = \frac{pR_1^2}{4D} + A} \tag{8.34}$$

or

$$\frac{1}{r}\frac{d}{dr}\left(r\frac{dw}{dr}\right) = \frac{pr^2}{4D} + A \qquad = \frac{pR_1^2}{4D} + A$$

or

$$\frac{d}{dr}\left(r\frac{dw}{dr}\right) = \frac{pr^3}{4D} + Ar \qquad = \frac{pR_1^2 r}{4D} + Ar$$

which on integrating becomes,

$$r\frac{dw}{dr} = \frac{pr^4}{16D} + \frac{Ar^2}{2} + C \qquad = \frac{pR_1^2 r^2}{8D} + \frac{Ar^2}{2} + F$$

or

$$\frac{dw}{dr} = \frac{pr^3}{16D} + \frac{Ar}{2} + \frac{C}{r} \qquad = \frac{pR_1^2 r}{8D} + \frac{Ar}{2} + \frac{F}{r} \tag{8.35}$$

@ $\quad r = 0, \quad \frac{dw}{dr} \neq \infty \quad$ therefore $\quad \underline{C = 0}$

For continuity @ $r = R_1$, the value of the slope must be the same from both expressions on the right of equation (8.35), i.e.

$$\frac{pR_1^3}{16D}+\frac{AR_1}{2}=\frac{pR_1^3}{8D}+\frac{AR_1}{2}+\frac{F}{R_1}$$

therefore

$$\underline{F=-pR_1^4/(16D)} \tag{8.36}$$

therefore

$$\frac{\mathrm{d}w}{\mathrm{d}r}=\frac{pr^3}{16D}+\frac{Ar}{2}\;\Bigg\vdots\;=\frac{pR_1^2r}{8D}+\frac{Ar}{2}-\frac{pR_1^4}{16Dr} \tag{8.37}$$

which on integrating becomes

$$w=\frac{pr^4}{64D}+\frac{Ar^2}{4}+G\;\Bigg\vdots\;=\frac{pR_1^2r^2}{16D}+\frac{Ar^2}{4}-\frac{pR_1^4}{16D}\ln r+H \tag{8.38}$$

Now, there are three unknowns in equation (8.38), namely A, G and H, and, therefore, three simultaneous equations are required to determine these unknowns. One equation can be obtained by considering the continuity of w at $r=R_1$ in (8.38), and the other two equations can be obtained by considering boundary conditions.

One suitable boundary condition is that at $r=R_2$, $M_r=0$, which can be obtained by considering that portion of the plate where $R_2>r>R_1$, as follows:

$$\frac{\mathrm{d}w}{\mathrm{d}r}=\frac{pR_1^2r}{8D}+\frac{Ar}{2}-\frac{pR_1^4}{16Dr}$$

$$\frac{\mathrm{d}^2w}{\mathrm{d}r^2}=\frac{pR_1^2}{8D}+\frac{A}{2}+\frac{pR_1^4}{16Dr^2}$$

Now,

$$M_r=D\left(\frac{\mathrm{d}^2w}{\mathrm{d}r^2}+\frac{v}{r}\frac{\mathrm{d}w}{\mathrm{d}r}\right)$$

$$=D\left\{\left(\frac{pR_1^2}{8D}+\frac{A}{2}+\frac{pR_1^4}{16Dr^2}\right)+\frac{v}{r}\left(\frac{pR_1^2r}{8D}+\frac{Ar}{2}-\frac{pR_1^4}{16Dr}\right)\right\}$$

$$=D\left\{\frac{pR_1^2}{8D}(1+v)+\frac{A}{2}(1+v)+\frac{pR_1^4}{16Dr^2}(1-v)\right\} \tag{8.39}$$

Now, @ $r=R_2$, $M_r=0$ therefore

$$\frac{A}{2}(1+v)=-\frac{pR_1^2}{8D}(1+v)-\frac{pR_1^4}{16DR_2^2}(1-v)$$

or,

$$\underline{A=-\frac{pR_1^2}{4D}-\frac{pR_1^4}{8DR_2^2}\left(\frac{1-v}{1+v}\right)} \tag{8.40}$$

Another suitable boundary condition is that

@ $\qquad r = R_1, \quad w = 0$

In this case, it will be necessary to consider only that portion of the plate where $r < R_1$, as follows:

$$w = \frac{pr^4}{64D} + \frac{Ar^2}{4} + G$$

@ $r = R_1$, $w = 0$ therefore

$$0 = \frac{pR_1^4}{64D} + \frac{AR_1^2}{4} + G$$

or,

$$G = -\frac{pR_1^4}{64D} + \frac{pR_1^4}{16D} + \frac{pR_1^6}{32DR_2^2}\left(\frac{1-v}{1+v}\right)$$

$$= -\frac{pR_1^4}{64D} + \left\{\frac{pR_1^2}{4D} + \frac{pR_1^4}{8DR_2^2}\left(\frac{1-v}{1+v}\right)\right\}\frac{R_1^2}{4}$$

or,

$$\underline{G = \frac{pR_1^4}{64D}\left\{3 + 2\left(\frac{R_1}{R_2}\right)^2\left(\frac{1-v}{1+v}\right)\right\}} \tag{8.41}$$

The central deflection δ occurs at $r = 0$; hence, from (8.41),

$$\delta = G$$

$$\underline{\delta = \frac{pR_1^4}{64D}\left\{3 + 2\left(\frac{R_1}{R_2}\right)^2\left(\frac{1-v}{1+v}\right)\right\}} \tag{8.42}$$

8.7.1 EXAMPLE 8.5 PLATE UNDER AN ANNULAR LOAD

A flat circular plate of outer radius R_2 is clamped firmly around its outer circumference. If a load W is applied concentrically to the plate, through a tube of radius R_1, as shown in Fig. 8.5, show that the central deflection δ is

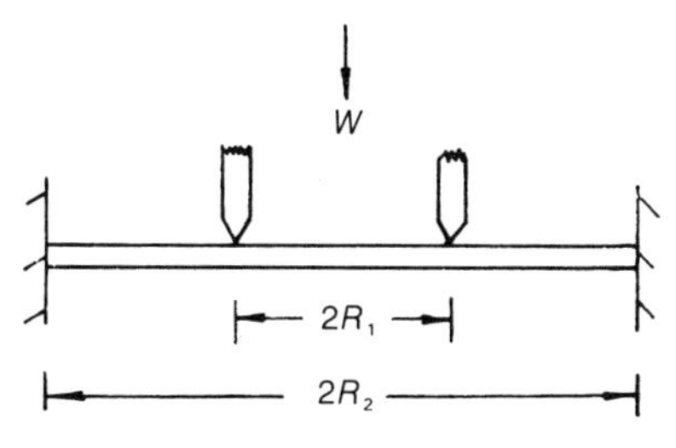

Fig. 8.5. Plate under an annular load.

given by:

$$\delta = \frac{W}{16\pi D}\left\{R_1^2 \ln\left(\frac{R_1}{R_2}\right)^2 + (R_2^2 - R_1^2)\right\}$$

8.7.2

Now when $r < R_1$, $F = 0$, and when $R_2 > r > R_1$, $F = W/2\pi r$, so that the plate differential equation becomes:

$$r < R_1 \quad \Big| \quad r > R_1$$

$$\frac{\mathrm{d}}{\mathrm{d}r}\left\{\frac{1}{r}\mathrm{d}\left(r\frac{\mathrm{d}w}{\mathrm{d}r}\right)\right\} = 0 \quad \Big| \quad = \frac{W}{2\pi r D}$$

or

$$\frac{1}{r}\mathrm{d}\left(r\frac{\mathrm{d}w}{\mathrm{d}r}\right) = A \quad \Big| \quad = \frac{W}{2\pi D}\ln r + B$$

or

$$\mathrm{d}\left(r\frac{\mathrm{d}w}{\mathrm{d}r}\right) = Ar \quad \Big| \quad = \frac{Wr\ln r}{2\pi D} + Br \tag{8.43}$$

From continuity considerations at $r = R_1$, the two expressions on the right of (8.43) must be equal, i.e.

$$A = \frac{W}{2\pi D}\ln R_1 + B \tag{8.44}$$

On integrating (8.43),

$$r\frac{\mathrm{d}w}{\mathrm{d}r} = \frac{Ar^2}{2} + C \quad \Big| \quad = \frac{W}{2\pi D}\left(\frac{r^2}{2}\ln r - \frac{r^2}{4}\right) + \frac{Br^2}{2} + F$$

or,

$$\frac{\mathrm{d}w}{\mathrm{d}r} = \frac{Ar}{2} + \frac{C}{r} \quad \Big| \quad = \frac{Wr}{4\pi D}\left(\ln r - \frac{r}{2}\right) + \frac{Br}{2} + \frac{F}{r} \tag{8.45}$$

@ $r = 0$, $\dfrac{\mathrm{d}w}{\mathrm{d}r} \neq \infty$ therefore $\underline{C = 0}$

From continuity considerations for $\mathrm{d}w/\mathrm{d}r$, at $r = R_1$,

$$\frac{AR_1}{2} = \frac{WR_1}{4\pi D}\left(\ln R_1 - \frac{R_1}{2}\right) + \frac{BR_1}{2} + \frac{F}{R_1} \tag{8.46}$$

On integrating (8.46),

$$w = \frac{Ar^2}{2} + G \quad \Big| \quad = \frac{W}{2\pi D}\left(\frac{r^2}{4}\ln r - \frac{r^2}{8} - \frac{r^2}{8}\right) + \frac{Br^2}{4} + F\ln r + H$$

or

$$w = \frac{Ar^2}{2} + G \quad \Big| \quad = \frac{Wr^2}{8\pi D}(\ln r - 1) + \frac{Br^2}{4} + F\ln r + H \tag{8.47}$$

From continuity considerations for w, @ $r = R_1$,

$$\frac{AR_1^2}{2} + G = \frac{WR_1^2}{8\pi D}(\ln R_1 - 1) + \frac{BR_1^2}{4} + F\ln R_1 + H \tag{8.48}$$

In order to obtain the necessary number of simultaneous equations to determine the arbitrary constants, it will be necessary to consider *boundary conditions.*

At $\quad r = R_2, \quad \dfrac{dw}{dr} = 0$

therefore

$$0 = \frac{WR_2}{4\pi D}\left(\ln R_2 - \frac{R_2}{2}\right) + \frac{BR_2}{2} + \frac{F}{R_2} \tag{8.49}$$

Also, @ $r = R_2$, $w = 0$

therefore

$$0 = \frac{WR_2^2}{8\pi D}(\ln R_2 - 1) + \frac{BR_2^2}{4} + F\ln(R_2) + H \tag{8.50}$$

Solving (8.46), (8.48), (8.49) and (8.50),

$$F = \frac{WR_1^2}{8\pi D} \tag{8.51}$$

$$H = -\frac{W}{8\pi D}\{-R_2^2/2 - R_1^2/2 + R_1^2\ln(R_2)\}$$

and

$$G = -\frac{WR_1^2}{8\pi D} + \frac{WR_1^2}{8\pi D}\ln(R_1) + H$$

$$= -\frac{WR_1^2}{8\pi D} + \frac{WR_1^2\ln(R_1)}{8\pi D} - \frac{W}{8\pi D}\left\{-\frac{R_2^2}{2} - \frac{R_1^2}{2} + R_1^2\ln(R_2)\right\}$$

$$= \frac{W}{16\pi D}\{-2R_1^2 + 2R_1^2\ln(R_1) + R_2^2 + R_1^2 - 2R_1^2\ln(R_2)\}$$

$$G = \frac{W}{16\pi D}\left\{R_1^2\ln\left(\frac{R_1}{R_2}\right)^2 + (R_2^2 - R_1^2)\right\}$$

δ occurs @ $r = 0$, i.e.

$$\delta = G = \frac{W}{16\pi D}\left\{R_1^2\ln\left(\frac{R_1}{R_2}\right)^2 + (R_2^2 - R_1^2)\right\}$$

8.8.1 LARGE DEFLECTIONS OF PLATES

If the maximum deflection of a plate exceeds half the plate thickness, the plate changes to a shallow shell, and withstands much of the lateral load as a membrane, rather than as a flexural structure.

For example, consider the membrane shown in Fig. 8.6, which is subjected to a uniform lateral pressure p.

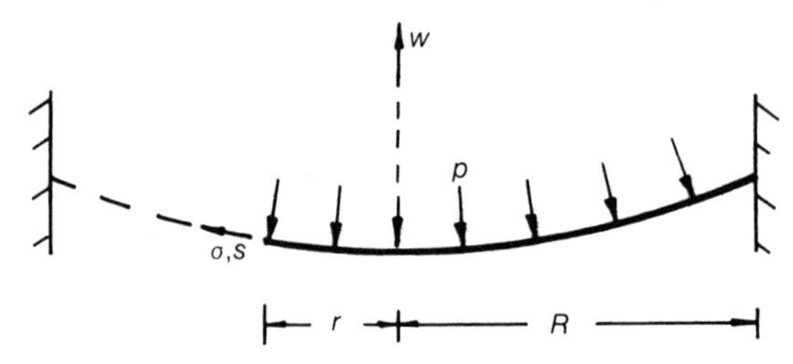

Fig. 8.6. Portion of circular membrane.

8.8.2

Let,

w = out-of-plane deflection at any radius r
σ = membrane tension at a radius r
t = thickness of membrane

Resolving vertically,

$$\sigma * t * 2\pi r * \frac{\mathrm{d}w}{\mathrm{d}r} = p * \pi r^2$$

or

$$\frac{\mathrm{d}w}{\mathrm{d}r} = \frac{pr}{2\sigma t} \tag{8.52}$$

or

$$w = \frac{pr^2}{4\sigma t} + A$$

@ $r = R$, $w = 0$ therefore

$$A = -\frac{pR^2}{4\sigma t}$$

i.e.

$\hat{w}$ = maximum deflection of membrane
$\underline{\hat{w} = -pR^2/(4\sigma t)}$

8.8.3

Now change of meridional (or radial) length =

$$\delta l = \int \mathrm{d}s - \int \mathrm{d}r$$

where,

s = any length along the meridian

Using Pythagoras' theorem,

$$\delta l = \int (\mathrm{d}w^2 + \mathrm{d}r^2)^{\frac{1}{2}} - \int \mathrm{d}r$$

$$= \int \left[1 + \left(\frac{\mathrm{d}w}{\mathrm{d}r}\right)^2\right]^{\frac{1}{2}} \mathrm{d}r - \int \mathrm{d}r$$

Expanding binomially, and neglecting higher order terms,

$$\delta l = \int \left[1 + \tfrac{1}{2}\left(\frac{dw}{dr}\right)^2\right] dr - \int dr$$

$$= \tfrac{1}{2} \int \left(\frac{dw}{dr}\right)^2 dr \tag{8.53}$$

Substituting the derivative of w, (8.52), into (8.53),

$$\delta l = \tfrac{1}{2} \int_0^R \left(\frac{pr}{2\sigma t}\right)^2 dr$$

$$= p^2R^3/(24\sigma^2t^2) \tag{8.54}$$

but,

$$\varepsilon = \text{strain} = \frac{\delta l}{R} = \frac{1}{E}(\sigma - \nu\sigma)$$

or,

$$\sigma^3 = \frac{E}{(1-\nu)}\left(\frac{p^2R^2}{24\sigma^2t^2}\right)$$

i.e.

$$\sigma = \sqrt[3]{\left\{\frac{E}{(1-\nu)}\left(\frac{p^2R^2}{24t^2}\right)\right\}} \tag{8.55}$$

but

$$\sigma = pR^2/(4t\hat{w}) \tag{8.56}$$

Equating (8.55) and (8.56),

$$p = \frac{8E}{3(1-\nu)}\left(\frac{t}{R}\right)\left(\frac{\hat{w}}{R}\right)^3 \tag{8.57}$$

According to small deflection theory of plates, (8.23),

$$p = \frac{64D}{R^3}\left(\frac{\hat{w}}{R}\right) \tag{8.58}$$

Thus, for the large deflections of clamped circular plates under lateral pressure, equations (8.57) and (8.58) should be added together, as follows:

$$p = \frac{64D}{R^3}\left(\frac{\hat{w}}{R}\right) + \frac{8}{3(1-\nu)}\left(\frac{t}{R}\right)\left(\frac{\hat{w}}{R}\right)^3 \tag{8.59}$$

If $\nu = 0.3$, then (8.59) becomes

$$\frac{pR^4}{64Dt} = \left(\frac{\hat{w}}{t}\right)\left\{1 + 0.65\left(\frac{\hat{w}}{t}\right)\right\} \tag{8.60}$$

where the second term in (8.60) represents the membrane effect, and the first term represents the flexural effect.

When $\hat{w}/t = 0.5$, the membrane effect is about 16.3% of the bending effect,

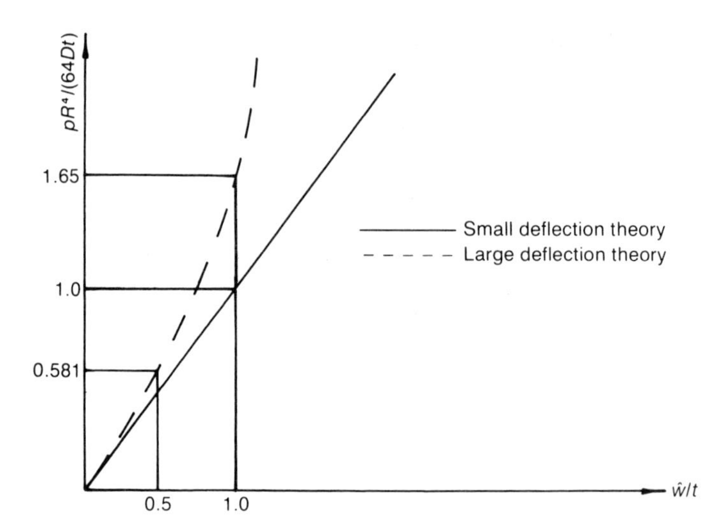

Fig. 8.7. Small and large deflection theory.

but when $\hat{w}/t = 1$, the membrane effect becomes about 65% of the bending effect. The bending and membrane effects are about the same when $\hat{w}/t = 1.24$.

A plot of the variation of $\hat{w}$ due to bending, and due to the combined effects of bending plus membrane stresses, is shown in Fig. 8.7.

8.8.4 Power Series Solution

This method of solution, which involves the use of data sheets, is based on a power series solution of the fundamental equations governing the large deflection theory of circular plates.

For a circular plate under a uniform lateral pressure p, the large deflection equations [24] are given by (8.61) to (8.63).

$$\mathrm{D}\frac{\mathrm{d}}{\mathrm{d}r}\left\{\frac{1}{r}\frac{\mathrm{d}}{\mathrm{d}r}\left(r\frac{\mathrm{d}w}{\mathrm{d}r}\right)\right\} = \sigma_r^t\frac{\mathrm{d}w}{\mathrm{d}r} + \frac{pr}{2} \tag{8.61}$$

$$\frac{\mathrm{d}}{\mathrm{d}r}(r\sigma_r) - \sigma_t = 0 \tag{8.62}$$

$$r\frac{\mathrm{d}}{\mathrm{d}r}(\sigma_r + \sigma_t) + \frac{E}{2}\left(\frac{\mathrm{d}w}{\mathrm{d}r}\right)^2 = 0 \tag{8.63}$$

Way [30] has shown that to assist in the solution of equations (8.61) to (8.63), by the power series method, it will be convenient to introduce the dimensionless ratio ζ, where,

$$\zeta = r/R$$

or

$$r = \zeta R$$

R = outer radius of disc
r = any value of radius between 0 and R

Substituting for r into (8.61):

$$\frac{1}{12(1-\nu^2)}\frac{\mathrm{d}}{\mathrm{d}(\zeta R)}\left\{\frac{1}{\zeta R}*\frac{\mathrm{d}(\zeta R\theta)}{\mathrm{d}(\zeta R)}\right\}=\frac{\sigma_{\mathrm{r}}\theta}{Et^2}+\frac{p\zeta R}{2Et^3}$$

or,

$$\frac{1}{12(1-\nu^2)}\frac{\mathrm{d}}{\mathrm{d}\zeta}\left\{\frac{1}{\zeta}*\frac{\mathrm{d}(\zeta\theta)}{\mathrm{d}\zeta}\right\}=\frac{\sigma_{\mathrm{r}}R^2\theta}{Et^2}+\frac{pR^3\zeta}{2Et^3} \tag{8.64}$$

Inspecting (8.64), it can be seen that the LHS is dependent only on the slope θ.

Now,

$$\theta=\frac{\mathrm{d}w}{\mathrm{d}r}=\frac{\mathrm{d}w}{\mathrm{d}(\zeta R)}$$

which on substituting into (8.64) gives:

$$\frac{1}{12(1-\nu^2)}\frac{\mathrm{d}}{\mathrm{d}\zeta}\left\{\frac{1}{\zeta}\frac{\mathrm{d}}{\mathrm{d}\zeta}\left[\frac{\zeta\mathrm{d}(w/t)}{\mathrm{d}\zeta}\right]\right\}=\frac{\sigma_{\mathrm{r}}}{E}\left(\frac{R}{t}\right)^2\frac{\mathrm{d}(w/t)}{\mathrm{d}\zeta}+\frac{p}{E}\left(\frac{R}{t}\right)^4\frac{\zeta}{2} \tag{8.65}$$

but,

$$\left(\frac{w}{t}\right)\cdot\frac{\sigma_{\mathrm{r}}}{E}\left(\frac{R}{t}\right)^2 \quad \text{and} \quad \frac{p}{E}\left(\frac{R}{t}\right)^4$$

are all dimensionless, and this feature will be used later on in the present chapter.

Now substituting r, in terms of ζ into (8.62), equation (8.66) is obtained:

$$\frac{\mathrm{d}}{\mathrm{d}\zeta}\left\{\frac{\zeta\sigma_{\mathrm{r}}}{E}\left(\frac{R}{t}\right)^2\right\}=\frac{\sigma_{\mathrm{t}}}{E}\left(\frac{R}{t}\right)^2 \tag{8.66}$$

Similarly, substituting r in terms of ζ into (8.63), equation (8.67) is obtained:

$$\zeta\frac{\mathrm{d}}{\mathrm{d}\zeta}\left\{\frac{\sigma_{\mathrm{r}}}{E}\left(\frac{R}{t}\right)^2+\frac{\sigma_{\mathrm{t}}}{E}\left(\frac{R}{t}\right)^2\right\}+\frac{E}{2}\theta^2=0 \tag{8.67}$$

Equation (8.67) can be seen to be dependent only on the deflected form of the plate.

The fundamental equations, which now appear as equations (8.65) to (8.67), can be put into dimensionless form by introducing the following dimensionless variables:

$$\begin{aligned}
X&=r/t=\zeta R/t\\
W&=w/t\\
U&=u/t\\
M'_{\mathrm{r}}&=M_{\mathrm{r}}t/D\\
S_{\mathrm{r}}&=\sigma_{\mathrm{r}}/E\\
S_{\mathrm{t}}&=\sigma_{\mathrm{t}}/E\\
S'_{\mathrm{r}}&=\sigma'_{\mathrm{r}}/E\\
S'_{\mathrm{t}}&=\sigma'_{\mathrm{t}}/E\\
S_{\mathrm{p}}&=p/E
\end{aligned} \tag{8.68}$$

So that,

$$\theta = \frac{dw}{dr} = \frac{dW}{dX} \tag{8.69}$$

or,

$$W = \int \theta \, dX \tag{8.70}$$

Now from standard circular plate theory,

$$\sigma_r' = \frac{6D}{t^2}\left(\frac{d\theta}{dr} + \frac{\nu\theta}{r}\right)$$

and,

$$\sigma_t' = \frac{6D}{t^2}\left(\frac{\theta}{r} + \nu\frac{d\theta}{dr}\right)$$

Hence,

$$S_r' = \frac{1}{2(1-\nu^2)}\left(\frac{d\theta}{dX} + \frac{\nu\theta}{X}\right) \tag{8.71}$$

and,

$$S_t' = \frac{1}{2(1-\nu^2)}\left(\frac{\theta}{X} + \nu\frac{d\theta}{dX}\right) \tag{8.72}$$

Now from elementary two-dimensional stress theory,

$$\frac{uE}{r} = \sigma_t - \nu\sigma_r$$

or,

$$U = X(S_t - \nu S_r) \tag{8.73}$$

where,

u = in-plane radial deflection at r

Substituting (8.68) to (8.73) into (8.65) to (8.67), the fundamental equations take the form of (8.74) to (8.76):

$$\frac{1}{12(1-\nu^2)}\frac{d}{dX}\left\{\frac{1}{X}\frac{d(X\theta)}{dX}\right\} = S_p\frac{X}{2} + S_r\theta \tag{8.74}$$

$$\frac{d(XS_r)}{dX} - S_t = 0 \tag{8.75}$$

$$X\frac{d}{dX}(S_r + S_t) + \frac{\theta^2}{2} = 0 \tag{8.76}$$

8.8.5

Solution of (8.74) to (8.76) can be achieved through a power series solution. Now S_r is a symmetrical function, i.e. $S_r(X) = S_r(-X)$, so that it can be approximated in an even series powers of X.

Furthermore, since θ is anti-symmetrical, i.e. $\theta(X) = -\theta(X)$, it can be expanded in an odd series power of X. Let,

$$S_r = B_1 + B_2X^2 + B_3X^4 + \cdots\cdots$$

and,

$$\theta = C_1X + C_2X^3 + C_3X^5 + \cdots\cdots$$

or,

$$S_r = \sum_{i=1}^{\infty} B_iX^{2i-2} \tag{8.77}$$

and,

$$\theta = \sum_{i=1}^{\infty} C_iX^{2i-1} \tag{8.78}$$

Now from (8.75):

$$S_t = \frac{d(XS_r)}{dX} = \sum_{i=1}^{\infty} (2i-1)B_iX^{2i-2} \tag{8.79}$$

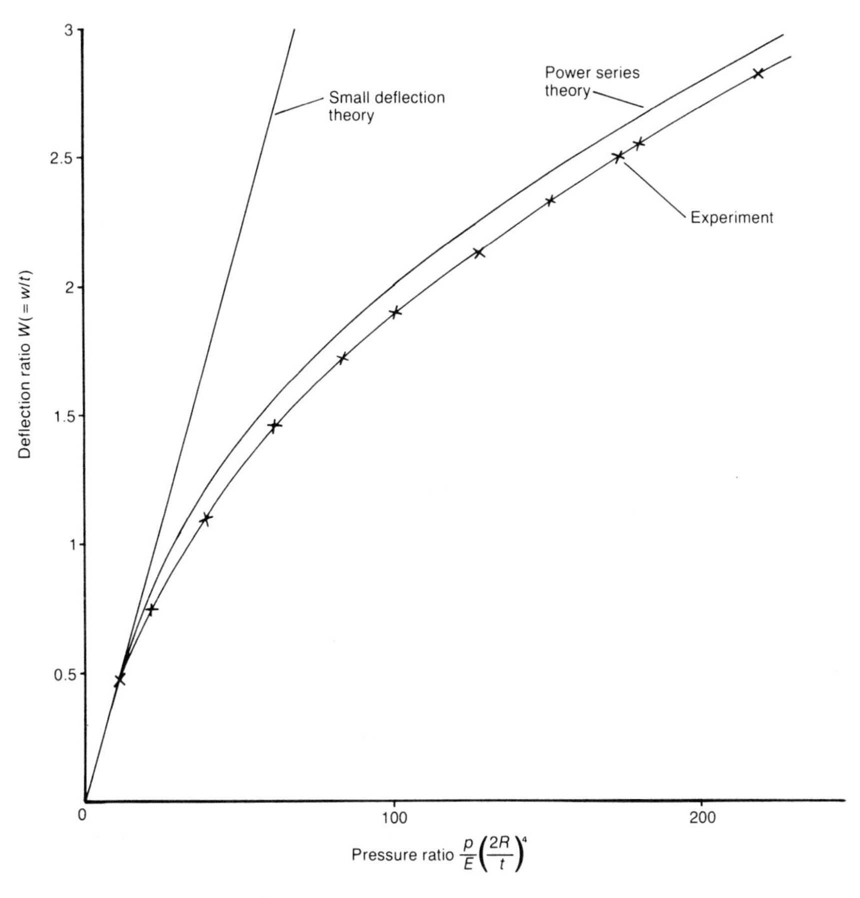

Fig. 8.8. Central deflection versus pressure for a simply-supported plate.

From (8.70):

$$W = \int \theta \, dX = \sum_{i=1}^{\infty} \left(\frac{1}{2i}\right) C_i X^{2i} \tag{8.80}$$

Hence,

$$S_r' = \sum_{i=1}^{\infty} \frac{(2i + \nu - 1)}{2(1 - \nu^2)} C_i X^{2i-2} \tag{8.81}$$

$$S_t' = \sum_{i=1}^{\infty} \frac{\{1 + \nu(2i - 1)\}}{2(1 - \nu^2)} C_i X^{2i-2} \tag{8.82}$$

Now,

$$\begin{aligned} U &= X(S_t - \nu S_r) \\ &= \sum_{i=1}^{\infty} (2i - 1 - \nu) B_i X^{2i-1} \end{aligned} \tag{8.83}$$

for $i = 1, 2, 3, 4 \to \infty$.

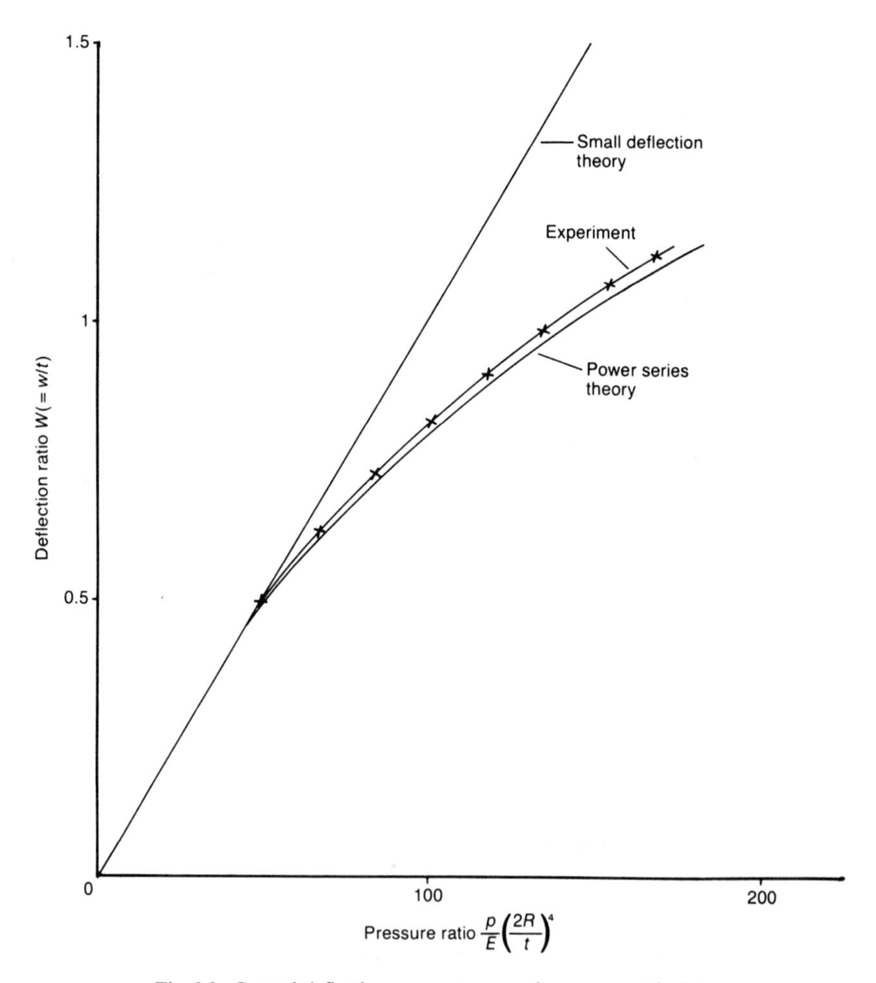

Fig. 8.9. Central deflection versus pressure for an encastré plate.

From (8.77) to (8.83), it can be seen that if B_i and C_i are known, all quantities of interest can readily be determined.

Way [30] has shown that:

$$B_k = \frac{-\sum_{m=1}^{k-1} C_m C_{k-m}}{8k(k-1)}$$

for $k = 2, 3, 4$, etc., and,

$$C_k = \frac{3(1-\nu^2)}{k(k-1)} \sum_{m=1}^{k-1} B_m C_{k-m}$$

for $k = 3, 4, 5$, etc., and,

$$C_2 = \frac{3(1-\nu^2)}{2}\left(\frac{S_p}{2} + B_1 C_1\right)$$

Once B_1 and C_1 are known, the other constants can be found. In fact, using this approach, Hewitt and Tannent [31] have produced a set of curves which

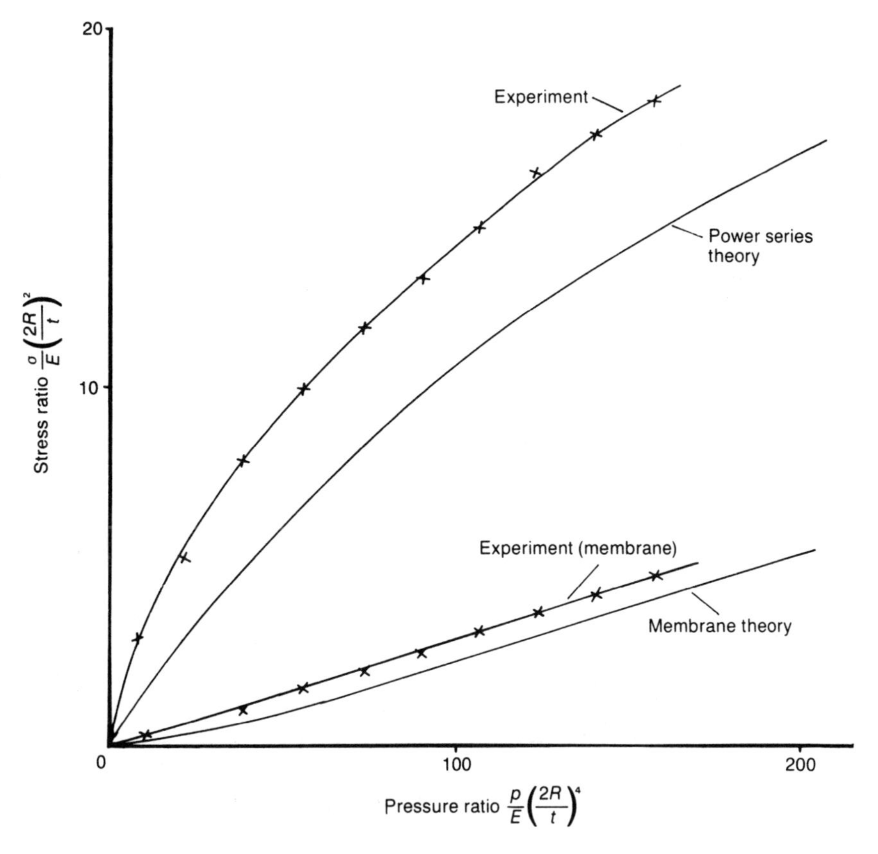

Fig. 8.10. Central stresses versus pressure for an encastré plate.

can be used directly for obtaining deflection and stresses in circular plates under uniform lateral pressure, as shown in Figs. 8.8 to 8.12. Hewitt and Tannent have also compared experiment and small deflection theory with these curves.

8.9.1 SHEAR DEFLECTIONS OF VERY THICK PLATES

If a plate is very thick, so that membrane effects are insignificant, then it is possible that shear deflections can become important.

For such cases, the bending effects and shear effects must be added together, as shown by equation (8.84), which is rather similar to the method

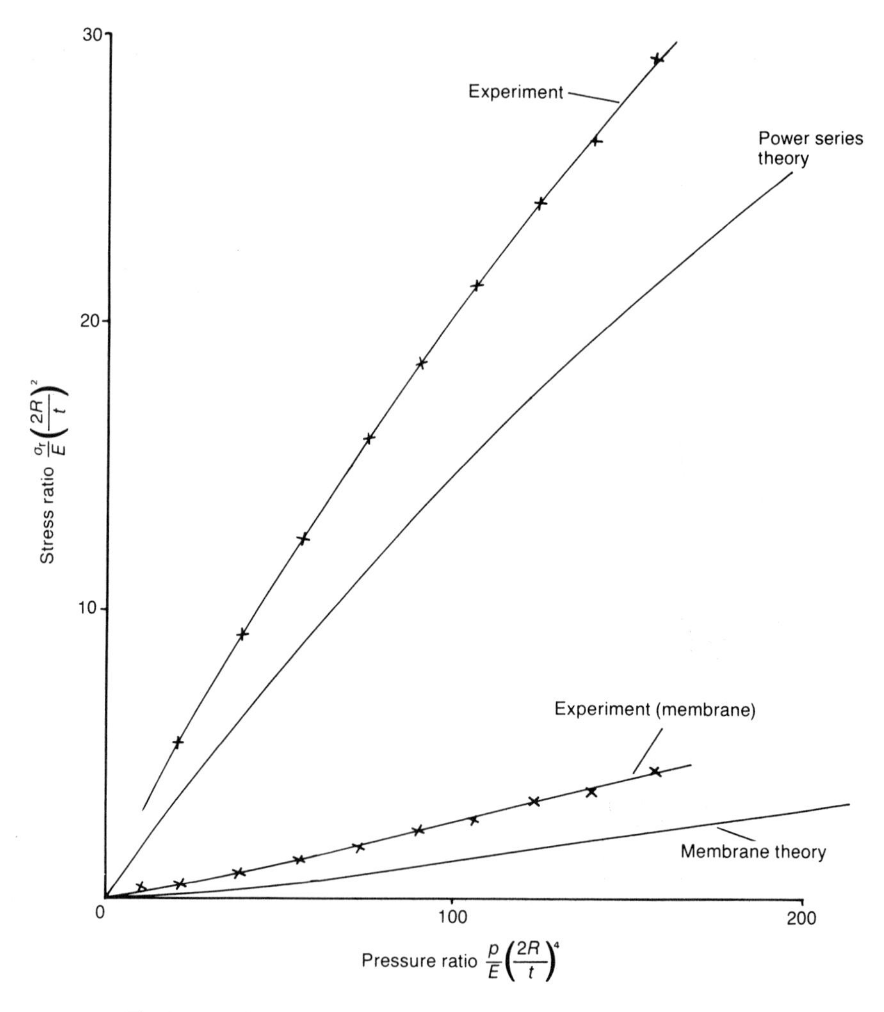

Fig. 8.11. Radial stresses near edge versus pressure for an encastré plate.

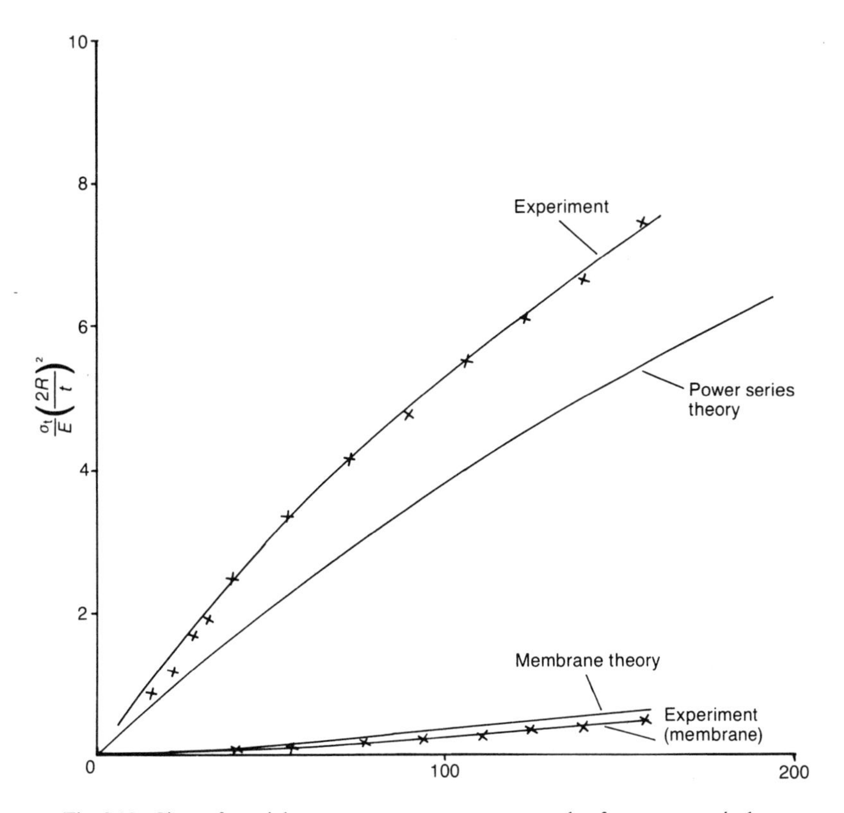

Fig. 8.12. Circumferential stresses versus pressure near edge for an encastré plate.

used for beams in Section 2.14.1.

$$\delta = \delta_{\text{bending}} + \delta_{\text{shear}}$$

which for a plate under a uniform pressure $p =$

$$\delta = pR\left\{k_1\left(\frac{R}{t}\right)^3 + k_2\left(\frac{t}{R}\right)^2\right\} \tag{8.84}$$

where, k_1 and k_2 = constants.

From (8.84), it can be seen that δ_{shear} becomes important for large values of (t/R).

EXAMPLES FOR PRACTICE 8

1. Determine an expression for the deflection of a circular plate of radius R, simply-supported around its edges, and subjected to a centrally placed concentrated load W.

$$\left\{w = \frac{W}{8\pi D}\left[\frac{(3+\nu)}{2(1+\nu)}(R^2 - r^2) + r^2 \ln\left(\frac{r}{R}\right)\right]\right\}$$

2. Determine expressions for the deflection and circumferential bending moments for a circular plate of radius R, simply-supported around its edges, and subjected to a uniform pressure p.

$$\left\{\hat{w} = \frac{pR^4}{64D}\left[\frac{(5+\nu)}{(1+\nu)} - \frac{(6+2\nu)}{(1+\nu)}\left(\frac{r}{R}\right)^2 + \left(\frac{r}{R}\right)^4\right];\right.$$

$$M_r = -\frac{(3+\nu)}{16}pR^2\left[1-\left(\frac{r}{R}\right)^2\right];$$

$$\left. M_t = \frac{pR^2}{16}\left[-(3+\nu)+(1+3\nu)\left(\frac{r}{R}\right)^2\right]\right\}$$

3. Determine an expression for the maximum deflection of a simply-supported circular plate, subjected to the loading shown in Fig. Q.8.3.

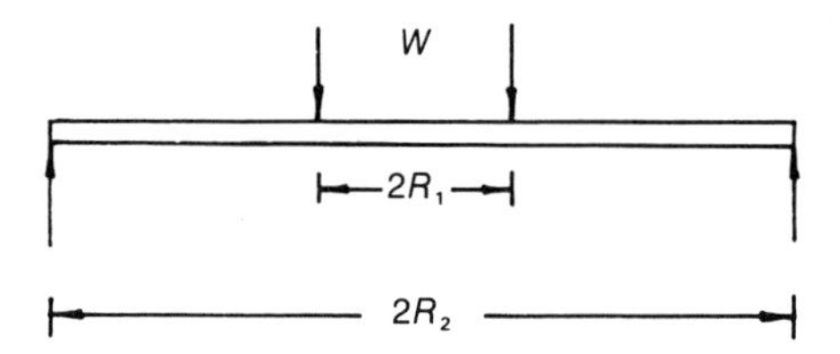

Fig. Q.8.3. Simply-supported plate.

$$\left\{\hat{w} = \frac{W}{16\pi D}\left[\frac{(3+\nu)}{(1+\nu)}(R_2^2 - R_1^2) + 2R_1^2 \ln\left(\frac{R_1}{R_2}\right)\right]\right\}$$

4. Determine expressions for the maximum deflection and bending moments for the concentrically loaded circular plates of Fig. Q.8.4(a) and (b).

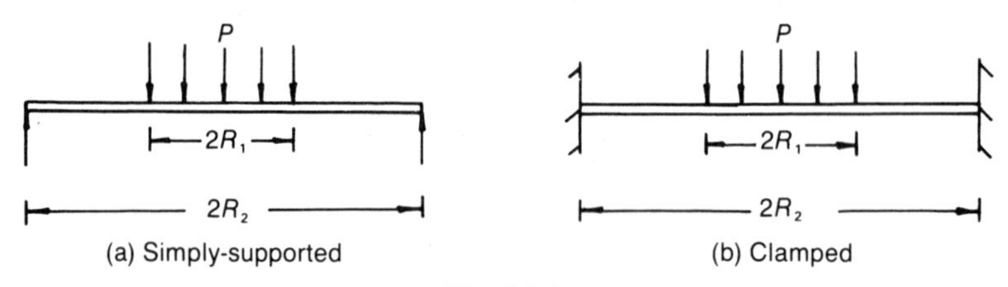

Fig. Q.8.4

$$\left\{(a)\ \hat{w} = \frac{P}{16\pi D}\left[\frac{(3+\nu)}{(1+\nu)}R_2^2 - \frac{(7+3\nu)}{4(1+\nu)}R_1^2 + R_1^2\ln\left(\frac{R_1}{R_2}\right)\right];\right.$$

$$\hat{M} = \frac{P}{4\pi}\left[1 - \frac{(1-\nu)}{4}\left(\frac{R_1}{R_2}\right)^2 - (1+\nu)\ln\left(\frac{R_1}{R_2}\right)\right];$$

$$(b)\ \hat{w} = \frac{P}{16\pi D}\left[R_2^2 - 0.75R_1^2 + R_1^2\ln\left(\frac{R_1}{R_2}\right)\right];$$

$$\hat{M} = \frac{P}{4\pi}\left[1 - 0.5\left(\frac{R_1}{R_2}\right)^2\right] \quad \text{for } R_1/R_2 > 0.57$$

and,

$$\hat{M} = \frac{P}{4\pi}(1 + \nu)\left[0.25\left(\frac{R_1}{R_2}\right)^2 - \ln\left(\frac{R_1}{R_2}\right)\right] \quad \text{for } R_1/R_2 < 0.57\text{; where}$$

$$P = p * \pi R_1^2 \}$$

5. A flat circular plate of radius R is firmly clamped around its boundary. The plate has stepped variation in its thickness, where the thickness inside a radius of $(R/5)$ is so large that its flexural stiffness may be considered to approach infinity. When the plate is subjected to a pressure p over its entire surface, determine the maximum central deflection and the maximum surface stress at any radius r. $\nu = 0.3$.

$$\left\{0.126pR^4/(Et^3); p\left(\frac{R}{t}\right)^2\left[-1.238\left(\frac{r}{R}\right)^2 + 0.507 + 0.0105\left(\frac{R}{r}\right)^2\right]\right\}$$

6. If the loading of Example 4 were replaced by a centrally applied concentrated load W, determine expressions for the central deflection and the maximum surface stress at any radius r.

$$\left\{0.115WR^2/(Et^3); \frac{W}{t^2}\left[0.621\ln\left(\frac{R}{r}\right) - 0.436 + 0.0224\left(\frac{R}{r}\right)^2\right]\right\}$$

9

Thick Cylinders and Spheres

9.1.1 THICK AND THIN SHELLS

If the thickness:radius ratio of a shell exceeds 1:30, the theory for thin shell starts to break down. The reason for this is that for thicker shells, the radius of the shell changes appreciably over its thickness, so that the strain can no longer be assumed to be constant over the thickness. Thick shells are of much importance in ocean egineering, civil engineering, nuclear engineering, etc.

In this chapter, in addition to considering thick shells under pressure, considerations will also be made of the plastic collapse of thick cylinders and discs, and thermal stresses in the structures. The collapse of discs and rings due to high speed rotation will also be discussed.

9.2.1 DERIVATION OF THE HOOP AND RADIAL STRESS EQUATIONS FOR A THICK-WALLED CYLINDER

The following convention will be used, where all the stresses and strains are assumed to be positive if they are tensile:

At any radius r,

σ_θ = hoop stress

σ_r = radial stress

σ_x = longitudinal stress

ε_θ = hoop strain

ε_r = radial strain

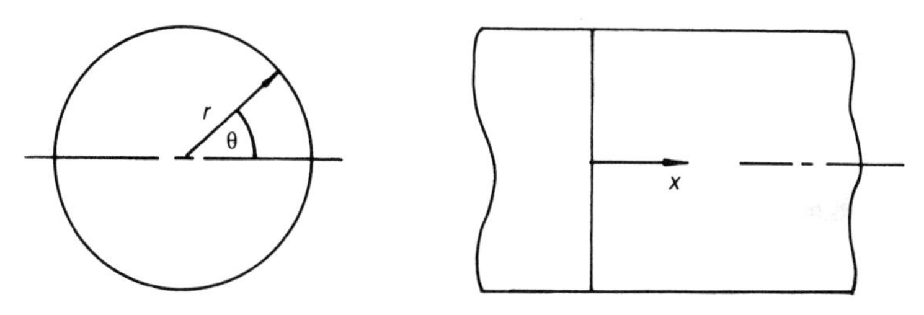

Fig. 9.1. Thick cylinder.

ε_x = longitudinal strain (assumed to be constant)

w = radial deflection

From Fig. 9.2, it can be seen that at any radius r,

$$\varepsilon_\theta = \frac{2\pi(r+w) - 2\pi r}{2\pi r}$$

or

$$\varepsilon_\theta = w/r \tag{9.1}$$

Similarly,

$$\varepsilon_\mathrm{r} = \frac{\delta w}{\delta r} = \frac{\mathrm{d}w}{\mathrm{d}r} \tag{9.2}$$

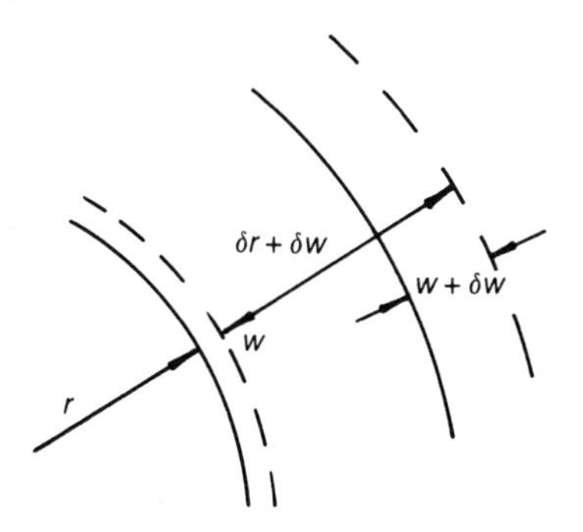

Fig. 9.2. Deformation at any radius r.

From the standard stress–strain relationships [3–7],

$$E\varepsilon_x = \sigma_x - \nu\sigma_\theta - \nu\sigma_\mathrm{r} = \text{a constant}$$

$$E\varepsilon_\theta = \frac{Ew}{r} = \sigma_\theta - \nu\sigma_x - \nu\sigma_\mathrm{r} \tag{9.3}$$

$$E\varepsilon_\mathrm{r} = E\frac{\mathrm{d}w}{\mathrm{d}r} = \sigma_\mathrm{r} - \nu\sigma_\theta - \nu\sigma_x \tag{9.4}$$

Multiplying (9.3) by r,

$$Ew = \sigma_\theta * r - \nu\sigma_x * r - \nu\sigma_\mathrm{r} * r \tag{9.5}$$

and differentiating (9.5) w.r.t r,

$$E\frac{dw}{dr} = \sigma_\theta - \nu\sigma_x - \nu\sigma_r + r\left(\frac{d\sigma_\theta}{dr} - \nu\frac{d\sigma_x}{dr} - \nu\frac{d\sigma_r}{dr}\right) \tag{9.6}$$

Subtracting (9.4) from (9.6),

$$(\sigma_\theta - \sigma_r)(1+\nu) + r\frac{d\sigma_\theta}{dr} - \nu r\frac{d\sigma_x}{dr} - \nu r\frac{d\sigma_r}{dr} = 0 \tag{9.7}$$

Since ε_x is constant

$$\sigma_x - \nu\sigma_\theta - \nu\sigma_r = \text{constant} \tag{9.8}$$

Differentiating (9.8) w.r.t. r,

$$\frac{d\sigma_x}{dr} - \nu\frac{d\sigma_\theta}{dr} - \nu\frac{d\sigma_r}{dr} = 0$$

or

$$\frac{d\sigma_x}{dr} = \nu\left(\frac{d\sigma_\theta}{dr} + \frac{d\sigma_r}{dr}\right) \tag{9.9}$$

Substituting (9.9) into (9.7),

$$(\sigma_\theta - \sigma_r)(1+\nu) + r(1-\nu^2)\frac{d\sigma_\theta}{dr} - \nu r(1+\nu)\frac{d\sigma_r}{dr} = 0 \tag{9.10}$$

and dividing through by $(1+\nu)$

$$\sigma_\theta - \sigma_r + r(1-\nu)\frac{d\sigma_\theta}{dr} - \nu r\frac{d\sigma_r}{dr} = 0 \tag{9.11}$$

Considering now the equilibrium of an element of the shell, as shown in Fig. 9.3,

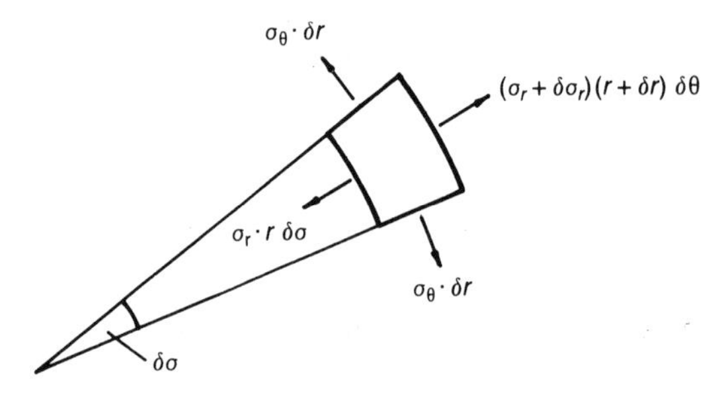

Fig. 9.3. Shell element.

$$2\sigma_\theta\cdot\delta r \sin\left(\frac{\delta\theta}{2}\right) + \sigma_r\cdot r\cdot\delta\theta - (\sigma_r + \delta\sigma_r)(r+\delta r)\delta\theta = 0$$

In the limit,

$$\sigma_\theta - \sigma_r - r\frac{d\sigma_r}{dr} = 0 \tag{9.12}$$

Subtracting (9.11) from (9.12)

$$\frac{d\sigma_\theta}{dr} + \frac{d\sigma_r}{dr} = 0 \qquad (9.13)$$

therefore

$$\sigma_\theta + \sigma_r = \text{constant} = 2A \qquad (9.14)$$

Subtracting (9.12) from (9.14),

$$2\sigma_r + r\frac{d\sigma_r}{dr} = 2A$$

or

$$\frac{1}{r}\frac{d(\sigma_r r^2)}{dr} = 2A$$

$$\frac{d(\sigma_r \cdot r^2)}{dr} = 2Ar$$

Integrating,

$$\sigma_r \cdot r^2 = Ar^2 - B$$

$$\sigma_r = A - \frac{B}{r^2} \qquad (9.15)$$

From (9.14),

$$\sigma_\theta = A + \frac{B}{r^2} \qquad (9.16)$$

9.3.1 LAMÉ LINE

Equations (9.15) and (9.16) can be represented by a single straight line, if they are plotted against $1/r^2$, as shown in Fig. 9.4, where σ_r is on the left of the diagram and σ_θ is on the right.

In Fig. 9.4, we see the case of a thick-walled cylinder of internal radius R_1 and external radius R_2, subjected to an internal pressure P. Now two points are known on this straight line: the radial stresses on the internal and external surfaces, which are $-P$ and zero, respectively (shown by a "X"). Hence, the straight line can be drawn, and any stress calculated throughout the thickness of the wall by equating similar triangles.

In Fig. 9.4,

$\sigma_{\theta 1}$ = internal hoop stress, which can be seen to be a maximum stress

$\sigma_{\theta 2}$ = external hoop stress

Furthermore, from Fig. 9.4, it can be seen that both $\sigma_{\theta 1}$ and $\sigma_{\theta 2}$ are tensile, and that σ_r is compressive.

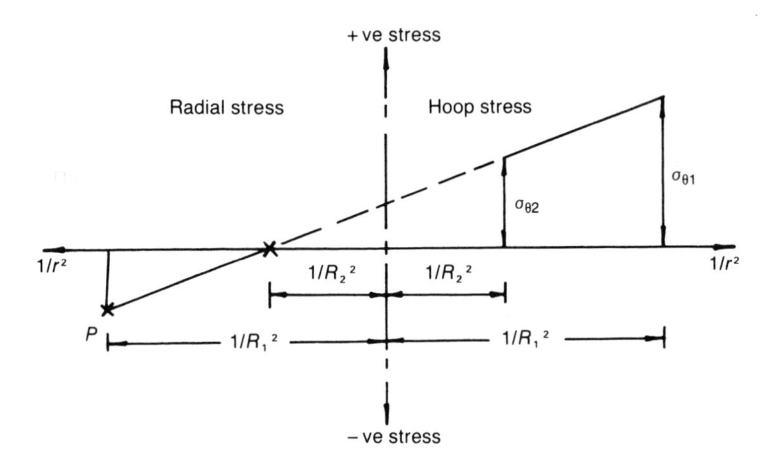

Fig. 9.4. Lamé line for the case of internal pressure.

9.3.2 To Calculate $\sigma_{\theta 1}$ and $\sigma_{\theta 2}$

Equating similar triangles in Fig. 9.4,

$$\frac{\sigma_{\theta 1}}{\left(\dfrac{1}{R_1^2}+\dfrac{1}{R_2^2}\right)}=\frac{P}{\left(\dfrac{1}{R_1^2}-\dfrac{1}{R_2^2}\right)}$$

or,

$$\sigma_{\theta 1}=\frac{P\left(\dfrac{1}{R_1^2}+\dfrac{1}{R_2^2}\right)}{\left(\dfrac{1}{R_1^2}-\dfrac{1}{R_2^2}\right)}*\frac{R_1^2R_2^2}{R_1^2R_2^2}$$

$$\underline{\sigma_{\theta 1}=\frac{P(R_1^2+R_2^2)}{(R_2^2-R_1^2)}}$$

Similarly,

$$\frac{\sigma_{\theta 2}}{\left(\dfrac{1}{R_2^2}+\dfrac{1}{R_2^2}\right)}=\frac{P}{\left(\dfrac{1}{R_1^2}-\dfrac{1}{R_2^2}\right)}$$

or,

$$\sigma_{\theta 2}=\frac{P\left(\dfrac{1}{R_2^2}\right)*2}{\left(\dfrac{1}{R_1^2}-\dfrac{1}{R_2^2}\right)}*\frac{R_1^2R_2^2}{R_1^2R_2^2}$$

$$\underline{\sigma_{\theta 2}=\frac{2PR_1^2}{(R_2^2-R_1^2)}}$$

9.4.1 EXAMPLE 9.1 THICK CYLINDER UNDER INTERNAL PRESSURE

A thick-walled cylinder of internal diameter 0.1 m is subjected to an internal pressure of 50 MPa. If the maximum permissible stress in the cylinder is limited to 200 MPa, determine the minimum possible external diameter.

9.4.2 Let d_2 = external diameter—to be determined

The Lamé line for this case would appear as shown in Fig. 9.5, where it can be seen that the Lamé line is obtained by knowing two values of radial stress. These are the radial stresses on the internal and external surfaces of the cylinder, which are -50 MPa and zero, respectively. However, as d_2 is unknown, a third point is required, which in this case is the maximum stress, (i.e. $\sigma_{\theta 1}$); this is 200 MPa.

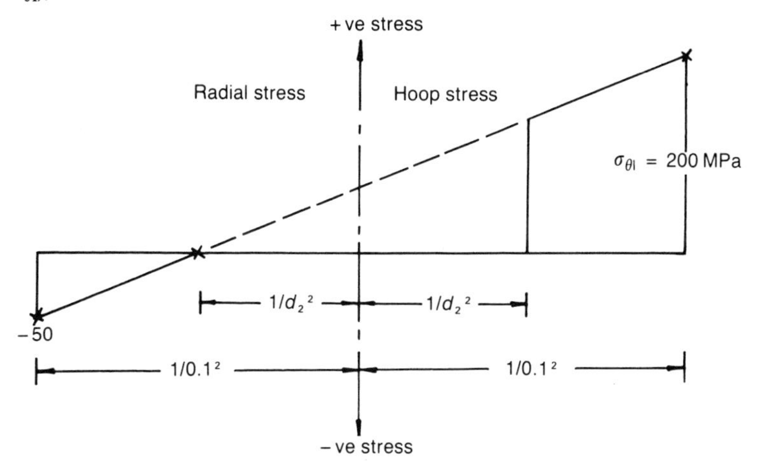

Fig. 9.5. Lamé line for thick cylinder.

Equating similar triangles in Fig. 9.5,

$$\frac{50}{\left(\frac{1}{0.1^2}-\frac{1}{d_2^2}\right)}=\frac{200}{\left(\frac{1}{0.1^2}+\frac{1}{d_2^2}\right)}$$

or,

$$\frac{\left(\frac{1}{0.1^2}+\frac{1}{d_2^2}\right)}{\left(\frac{1}{0.1^2}-\frac{1}{d_2^2}\right)}*\left(\frac{0.1^2*d_2^2}{0.1^2*d_2^2}\right)=4$$

$$\left(\frac{d_2^2+0.1^2}{d_2^2-0.1^2}\right)=4$$

$$d_2^2+0.1^2=4d_2^2-4*0.1^2$$

$$3d_2^2=0.1^2*5$$

$$\underline{d_2=0.129\,\text{m}}$$

9.5.1 EXAMPLE 9.2 THICK CYLINDER UNDER EXTERNAL PRESSURE

If the vessel of Example 9.1 is subjected to an external pressure of 50 MPa, determine the maximum value of stress that would occur in this vessel.

9.5.2

The Lamé line for this case is shown in Fig. 9.6, where it can be seen that two values of radial stress are known. These are the radial stresses on the internal and external surfaces, which are zero and -50 MPa, respectively. Let $\sigma_{\theta I}$ = hoop stress on the internal surface, which from Fig. 9.6 can be seen to have the largest magnitude.

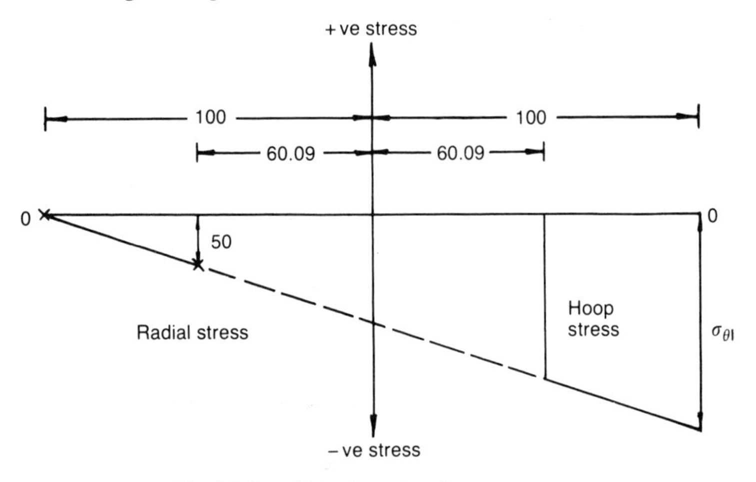

Fig. 9.6. Lamé line for external pressure case.

By equating similar triangles,

$$\frac{\sigma_{\theta I}}{100+100} = \frac{-50}{100-60.09}$$

$$\sigma_{\theta I} = \frac{-50 \times 200}{39.91}$$

$$\underline{\sigma_{\theta I} = -250.6\,\text{MPa}}$$

N.B. It should be noted in Examples 9.1 and 9.2 that the maximum stress for both cases was the internal hoop stress.

9.6.1 EXAMPLE 9.3 A STEEL RING SHRUNK ONTO A SOLID SHAFT

A steel ring of external diameter 10 cm and internal diameter 5 cm, is to be shrunk into a solid steel shaft of diameter 5 cm, where all the dimensions are nominal. If the interference fit at the common surface between the ring and the shaft is 0.01 cm, based on a diameter, determine the maximum stress in the material.

$$E = 2\text{E}11\ \text{N/m}^2 \quad \nu = 0.3$$

9.6.2 Consider First the Steel Ring

The Lamé line for the steel ring will be as shown in Fig. 9.7, where the radial stress on its outer surface will be zero, and that on its internal surface will be P_c (the radial pressure at the common surface between the steel ring and the shaft).

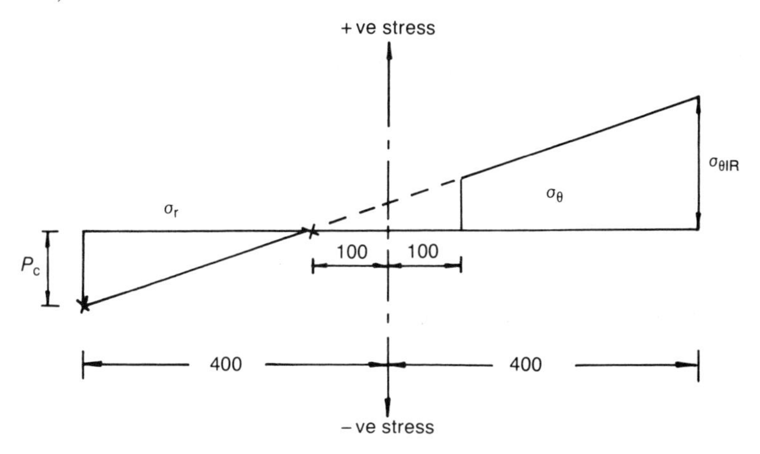

Fig. 9.7. Lamé line for steel ring.

Let,

$\sigma_{\theta IR}$ = hoop stress (maximum stress) on the internal surface of the ring.

σ_{rIR} = radial stress on the internal surface of the ring.

Equating similar triangles,

$$\frac{\sigma_{\theta IR}}{400 + 100} = \frac{P_c}{400 - 100}$$

therefore

$$\underline{\sigma_{\theta IR} = 1.667 P_c} \tag{9.17}$$

9.6.3 Consider Now the Shaft

For this case, the Lamé line will be horizontal, because if it had any slope at all, the stresses at the centre will be infinity, for a finite value of P_c, which is impossible.

From Fig. 9.8, it can be seen that for a solid shaft, under an axisymmetric

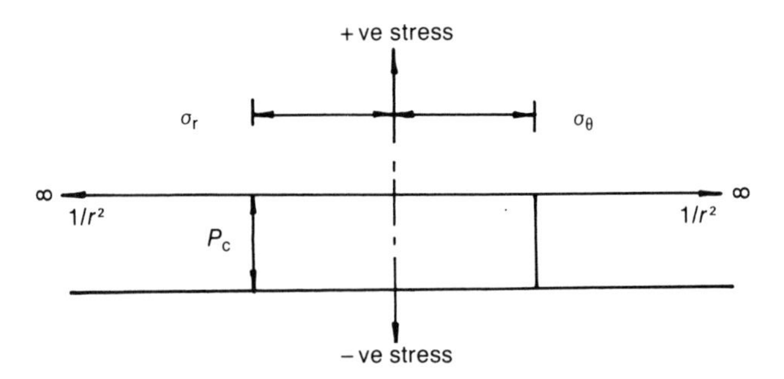

Fig. 9.8. Lamé line for a solid shaft.

exteral pressure of P_c,

$$\sigma_r = \sigma_\theta = -P_c \text{ (everywhere)} \tag{9.18}$$

9.6.4

Let,

w_R = increase in the radius of the ring at its inner surface

w_s = increase in the radius of the shaft at its outer surface

Now, applying the expression

$$E\varepsilon_\theta = \frac{w}{r} = \sigma_\theta - \nu\sigma_r - \nu\sigma_x$$

to the inner surface of the ring

$$\frac{Ew_R}{2.5\text{E-}2} = \sigma_{\theta IR} - \sigma_{rIR}$$

but,

$$\sigma_{rIR} = -P_c$$

therefore

$$\frac{2\text{E}11 * w_R}{2.5\text{E-}2} = 1.667\,P_c + 0.3P_c$$

$$\underline{w_R = 2.459\text{E-}13P_c} \tag{9.19}$$

Similarly for the shaft,

$$\frac{Ew_s}{2.5\text{E-}2} = \sigma_{\theta s} - \nu\sigma_{rs}$$

but,

$$\sigma_{\theta s} = \sigma_{rs} = -P_c$$

therefore

$$\frac{2\text{E}11 * w_s}{2.5\text{E-}2} = -P_c(1-\nu)$$

or,

$$\underline{w_s = -8.75\text{E-}14 P_c} \tag{9.20}$$

9.6.5

Now the interference fit on the diameters = 0.01E-2 m.

Therefore the interference fit on the radii = 5E-5 m = $w_R - w_s$
or,

$$(2.459\text{E-}13 + 8.75\text{E-}14) P_c = 5\text{E-}5$$

$$\underline{P_c = 149.97\ \text{MPa}}$$

From (9.17), the maximum stress =

$$\sigma_{\theta IR} = 1.667 * P_c$$

$$\underline{\sigma_{\theta IR} = 250\ \text{MPa}}$$

9.7.1 COMPOUND CYLINDERS

A cylinder made from two different materials is sometimes found useful in engineering, when one material is suitable for resisting corrosion in a certain environment, but because this material is expensive or weak, another material is used to strengthen it.

9.8.1 EXAMPLE 9.4 AN ALUMINIUM DISC, SHRUNK ONTO A STEEL SHAFT

An aluminium alloy disc of constant thickness, and of internal and external radii R_1 and R_2, respectively, is shrunk onto a solid steel shaft of external radius $R_1 + \delta$. Show that the maximum stress ($\hat{\sigma}$) in the disc is given by:

$$\hat{\sigma} = \delta \Bigg/ \left\{ R_1 \left[\left(\frac{1 - \nu_s}{E_s} + \frac{\nu_a}{E_a} \right) \left(\frac{R_2^2 - R_1^2}{R_1^2 + R_2^2} \right) + \frac{1}{E_a} \right] \right\}$$

where,

E_s = elastic modulus of steel

E_a = elastic modulus of aluminium alloy

ν_s = Poisson's ratio for steel

ν_a = Poisson's ratio for aluminium alloy

9.8.2

Let,

P_c = radial pressure at the common surface—to be determined

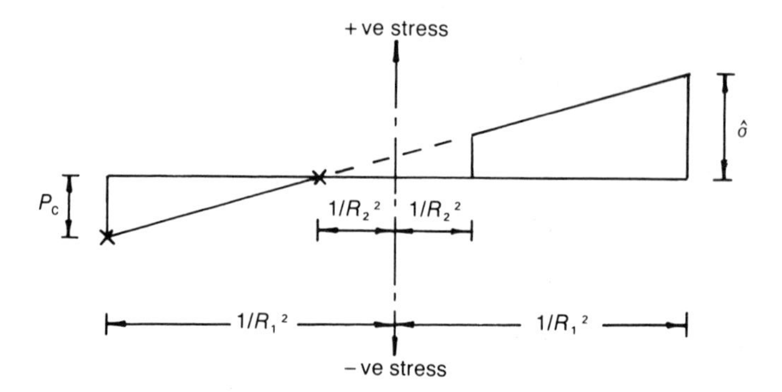

Fig. 9.9. Lamé line for the disc.

Now the radial stress for the disc, on the external surface, is zero, and the radial stress on the internal surface of the disc is $-P_c$; hence, the Lamé line will take the form shown in Fig. 9.9.

Equating similar triangles in Fig. 9.99,

$$\frac{P_c}{\left(\frac{1}{R_1^2}-\frac{1}{R_2^2}\right)}=\frac{\hat{\sigma}}{\left(\frac{1}{R_1^2}+\frac{1}{R_2^2}\right)}$$

therefore

$$P_c=\hat{\sigma}*\frac{(R_2^2-R_1^2)}{(R_2^2+R_1^2)} \tag{9.21}$$

9.8.3

Now for the steel shaft, the Lamé line must be horizontal, otherwise the stresses at the centre of the shaft will be infinite for a finite value of P_c, i.e.

$$\begin{aligned}\sigma_{\theta s} &= \text{hoop stress in the shaft}\\ &= \text{radial stress in the shaft } (\sigma_{rs})\\ &= -P_c \text{ (everywhere in the shaft)}\end{aligned}$$

9.8.4

Let,

w_a = increase in radius of the aluminium disc at its internal surface

w_s = increase in radius of the steel shaft on its external surface

Applying the expression

$$E\varepsilon_\theta=\frac{Ew}{r}=\sigma_\theta-\nu\sigma_r-\nu\sigma_x$$

to the internal surface of the aluminium disc,

$$\frac{E_a w_a}{R_1} = \hat{\sigma} + \nu_a P_c$$

or,

$$w_a = \frac{R_1}{E_a}(\hat{\sigma} + \nu_a P_c) \tag{9.22}$$

Similarly, for the steel shaft,

$$\frac{E_s w_s}{R_1} = -P_c + \nu_s P_c$$

or,

$$w_s = \frac{-R_1}{E_s} * P_c(1 - \nu_s) \tag{9.23}$$

but,

$$\delta = w_a - w_s$$

$$= \frac{R_1}{E_a}(\hat{\sigma} + \nu_a P_c) + \frac{R_1}{E_s} P_c(1 - \nu_s) \tag{9.24}$$

Substituting (9.21) into (9.24),

$$\underline{\hat{\sigma} = \delta \Bigg/ \left\{ R_1 \left[\left(\frac{1 - \nu_s}{E_s} + \frac{\nu_a}{E_a} \right) \left(\frac{R_2^2 - R_1^2}{R_1^2 + R_2^2} \right) + \frac{1}{E_a} \right] \right\}}$$

9.9.1 EXAMPLE 9.5 COMPOUND TUBE

A steel cylinder with external and internal diameters of 10 cm and 8 cm, respectively, is shrunk onto an aluminium alloy cylinder with internal and external diameters of 5 cm and 8 cm, respectively, where all the dimensions are nominal.

Find the radial pressure at the common surface, due to shrinkage alone, so that when there is an internal pressure of 150 MPa, the maximum hoop in the inner cylinder is 110 MPa. Determine, also, the maximum hoop stress in the outer cylinder, and plot the distributions across the sections.

For steel,

$$E_s = 2\text{E}11\ \text{N/m}^2; \quad \nu_s = 0.3$$

For aluminium alloy,

$$E_a = 6.7\text{E}10\ \text{N/m}^2; \quad \nu_a = 0.32$$

Let,

P_c^s = the radial pressure at the common surface due to shrinkage alone

σ_θ^s = the hoop stress due to shrinkage alone

σ_{θ}^{p} = the hoop stress due to pressure alone
$\sigma_{\theta,10s}$ = hoop stress in the steel at its 10 cm diameter
$\sigma_{\theta,8s}$ = hoop stress in the steel at its 8 cm diameter
$\sigma_{r,10s}$ = radial stress in the steel at its 10 cm diameter
$\sigma_{r,8s}$ = radial stress in the steel at its 8 cm diameter
$\sigma_{\theta,8a}$ = hoop stress in the aluminium alloy at its 8 cm diameter
$\sigma_{r,8a}$ = radial stress in the aluminium alloy at its 8 cm diameter
$\sigma_{\theta,5a}$ = hoop stress in the aluminium alloy at its 5 cm diameter
$\sigma_{r,5a}$ = radial stress in the aluminium alloy at its 5 cm diameter

9.9.2 Consider first the stress due to *shrinkage alone*

For the aluminium alloy tube, the Lamé line due to shrinkage alone is shown in Fig. 9.10.

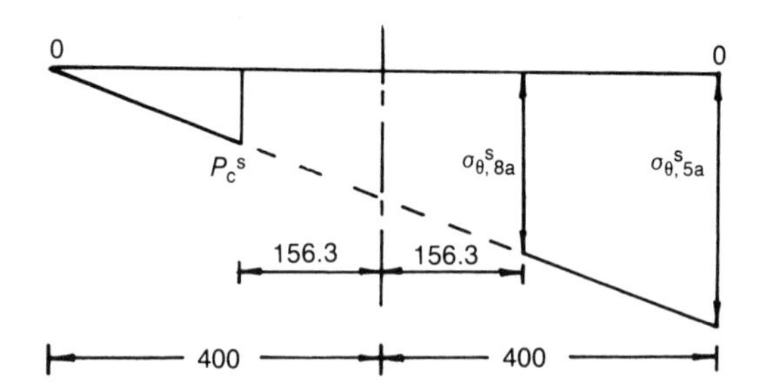

Fig. 9.10. Lamé line for aluminium alloy tube.

Equating similar triangles in Fig. 9.10,

$$\frac{\sigma_{\theta,5a}^{s}}{400+400}=\frac{-P_c^s}{400-156.3}$$

therefore

$$\underline{\sigma_{\theta,5a}^{s}=-3.282P_c^s} \qquad (9.25$$

Similarly,

$$\frac{\sigma_{\theta,8a}^{s}}{400+156.3}=\frac{-P_c^s}{400-156.3}$$

therefore

$$\underline{\sigma_{\theta,8a}^{s}=-2.282P_c^s} \qquad (9.26)$$

For the steel tube, the Lamé line due to shrinkage alone will be as shown in Fig. 9.11.

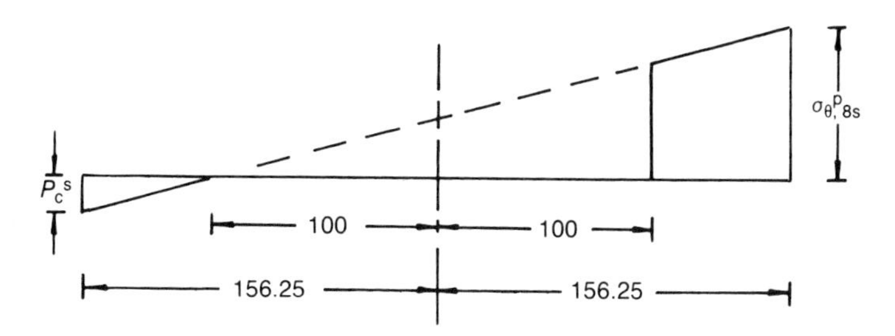

Fig. 9.11. Lamé line for steel tube, due to shrinkage.

Equating similar triangles in Fig. 9.11,

$$\frac{\sigma_{\theta,8s}^{s}}{156.25+100}=\frac{P_c^s}{156.25-100}$$

$$\underline{\sigma_{\theta,8s}^{s}=4.556P_c^s} \tag{9.27}$$

9.9.3 Consider now the stresses due to *pressure* alone.

Let,

P = internal pressure

P_c^p = pressure at the common surface, due to pressure alone

For the aluminium alloy tube, the Lamé line due to pressure alone will be as shown in Fig. 9.12.

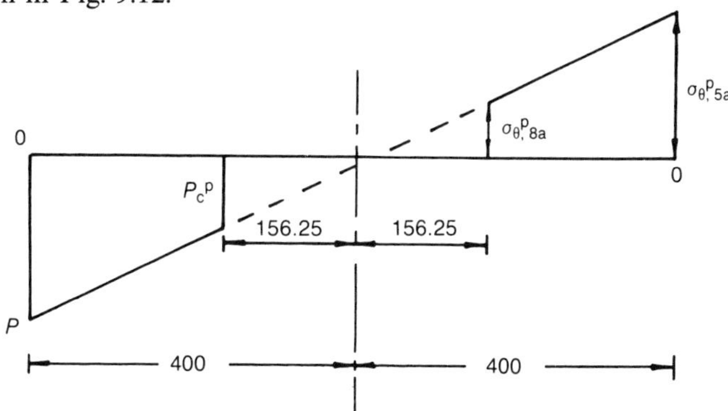

Fig. 9.12. Lamé line in aluminium alloy, due to pressure alone.

Equating similar triangles in Fig. 9.12,

$$\frac{P-P_c^p}{400-156.25}=\frac{\sigma_{\theta,8a}^{p}+P}{400+156.25}$$

or,

$$\frac{150-P_c^p}{243.75}=\frac{\sigma_{\theta,8a}^{p}+150}{556.25}$$

$$\underline{\sigma_{\theta,8a}^{p}=192.3-2.282P_c^p} \tag{9.28}$$

Similarly,

$$\frac{P - P_c^p}{243.75} = \frac{\sigma_{\theta,5a}^p + P}{800}$$

$$\frac{150 - P_c^p}{243.75} = \frac{\sigma_{\theta,5a}^p + 150}{800}$$

therefore

$$\underline{\sigma_{\theta,5a}^p = 342.5 - 3.282 P_c^p} \tag{9.29}$$

For the steel tube, due to pressure alone, the Lamé line is as shown in Fig. 9.13.

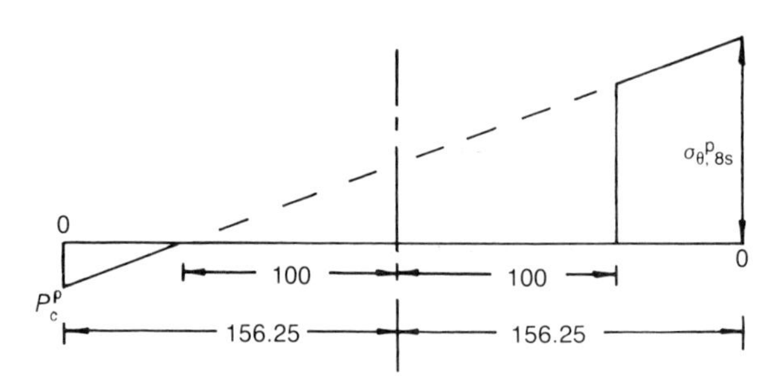

Fig. 9.13. Lamé line for steel, due to pressure alone.

Equating similar triangles in Fig. 9.13,

$$\frac{\sigma_{\theta,8s}^p}{156.25 + 100} = \frac{P_c^p}{156.25 - 100}$$

$$\underline{\sigma_{\theta,8s}^p = 4.556 P_c^p} \tag{9.30}$$

Owing to pressure alone, there is no interference fit, so that,

$$w_s^p = w_a^p$$

Now,

$$\frac{E_s * w_s^p}{4\text{E-2}} = \sigma_{\theta,8s}^p + \nu_s P_c^p$$

therefore

$$w_s^p = \frac{4\text{E-2}}{2\text{E11}}(4.556 + 0.3) P_c^p$$

$$\underline{w_s^p = 9.712\text{E-13} P_c^p} \tag{9.31}$$

Similarly,

$$\frac{E_a * w_a^p}{4\text{E-2}} = \sigma_{\theta,8a}^p + \nu_a P_c^p$$

therefore

$$w_a^p = \frac{4E\text{-}2}{6.7E10}(\sigma_{\theta,8a}^p + \nu_a P_c^p)$$

$$= \frac{4E\text{-}2}{6.7E10}\{192.3 + (-2.282 + 0.32)P_c^p\}$$

$$\underline{w_a^p = 1.148E\text{-}10 - 1.171E\text{-}12P_c^p} \qquad (9.32)$$

Equating (9.31) and (9.32),

$$9.712E\text{-}13P_c^p = 1.148E\text{-}10 - 1.171E\text{-}12P_c^p$$

therefore

$$\underline{P_c^p = 53.59\ \text{MPa}} \qquad (9.33)$$

Substituting (9.33) into (9.28) and (9.29),

$$\sigma_{\theta,5a}^p = 342.5 - 3.282 \times 53.59 = \underline{166.6\ \text{MPa}}$$

$$\sigma_{\theta,8a}^p = 192.3 - 2.282 \times 53.59 = \underline{70\ \text{MPa}}$$

Now the maximum hoop stress in the inner tube lies either on its outer surface or on its inner surface, so that,

$$\text{either} \quad \sigma_{\theta,8a}^p + \sigma_{\theta,8a}^s = 110 \qquad (9.34)$$

$$\text{or} \quad \sigma_{\theta,5a}^p + \sigma_{\theta,5a}^s = 110 \qquad (9.35)$$

From (9.34),

$$P_c^s = -17.5\ \text{MPa}$$

and from (9.35),

$$P_c^s = 17.27\ \text{MPa}$$

i.e. the required $\underline{P_c^s = 17.27\ \text{MPa}}$

Hence, owing to the *combined effects of pressure + shrinkage*,

$$\underline{P_c = P_c^p + P_c^s = 17.27 + 53.59 = 70.86\ \text{MPa}}$$

The resultant hoop stress in the steel tube on its 8 cm diameter =

$$\underline{\sigma_{\theta,8s} = 4.556 * (P_c^s + P_c^p) = 322.8\ \text{MPa}}$$

Similarly, the resultant hoop stress in the aluminium alloy tube on its 8 cm diameter =

$$\sigma_{\theta,8a} = 192.3 - 2.282(P_c^p + p_c^s) = \underline{30.6\ \text{MPa}}$$

and,

$$\sigma_{\theta,5a} = 342.5 - 3.282(P_c^p + p_c^s) = \underline{110\ \text{MPa}}$$

The hoop stress distribution through the two walls is shown in Fig. 9.14.

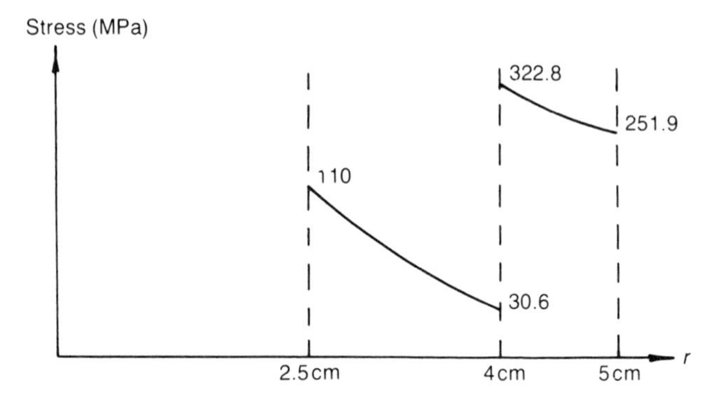

Fig. 9.14. Stress distribution across the compound cylinder.

9.10.1 THICK CYLINDER WITH TEMPERATURE VARIATION

Consider a thick cylinder to be subjected to a temperature rise T, at any radius r, so that the three strains ε_x, ε_θ and ε_r will be increased by an amount αT, as shown by equations (9.36) to (9.38)

$$E\varepsilon_x = \sigma_x - \nu\sigma_\theta - \nu\sigma_r + E\alpha T \tag{9.36}$$

$$E\varepsilon_\theta = \frac{Ew}{r} = \sigma_\theta - \nu\sigma_x - \nu\sigma_r + E\alpha T \tag{9.37}$$

$$E\varepsilon_r = E\frac{dw}{dr} = \sigma_r - \nu\sigma_x - \nu\sigma_\theta + E\alpha T \tag{9.38}$$

Equation (9.37) can be rewritten as follows:

$$Ew = \sigma_\theta r - \nu\sigma_x r - \nu\sigma_r r + Er\alpha T \tag{9.39}$$

Differentiating (9.39) w.r.t. r,

$$E\frac{dw}{dr} = \sigma_\theta + r\frac{d\sigma_\theta}{dr} - \nu\sigma_x - \nu r\frac{d\sigma_x}{dr} - \nu\sigma_r$$

$$-\nu r\frac{d\sigma_r}{dr} + E\alpha T + rE\alpha\frac{dT}{dr} \tag{9.40}$$

Taking (9.38) from (9.40),

$$0 = \sigma_\theta(1+\nu) - \sigma_r(1+\nu) - \nu r\frac{d\sigma_x}{dx}$$

$$+r\frac{d\sigma_\theta}{dr} - \nu r\frac{d\sigma_r}{dr} + rE\alpha\frac{dT}{dr}$$

or

$$0 = (\sigma_\theta - \sigma_r)(1+\nu) - \nu r\frac{d\sigma_x}{dr} + r\frac{d\sigma_\theta}{dr}$$

$$-\nu r\frac{d\sigma_r}{dr} + rE\alpha\frac{dT}{dr} \tag{9.41}$$

Since $\varepsilon_x = \text{constant}$, then from (9.36),

$$0 = \frac{d\sigma_x}{dr} - \nu\frac{d\sigma_\theta}{dr} - \nu\frac{d\sigma_r}{dr} + E\alpha\frac{dT}{dr}$$

or,

$$\frac{d\sigma_x}{dr} = \nu\frac{d\sigma_\theta}{dr} + \nu\frac{d\sigma_r}{dr} - E\alpha\frac{dT}{dr} \tag{9.42}$$

Substituting (9.42) into (9.41),

$$0 = (\sigma_\theta - \sigma_r)(1+\nu) + r\frac{d\sigma_\theta}{dr} - \nu r\frac{d\sigma_r}{dr}$$

$$-\nu^2 r\frac{d\sigma_\theta}{dr} - \nu^2 r\frac{d\sigma_r}{dr} + \nu rE\alpha\frac{dT}{dr} + rE\alpha\frac{dT}{dr}$$

$$0 = (\sigma_\theta - \sigma_r)(1+\nu) + r\frac{d\sigma_\theta}{dr}(1-\nu^2) - \nu r\frac{d\sigma_r}{dr}(1+\nu)$$

$$+ rE\alpha\frac{dT}{dr}(1+\nu)$$

$$(\sigma_\theta - \sigma_r) + r(1-\nu)\frac{d\sigma_\theta}{dr} - \nu r\frac{d\sigma_r}{dr} + rE\alpha\frac{dT}{dr} = 0 \tag{9.43}$$

For this case, the equilibrium considerations are the same as (9.12), i.e.

$$\sigma_\theta - \sigma_r - r\frac{d\sigma_r}{dr} = 0 \tag{9.44}$$

Subtracting (9.44) from (9.43),

$$r(1-\nu)\frac{d\sigma_\theta}{dr} - \nu r\frac{d\sigma_r}{dr} + r\frac{d\sigma_r}{dr} + rE\alpha\frac{dT}{dr} = 0$$

$$r(1-\nu)\frac{d\sigma_\theta}{dr} + r(1-\nu)\frac{d\sigma_r}{dr} + rE\alpha\frac{dT}{dr} = 0$$

$$\frac{d\sigma_\theta}{dr} + \frac{d\sigma_r}{dr} + \frac{E\alpha}{(1-\nu)}\frac{dT}{dr} = 0$$

Integrating the above,

$$\sigma_\theta + \sigma_r + \frac{E\alpha T}{(1-\nu)} = 2A \tag{9.45}$$

Subtracting (9.45) from (9.44),

$$-2\sigma_r - r\frac{d\sigma_r}{dr} - \frac{E\alpha T}{(1-\nu)} = -2A$$

or

$$2\sigma_r + r\frac{d\sigma_r}{dr} + \frac{E\alpha T}{(1-\nu)} = 2A$$

or

$$\frac{1}{r}\frac{d(\sigma_r r^2)}{dr} = 2A - \frac{E\alpha T}{(1-\nu)}$$

$$\frac{d(\sigma_r r^2)}{dr} = 2Ar - \frac{E\alpha Tr}{(1-\nu)}$$

$$\sigma_r r^2 = Ar^2 + B - \frac{E\alpha}{(1-\nu)}\int Tr\,dr$$

therefore

$$\underline{\underline{\sigma_r = A + \frac{B}{r^2} - \frac{E\alpha}{(1-\nu)r^2}\int Tr\,dr}} \tag{9.46}$$

From (9.45),

$$\sigma_\theta = 2A - \frac{E\alpha T}{(1-\nu)} - A - \frac{B}{r^2} + \frac{E\alpha}{(1-\nu)r^2}\int Tr\,dr$$

therefore

$$\sigma_\theta = A - \frac{B}{r^2} - \frac{E\alpha T}{(1-\nu)} + \frac{E\alpha}{(1-\nu)r^2}\int Tr\,dr \tag{9.47}$$

9.11.1 EXAMPLE 9.6 THERMAL STRESSES IN A THICK-WALLED PIPE

A thick-walled pipe of bore 18 mm and of wall thickness 3 mm is subjected to an internal pressure of 20 MPa. The pipe is also subjected to a thermal gradient through its wall, which is given by the equation:

$$T = 180 - 1.5r^2\,°\mathrm{C}$$

where,

$$r = \text{the radius in mm}$$

Determine the hoop stresses on the internal and external surfaces of the pipe.

$$E = 2\mathrm{E}11\ \mathrm{N/m^2} \quad \nu = 0.32 \quad \alpha = 15\mathrm{E}\text{-}6/°\mathrm{C}$$

9.11.2.

@ $r = 9\,\mathrm{mm}, \quad \sigma_r = -20\,\mathrm{N/mm^2}$

Substituting this boundary value into (9.46),

$$-20 = A + B/81 - 0$$

therefore

$$\underline{A = -B/81 - 20} \tag{9.48}$$

@ $r = 12\,\mathrm{mm}, \quad \sigma_r = 0$

Substituting this boundary value into (9.46),

$$0 = A + \frac{B}{144} - \frac{2E5 * 15E\text{-}6}{(1 - 0.32) * 144}\int_{9}^{12}(180r - 1.5r^3)\mathrm{d}r$$

$$= A + \frac{B}{144} - 3.064E\text{-}2\left[\frac{180r^2}{2} - \frac{1.5r^4}{4}\right]_{9}^{12}$$

therefore

$$0 = A + \frac{B}{144} - 3.064E\text{-}2\{(12\,960 - 7776) - (7290 - 2460)\}$$

$$0 = A + \frac{B}{144} - 10.85 \qquad (9.49)$$

Substituting (9.48) into (9.49),

$$0 = -\frac{B}{81} + \frac{B}{144} - 20 - 10.85$$

or,

$$5.40E\text{-}3B = -30.85$$

or,

$$\underline{B = -5712}$$

and from (9.48),

$$\underline{A = \frac{5712}{81} - 20 = 50.5}$$

Hence,

$$\underline{\sigma_\theta = 50.5 + \frac{5712}{r^2} - \frac{E\alpha T}{(1-\nu)} + \frac{E\alpha}{(1-\nu)r^2}\int Tr\,\mathrm{d}r} \qquad (9.50)$$

From (9.50), the hoop stress on the internal surface =

$$\sigma_{\theta,9} = 50.5 + \frac{5712}{81} - \frac{2E5 * 15E\text{-}6 * (180 - 1.5 * 81)}{0.68} + 0$$

$$\underline{\underline{\sigma_{\theta,9} = -137.1\,\mathrm{MPa}}}$$

From (9.50), the hoop stress on the external surface =

$$\sigma_{\theta,12} = 50.5 + \frac{5712}{144} - \frac{2E5 * 15E\text{-}6 * (180 - 1.5 * 144)}{0.68}$$

$$+ \frac{2E5 * 15E\text{-}6}{0.68 * 144}\int_{9}^{12}(180 - 1.5r^3)\mathrm{d}r$$

$$= 50.5 + 39.67 + 158.8 + 3.063E\text{-}2 * 354$$

$$\underline{\sigma_{\theta,12} = 260.8\,\mathrm{MPa}}$$

9.12.1 PLASTIC YIELDING OF THICK TUBES

The following assumptions are made in this theory.

1. The tube is constructed from an ideally elastic–plastic material.
2. The longitudinal stress is the "minimax" stress.
3. Yield occurs according to Tresca's criterion.

For this case, the equilibrium considerations of (9.12) apply, i.e.

$$\sigma_\theta - \sigma_r - r\frac{d\sigma_r}{dr} = 0 \tag{9.51}$$

Now, Tresca's criterion is that

$$\sigma_\theta - \sigma_r = \sigma_{yp}$$

$$\sigma_\theta = \sigma_{yp} + \sigma_r \tag{9.52}$$

Substituting (9.52) into (9.51),

$$\sigma_{yp} + \sigma_r - \sigma_r - r\frac{d\sigma_r}{dr} = 0$$

$$d\sigma_r = \sigma_{yp}\frac{dr}{r}$$

$$\sigma_r = \sigma_{yp}\ln r + C \tag{9.53}$$

For the case of a partially plastic cylinder, as shown in Fig. 9.15,

@ $r = r_2, \quad \sigma_r = -P_2$

Substituting the above boundary value into (9.53),

$$-P_2 = \sigma_{yp}\ln r_2 + C$$

therefore

$$C = -\sigma_{yp}\ln r_2 - P_2$$

and,

$$\underline{\sigma_r = \sigma_{yp}\ln\left(\frac{r}{r_2}\right) - P_2} \tag{9.54}$$

Similarly, from (9.52),

$$\underline{\sigma_\theta = \sigma_{yp}\left\{1 + \ln\left(\frac{r}{r_2}\right)\right\} - P_2} \tag{9.55}$$

where,

r_1 = internal radius

r_2 = outer radius of plastic section of cylinder

r_3 = external radius

P_1 = internal pressure

P_2 = radial pressure at outer radius of plastic zone

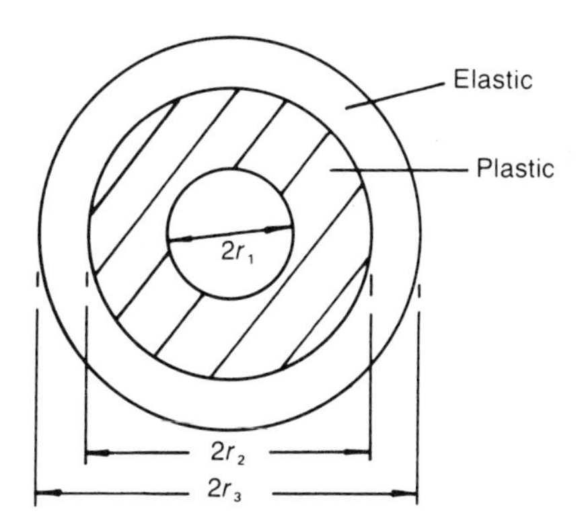

Fig. 9.15. Partially plastic cylinder.

The vessel can be assumed to behave as a compound cylinder, with the internal portion behaving plastically, and the external portion elastically.

The Lamé line for the elastic portion of the cylinder is shown in Fig. 9.16.

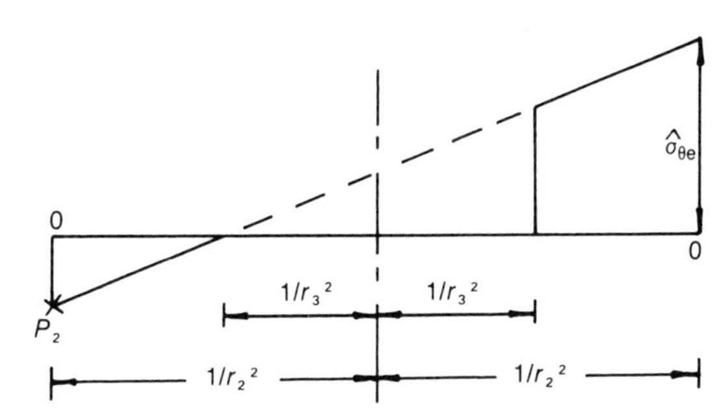

Fig. 9.16. Lamé line for elastic zone.

In Fig. 9.16,

$$\hat{\sigma}_{\theta e} = \text{elastic hoop stress @ } r = r_2$$

so that according to Tresca's criterion, on this radius,

$$\sigma_{yp} = \hat{\sigma}_{\theta e} + P_2 \tag{9.56}$$

From Fig. 9.16,

$$\frac{P_2}{\left(\dfrac{1}{r_2^2} - \dfrac{1}{r_3^2}\right)} = \frac{\hat{\sigma}_{\theta e}}{\left(\dfrac{1}{r_2^2} + \dfrac{1}{r_3^2}\right)}$$

therefore

$$\hat{\sigma}_{\theta e} = \frac{P_2(r_3^2 + r_2^2)}{(r_3^2 - r_2^2)} \tag{9.57}$$

Substituting (9.57) into (9.56),

$$P_2 = \sigma_{yp}(r_3^2 - r_2^2)/(2r_3^2) \tag{9.58}$$

9.12.2

Consider now the portion of the cylinder that is plastic.

Substituting (9.58) into (9.54) and (9.55), the stress distributions in the plastic zone are given by:

$$\sigma_r = -\sigma_{yp}\left\{\ln\left(\frac{r_2}{r}\right) + \frac{(r_3^2 - r_2^2)}{2r_3^2}\right\} \tag{9.59}$$

$$\sigma_\theta = \sigma_{yp}\left\{\frac{(r_3^2 + r_2^2)}{2r_3^2} - \ln\left(\frac{r_2}{r}\right)\right\} \tag{9.60}$$

To find the *pressure to just cause yield*, put

$$\sigma_r = -P_1 \quad @ \quad r = r_1$$

where,

P_1 = internal pressure that causes the onset of yield

therefore

$$P_1 = \sigma_{yp}\left\{\ln\left(\frac{r_2}{r_1}\right) + \left(\frac{r_3^3 - r_2^2}{2r_3^2}\right)\right\} \tag{9.61}$$

but if yield is only on the inside surface,

$$r_1 = r_2$$

in (9.61), so that,

$$P_1 = \sigma_{yp}\{(r_3^2 - r_1^2)/(2r_3^2)\} \tag{9.62}$$

To determine the *plastic collapse pressure* P_p, put $r_2 = r_3$ in (9.61). Therefore

$$P_p = \sigma_{yp}\ln\left(\frac{r_3}{r_1}\right) \tag{9.63}$$

To determine the hoop stress distribution in the plastic zone, $\sigma_{\theta p}$, it must be remembered that

$$\sigma_{yp} = \sigma_\theta - \sigma_r$$

therefore

$$\sigma_{\theta p} = \sigma_{yp}\{1 + \ln(r_3/r_1)\} \tag{9.64}$$

Plots of the stress distributions in a partially plastic cylinder, under internal pressure, are shown in Fig. 9.17.

9.13.1 EXAMPLE 9.7 ELASTO-PLASTIC COMPOUND CYLINDER

A high tensile steel cylinder of 0.5 m outer diameter, 0.4 m inner diameter, is shrunk onto a mild steel cylinder of 0.3 m bore. If the interference fit is such

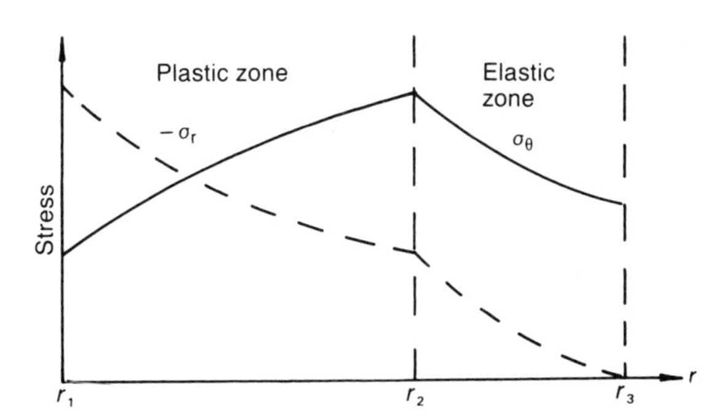

Fig. 9.17. Stress distributions in a partially plastic cylinder.

that when the internal pressure is $100\,\mathrm{MN/m^2}$, the inner face of the inner cylinder is on the point of yielding, determine the internal pressure which will cause plastic penetration through half the thickness of the inner cylinder. The material of the outer cylinder may be assumed to be of a higher quality, so that it does not yield and the inner cylinder material is perfectly elastic–plastic, yielding at a constant shear stress of $140\,\mathrm{MN/m^2}$. Both materials may be assumed to have the same elastic modulus and Poisson's ratio.

{Portsmouth Polytechnic, 1983}

9.13.2

The Lamé line for the compound cylinder at the onset of yield is shown in Fig. 9.18.

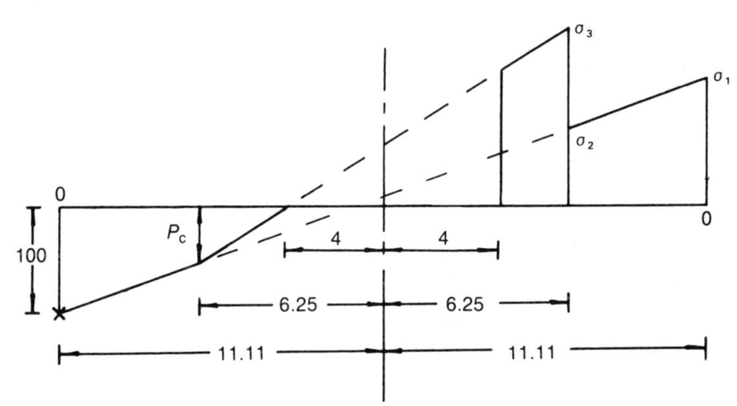

Fig. 9.18. Lamé line for compound cylinder.

In Fig. 9.18,

σ_1 = hoop stress on inner surface of inner cylinder

σ_2 = hoop stress on outer surface of inner cylinder

σ_3 = hoop stress on inner surface of outer cylinder

As yield occurs on the inner surface of the inner cylinder,

$$\frac{\sigma_1 - (-100)}{2} = \tau_{yp} = 140$$

therefore

$$\underline{\sigma_1 = 180\,\text{MPa}}$$

where,

$$\tau_{yp} = \text{yield stress in shear}$$

Equating similar triangles in Fig. 9.18.

$$\frac{\sigma_1 + 100}{11.11 + 11.11} = \frac{100 - P_c}{11.11 - 6.25}$$

or

$$\frac{280}{22.22} = \frac{100 - P_c}{4.86}$$

therefore

$$\underline{P_c = 38.76\,\text{MPa}}$$

Similarly, from Fig. 9.18,

$$\frac{\sigma_2 + 100}{11.11 + 6.25} = \frac{280}{22.22}$$

therefore

$$\underline{\underline{\sigma_2 = 118.8\,\text{MPa}}} \tag{9.65}$$

Also, from Fig. 9.18,

$$\frac{\sigma_3}{6.25 + 4} = \frac{P_c}{6.25 - 4}$$

but,

$$P_c = 38.76$$

therefore

$$\underline{\underline{\sigma_3 = 176.6\,\text{MPa}}} \tag{9.66}$$

9.13.3

Consider, now, plastic penetration of the inner cylinder to a diameter of 0.35 m.

The Lamé line in the elastic zones will be as shown in Fig. 9.19.

From Fig. 9.19,

$$\frac{\sigma_6 + P_3}{2} = 140$$

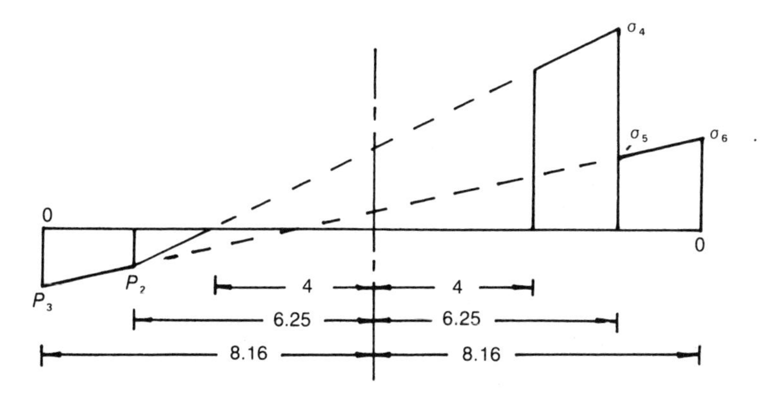

Fig. 9.19. Lamé line in elastic zones.

therefore

$$\underline{\sigma_6 = 280 - P_3} \tag{9.67}$$

Similarly,

$$\frac{P_3 - P_2}{8.16 - 6.25} = \frac{\sigma_6 + P_3}{8.16 \times 2} = \frac{280}{16.32}$$

therefore

$$P_3 - P_2 = \frac{280}{16.32} \times 1.91 = 32.77$$

or

$$P_2 = P_3 - 32.77 \tag{9.68}$$

Also from Fig. 9.19,

$$\frac{\sigma_4}{(6.25 + 4)} = \frac{P_2}{6.25 - 4}$$

or,

$$\underline{\sigma_4 = 4.56 P_2} \tag{9.69}$$

Substituting (9.68) into (9.69),

$$\underline{\sigma_4 = 4.56\, P_3 - 149.3} \tag{9.70}$$

and,

$$\frac{\sigma_5 + P_3}{8.16 + 6.25} = \frac{280}{16.32}$$

therefore

$$\underline{\sigma_5 = 247.2 - P_3} \tag{9.71}$$

9.13.4

Consider strains during the additional pressurisation.

Now,

$$w = \frac{r}{E}(\sigma_\theta - \nu\sigma_r)$$

which will be the same for both cylinders at the common surface, i.e.

$$\frac{1}{E}\{(\sigma_5 - \sigma_2) - \nu(P_2 - P_c)\} = \left\{\frac{1}{E}(\sigma_4 - \sigma_3) - \nu(P_2 - P_c)\right\}$$

or,

$$\sigma_5 - \sigma_2 = \sigma_4 - \sigma_3 \tag{9.72}$$

Substituting (9.65), (9.66), (9.70) and (9.71) into (9.72),

$$247.2 - P_3 - 118.8 = 4.56P_3 - 149.3 - 176.6$$

or

$$5.56P_3 = 454.3$$

therefore

$$\underline{P_3 = 81.71\ \text{MPa}}$$

9.13.5

Consider the *yielded portion.*

Now,

$$\sigma_r = \sigma_{yp} \ln r + C$$

$$\sigma_{yp} = 280\ \text{MPa}$$

and

@ $r = 0.175,$

$$\sigma_r = -P_3 = -81.71$$

therefore

$$\underline{C = 406.32}$$

@ $r = 0.15,$

$$\sigma_r = -P$$

therefore

$$-P = 280 \ln(0.15) + 406.3$$

$$= -531.2 + 406.3$$

$$\underline{P = 124.9\ \text{MPa}}$$

i.e. pressure to cause plastic penetration = 124.9 MPa.

9.14.1 EXAMPLE 9.8 PRESSURE TO CAUSE TOTAL PLASTIC COLLAPSE

Determine the internal pressure that will cause total plastic failure of the compound vessel of Example 9.7, given that σ_{yp} for the outer cylinder = 600 MPa and that E and ν are the same for both cylinders.

Now,

$$P_p = \sigma_{yp} \ln\left(\frac{r_3}{r_1}\right)$$

$$= 280 \ln\left(\frac{0.2}{0.15}\right) + 600 \ln\left(\frac{0.25}{0.2}\right)$$

$$= 80.55 + 133.89$$

therefore

$$\underline{P_p = 214.44\,\text{MPa}}$$

i.e. plastic collapse pressure = 214.44 MPa.

9.15.1 THICK SPHERICAL SHELLS

Consider a hemispherical element of a thick spherical shell at any radius r, under a compressive radial stress P, as shown in Fig. 9.20.

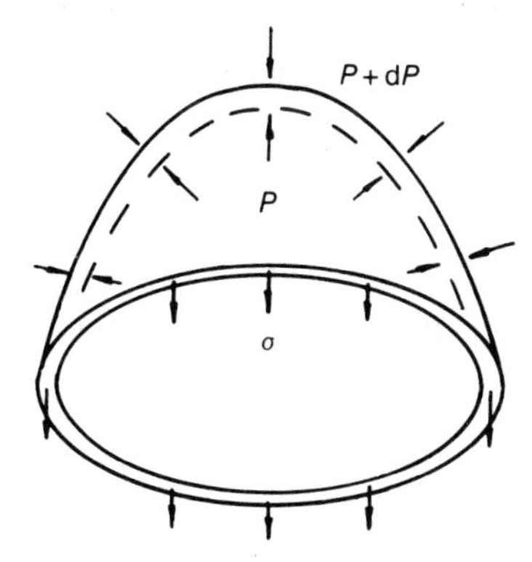

Fig. 9.20. Hemispherical shell element.

Let,

w = radial deflection at any radius r

so that

Hoop strain = w/r

and

Radial strain = $\mathrm{d}w/\mathrm{d}r$

From three-dimensional stress–strain relationships,

$$E\frac{w}{r} = \sigma - \nu\sigma + \nu P \tag{9.73}$$

and,

$$E\frac{\mathrm{d}w}{\mathrm{d}r} = -P - \nu\sigma - \nu\sigma$$

$$= -P - 2\nu\sigma \tag{9.74}$$

Multiplying (9.73) by r,

$$Ew = \sigma r - \nu\sigma r + \nu P r$$

and differentiating w.r.t. r,

$$E\frac{dw}{dr} = \sigma + r\frac{d\sigma}{dr} - \nu\sigma - \nu r\frac{d\sigma}{dr} + \nu P + \nu r\frac{dP}{dr}$$

$$= (1-\nu)\left(\sigma - r\frac{d\sigma}{dr}\right) + \nu\left(P + r\frac{dP}{dr}\right) \tag{9.75}$$

Equating (9.74) and (9.75),

$$-P - 2\nu\sigma = (1-\nu)\left(\sigma - r\frac{d\sigma}{dr}\right) + \nu\left(P + r\frac{dP}{dr}\right)$$

or,

$$(1+\nu)(\sigma + P) + r(1-\nu)\frac{d\sigma}{dr} + \nu r\frac{dP}{dr} = 0 \tag{9.76}$$

Considering now the equilibrium of the hemispherical shell element,

$$\sigma * 2\pi r * dr = P * \pi r^2 - (P + dP) * \pi * (r + dr)^2 \tag{9.77}$$

Neglecting higher order terms, (9.77) becomes

$$\sigma + P = (-r/2)\frac{dP}{dr} \tag{9.78}$$

Substituting (9.78) into (9.76),

$$-(r/2)(dP/dr)(1+\nu) + r(1-\nu)(d\sigma/dr) + \nu r(dP/dr) = 0$$

or,

$$\frac{d\sigma}{dr} - \tfrac{1}{2}\frac{dP}{dr} = 0 \tag{9.79}$$

which on integrating becomes,

$$\sigma - P/2 = A \tag{9.80}$$

Substituting (9.80) into (9.78),

$$3P/2 + A = (-r/2)(dP/dr)$$

or

$$-\frac{1}{r^2}\frac{d(P * r^3)}{dr} = 2A$$

or

$$\frac{d(P * r^3)}{dr} = -2Ar^2$$

which on integrating becomes,

$$P * r^3 = -2Ar^3/3 + B$$

or

$$\underline{P = -2A/3 + B/r^3} \tag{9.81}$$

and

$$\sigma = 2A/3 + B/(2r^3) \tag{9.82}$$

9.16.1 ROTATING DISCS

These are a common feature in engineering, which from time to time suffer failure, due to high speed rotation.

In this section, equations will be obtained for calculating hoop and radial stresses, and the theory will be extended for calculating angular velocities to cause plastic collapse of rotating discs and rings.

Consider a uniform thickness disc, of density ρ, rotating at a constant angular velocity ω.

From Section 9.2,

$$E\frac{dw}{dr} = \sigma_r - \nu\sigma_\theta \tag{9.83}$$

and,

$$E\frac{w}{r} = \sigma_\theta - \nu\sigma_r \tag{9.84}$$

or,

$$Ew = \sigma_\theta * r + \nu\sigma_r * r \tag{9.85}$$

Differentiating (9.85) w.r.t. r,

$$E\frac{dw}{dr} = \sigma_\theta + r\frac{d\sigma_\theta}{dr} - \nu\sigma_r - \nu r\frac{d\sigma_r}{dr} \tag{9.86}$$

Equating (9.83) and (9.86),

$$(\sigma_\theta - \sigma_r)(1+\nu) + r\frac{d\sigma_\theta}{dr} - \nu r\frac{d\sigma_r}{dr} = 0 \tag{9.87}$$

Considering equilibrium of an element of the disc, as shown in Fig. 9.21,

$$2\sigma_\theta * dr * \sin\left(\frac{d\theta}{2}\right) + \sigma_r * r * d\theta$$

$$-(\sigma_r + d\sigma_r)(r + dr)d\theta = \rho * \omega^2 * r^2 * dr * d\theta$$

In the limit, this reduces to

$$\sigma_\theta - \sigma_r - r\frac{d\sigma_r}{dr} = \rho\omega^2 r^2 \tag{9.88}$$

Substituting (9.88) into (9.87),

$$\left(r\frac{d\sigma_r}{dr} + \rho\omega^2 r^2\right)(1+\nu) + r\frac{d\sigma_\theta}{dr} - \nu r\frac{d\sigma_r}{dr} = 0$$

or,

$$\frac{d\sigma_\theta}{dr} + \frac{d\sigma_r}{dr} = -\rho\omega^2 r^2(1+\nu)$$

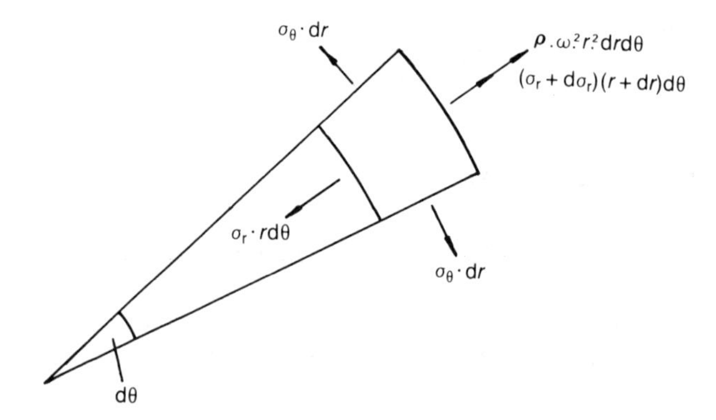

Fig. 9.21. Element of disc.

which on integrating becomes,

$$\sigma_\theta + \sigma_r = -(\rho\omega^2 r^2/2)(1+\nu) + 2A \tag{9.89}$$

Subtracting (9.88) from (9.89),

$$2\sigma_r + r\frac{d\sigma_r}{dr} = -(\rho\omega^2 r^2/2)(3+\nu) + 2A$$

or,

$$\frac{1}{r}\frac{d(\sigma_r * r^2)}{dr} = -\frac{\rho\omega^2 r^2(3+\nu)}{2} + 2A$$

which on integrating becomes,

$$\sigma_r * r^2 = -(\rho\omega^2 r^4/8)(3+\nu) + Ar^2 - B$$

or,

$$\underline{\sigma_r = A - B/r^2 - (3+\nu)(\rho\omega^2 r^2/8)} \tag{9.90}$$

and,

$$\underline{\sigma_\theta = A + B/r^2 - (1+3\nu)(\rho\omega^2 r^2/8)} \tag{9.91}$$

9.17.1 EXAMPLE 9.9 DISC OF UNIFORM STRENGTH

Obtain an expression for the radial variation in the thickness of a disc, so that it will be of constant strength, when it is rotated at an angular velocity ω.

9.17.2

Let,

$$t_0 = \text{thickness at centre}$$

$$t = \text{thickness at a radius } r$$

$$t + dt = \text{thickness at a radius } r + dr$$

$$\sigma = \text{stress} = \text{constant (everywhere)}$$

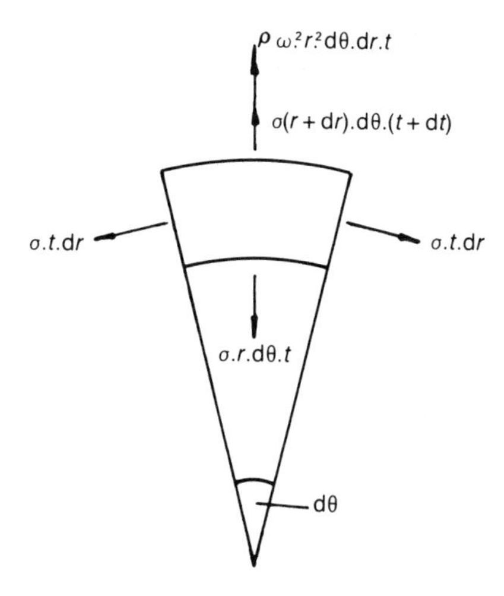

Fig. 9.22. Element of constant strength disc.

Consider the equilibrium of an element of this disc at any radius r, as shown in Fig. 9.22.

Resolving radially

$$2\sigma * t * \mathrm{d}r \sin\left(\frac{\mathrm{d}\theta}{2}\right) + \sigma tr\,\mathrm{d}\theta = \sigma(r + \mathrm{d}r)(t + \mathrm{d}t)\mathrm{d}\theta + \rho\omega^2 r^2 t\,\mathrm{d}\theta\,\mathrm{d}r$$

In the limit, this becomes,

$$\sigma t\,\mathrm{d}t = \sigma r\,\mathrm{d}t + \sigma t\,\mathrm{d}r + \rho\omega^2 rt\,\mathrm{d}r$$

or

$$\frac{\mathrm{d}t}{\mathrm{d}r} = -\rho\omega^2 rt/\sigma$$

which on integrating becomes,

$$\ln t = -\rho\omega^2 r^2 t/(2\sigma) + \ln C$$

or,

$$t = C\mathrm{e}^{(-\rho\omega^2 r^2/2\sigma)}$$

Now, @ $r = 0$, $t = t_0 = C$

therefore

$$\underline{t = t_0 \mathrm{e}^{(-\rho\omega^2 r^2/2\sigma)}} \tag{9.92}$$

9.18.1 PLASTIC COLLAPSE OF DISCS

Assume that $\sigma_\theta > \sigma_r$, and that plastic collapse occurs when

$$\sigma_\theta = \sigma_{yp}$$

Let,

R = external radius of disc

From *equilibrium considerations*

$$\sigma_{yp} - \sigma_r - r\frac{d\sigma_r}{dr} = \rho\omega^2 r^2$$

or,

$$\int r\,d\sigma_r = \int \{\sigma_{yp} - \sigma_r - \rho\omega^2 r^2\}\,dr$$

Integrating the LHS by parts,

$$r\cdot\sigma_r - \int \sigma_r\cdot dr = \sigma_{yp}\cdot r - \int \sigma_r\cdot dr - \rho\omega^2 r^3/3 + A$$

therefore

$$\underline{\sigma_r = \sigma_{yp} - \rho\omega^2 r^2/3 + A/r} \tag{9.93}$$

For a *solid disc*, @ $r = 0$, $\sigma_r \neq \infty$
therefore

$$\underline{A = 0}$$

therefore

$$\sigma_r = \sigma_{yp} - \rho\omega^2 r^2/3$$

@ $r = R$, $\sigma_r = 0$

therefore

$$0 = \sigma_{yp} - \rho\omega^2 R^2/3$$

therefore

$$\underline{\omega = \frac{1}{R}\sqrt{\frac{3\sigma_{yp}}{\rho}}} \tag{9.94}$$

where,

ω = the angular velocity of the disc, which causes plastic collapse

For an *annular disc*, of internal radius R_1 and external radius R_2, suitable boundary values for (9.93) are:

@ $r = R_1$, $\sigma_r = 0$

therefore

$$A = (\rho\omega^2 R_1^2/3 - \sigma_{yp})R_1$$

therefore

$$\sigma_r = \sigma_{yp} - \rho\omega^2 r^2/3 + (\rho\omega^2 R_1^2/3 - \sigma_{yp})(R_1/r) \tag{9.95}$$

@ $r = R_2, \quad \sigma_r = 0$

therefore

$$0 = \sigma_{yp} - \rho\omega^2 R_2^2/3 + (\rho\omega^2 R_1^2/3 - \sigma_{yp})(R_1/R_2)$$

i.e.

$$\omega = \sqrt{\left\{\left(\frac{3\sigma_{yp}}{\rho}\right)\frac{(R_2 - R_1)}{(R_2^3 - R_1^3)}\right\}} \tag{9.96}$$

9.19.1 ROTATING RINGS

Consider the equilibrium of the semi-circular ring element shown in Fig. 9.23.

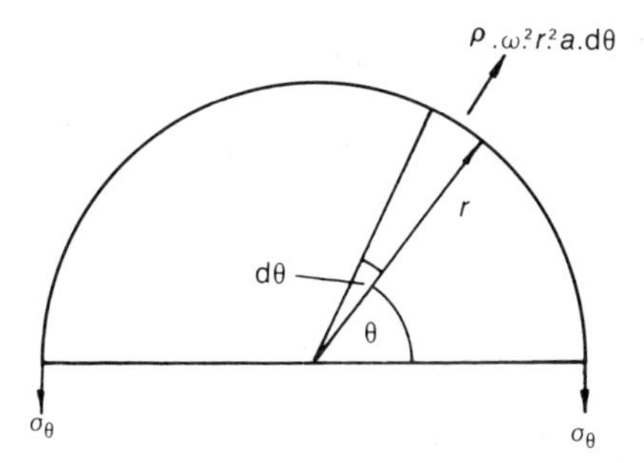

Fig. 9.23. Ring element.

Let,

a = cross-sectional area of ring

Resolving vertically

$$\sigma_\theta * a * 2 = \int_0^\pi \rho\omega^2 r^2 a \, d\theta \cdot \sin\theta$$

$$= \rho\omega^2 r^2 a[-\cos\theta]_0^\pi$$

$$= 2\rho\omega^2 r^2 a$$

therefore

$$\sigma_\theta = \rho\omega^2 r^2$$

@ collapse, $\sigma_\theta = \sigma_{yp}$
therefore

$$\omega = \frac{1}{r}\sqrt{\left(\frac{\sigma_{yp}}{\rho}\right)} \tag{9.97}$$

where,

ω = the angular velocity required to fracture the ring

9.20.1 THERMAL STRESSES IN DISCS

For discs subjected to a temperature rise of $T\,^\circ$C, at any radius r, it can readily be shown that the hoop and radial strains will be given by:

$$\varepsilon_\theta = \frac{w}{r} = \frac{1}{E}(\sigma_\theta - \nu\sigma_r) + \alpha T \tag{9.98}$$

$$\varepsilon_r = \frac{dw}{dr} = \frac{1}{E}(\sigma_r - \nu\sigma_\theta) + \alpha T \tag{9.99}$$

Multiplying (9.98) by r and differentiating it w.r.t. r,

$$E\frac{dw}{dr} = \sigma_\theta + r\frac{d\sigma_\theta}{dr} - \nu\sigma_r - \nu r\frac{d\sigma_r}{dr} + E\alpha T + E\alpha r\frac{dT}{dr} \tag{9.100}$$

Solving (9.99) and (9.100), the following is obtained:

$$(\sigma_\theta - \sigma_r)(1+\nu) + r\left(\frac{d\sigma_\theta}{dr}\right) - \nu r\left(\frac{d\sigma_r}{dr}\right) + E\alpha r\left(\frac{dT}{dr}\right) \tag{9.101}$$

From (9.88),

$$\sigma_\theta - \sigma_r + r\left(\frac{d\sigma_r}{dr}\right) = \rho\omega^2 r^2 \tag{9.102}$$

Substituting (9.102) into (9.101),

$$\frac{d\sigma_\theta}{dr} + \frac{d\sigma_r}{dr} = -(1+\nu)\rho\omega^2 r^2 - E\alpha\frac{dT}{dr}$$

which on integrating becomes

$$\sigma_\theta + \sigma_r = -(1+\nu)\rho\omega^2 r^2/2 - E\alpha T + 2A$$

Hence,

$$\underline{\sigma_\theta = A + \frac{B}{r^2} - \frac{(1+\nu)\rho\omega^2}{8}r^2 - E\alpha T + \frac{E\alpha}{r^2}\int Tr\,dr} \tag{9.103}$$

$$\underline{\sigma_r = A - \frac{B}{r^2} - \frac{(3+\nu)\rho\omega^2}{8}r^2 - \frac{E\alpha}{r^2}\int Tr\,dr} \tag{9.104}$$

9.21.1 Wire-wound Cylinders

On some occasions it is found cheaper to strengthen a cylinder by closely winding wire around it, rather than by thickening it, or by making it into a compound cylinder.

Such a vessel is shown diagrammatically in Fig. 9.24, where, very often, the wire is shrunk onto the external surface of the cylinder, or tensioned, so that the cylinder is initially under compression. In this condition, the cylinder can withstand a higher internal pressure, then it could if the wire were not

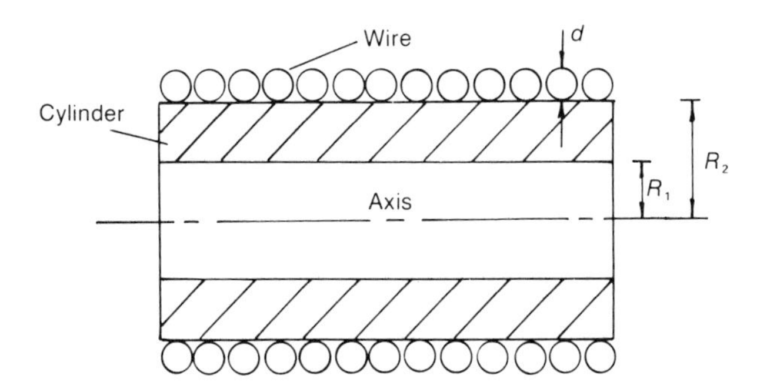

Fig. 9.24. Wire-wound cylinder.

shrunk onto it. Its analysis can be carried out, assuming that the vessel behaves as an equivalent compound cylinder, as shown in Fig. 9.25.

9.21.2

Consider a wire-wound cylinder, where the initial hoop stress in the wire $= \sigma^s_{\theta w}$. Let,

$\sigma^p_{\theta w}$ = hoop stress in the wire, due to the effects of pressure only

σ^s_r = initial radial stress in the cylinder at any radius r

σ^s_θ = initial hoop stress in the cylinder at any radius r

σ^p_r = radial stress in the cylinder, due to the effects of pressure alone

σ^p_θ = hoop stress in the cylinder, due to the effects of pressure alone

t_w = thickness of an imaginary outer cylinder, which is used to represent the effects of wire winding

$$= \frac{\pi d^2}{4} * \frac{1}{d} = \frac{\pi d}{4} \text{(see Fig. 9.25)}$$

d = wire diameter

From Fig. 9.25, it can be seen that the wire-wound cylinder can be represented by an equivalent compound cylinder, where the imaginary outer

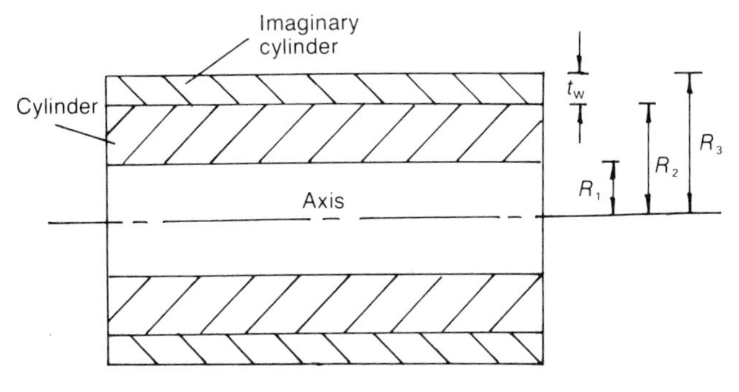

Fig. 9.25. Equivalent compound cylinder.

cylinder is made equivalent in terms of area. Thus, solution can take place as described in Section 9.9.1, and the resulting stresses obtained by superimposing those due to "shrinkage", together with those due to the effects of pressure alone, as follows:

$$\sigma_{\theta w} = \text{resultant hoop stress in the wire}$$

$$= \sigma_{\theta w}^{s} + \sigma_{\theta w}^{p}$$

$$\sigma_r = \sigma_r^s + \sigma_r^p$$

$$\sigma_\theta = \sigma_\theta^s + \sigma_\theta^p$$

For a more comprehensive analysis of thick-walled shells and discs, see reference [25].

EXAMPLES FOR PRACTICE 9

1. Determine the maximum permissible internal pressure that a thick-walled cylinder of internal diameter 0.2 m and of wall thickness 0.1 m can be subjected to, if the maximum permissible stress in this vessel is not to exceed 250 MPa.

   {150 MPa}

2. Determine the maximum permissible internal pressure that a thick-walled cylinder of internal diameter 0.2 m and of wall thickness 0.1 m can be subjected to, if the cylinder is also subjected to an external pressure of 20 MPa. $\sigma_{yp} = 300$ MPa.

   {212 MPa}

3. A steel ring of 9 cm external diameter and of 5 cm internal diameter is to be shrunk into a solid bronze shaft, where the interference fit is 0.005E-2 m, based on the diameter.

   Determine the maximum tensile stress that is set up in the material given that:

   For steel

   $$E_s = 2\text{E}11 \text{ N/m}^2 \quad \nu_s = 0.3$$

   For bronze

   $$E_b = 1\text{E}11 \text{ N/m}^2 \quad \nu_b = 0.35$$

   $\{P_c = 57.3$ MPa; $\hat{\sigma}_\theta = 108.3$ MPa$\}$

4. A compound cylinder is manufactured by shrinking a steel cylinder of external diameter 22 cm and of internal diameter 18 cm onto another steel cylinder of internal diameter 14 cm, the dimensions being nominal. If the maximum tensile stress in the outer cylinder is 100 MPa, determine the radial compressive stress at the common surface and the interference fit

at the common diameter. Determine, also, the maximum stress in the inner cylinder.

$$E_s = 2E11\ \text{N/m}^2 \quad \nu_s = 0.3$$

{19.8 MPa; $\delta = 0.16$ mm; -100 MPa}

5. If the inner cylinder of Example 4 were made from bronze, what would be the value of δ?

   For bronze

   $$E_b = 1E11\ \text{N/m}^2, \quad \nu_b = 0.4$$

   {$\delta = 0.226$ mm}

6. If the compound cylinder of Example 4 where subjected to an internal pressure of 50 MPa, what would be the value of the maximum resultant stress?

   {227.9 MPa}

7. If the compound cylinder of Example 5 were subjected to an internal pressure of 50 MPa, what would be the value of the maximum resultant stress?

   {241.1 MPa}

8. A thick compound cylinder consists of a brass cylinder of internal diameter 0.1 m and external diameter 0.2 m, surrounded by a steel cylinder of external diameter 0.3 m and of the same length.

   If the compound cylinder is subjected to a compressive axial load of 5 MN, and the axial strain is constant for both cylinders, determine the pressure at the common surface and the longitudinal stresses in the two cylinders, due to this load.

   The following assumptions may be made:

   (a) σ_{Ls} = a longitudinal stress in steel cylinder

   = a constant

   (b) σ_{LB} = longitudinal stress in brass cylinder

   = a constant

   For steel

   $$E_s = 2 \times 10^{11}\ \text{N/m}^2$$

   $$\nu_s = 0.3$$

   For brass

   $$E_B = 1 \times 10^{11}\ \text{N/m}^2$$

   $$\nu_B = 0.4$$

{Portsmouth Polytechnic, 1984}

{2.2 MPa; $\sigma_{Ls} = -96.52$ MPa; $\sigma_{LB} = -51.14$ MPa}

9. If, in Example 8, the brass cylinder is external and the steel cylinder is internal, what would be the pressure at the common surface due to a compressive axial load of 5 MN?

   {0}

10. A steel cylinder of 0.4 m outer diameter, 0.3 m inner diameter, is shrunk onto another steel cylinder of 0.2 m bore. If the interference fit is such that when the internal pressure is 100 MPa, the inner face of the inner cylinder is on the point of yielding, determine the value of internal pressure which will cause plastic penetration through half the thickness of the inner cylinder. The outer cylinder may be assumed not to yield.

    For the inner cylinder,

    $$\sigma_{yp} = 300 \text{ MN/m}^2$$

    {145.9 MPa}

11. Determine the plastic collapse pressure of Example 1, assuming that $\sigma_{yp} = 250$ MPa.

    {173.3 MPa}

12. Determine the plastic collapse pressure of Example 10, assuming that for the outer cylinder, $\sigma_{yps} = 500$ MPa.

    {265.4 MPa}

10

Finite Difference Methods

10.1.1 NUMERICAL METHODS

A numerical method, which is often used for structural problems, is the finite difference method. It is quite different to the finite element method, in that the appropriate differential equation is approximated by a finite difference equation. This difference equation is then applied to a number of discrete points in the structure, so that several simultaneous equations are obtained, and solution of these equations results in values of the unknown function at the pre-selected discrete points. Like the finite element method, the finite difference method is virtually useless without the aid of a computer.

10.2.1 BASIC THEORY OF CENTRAL DIFFERENCES

Prior to applying the finite difference method, it will be necessary to cover some of the basic theory on finite differences.

Consider a curve $v = f(x)$, where v is the unknown function, as shown in Fig. 10.1.

Let δx = the distance between any two values of v.

If δx is so chosen, that the portion of the curve between v_i and v_{i+1} is approximately straight, then at the station $(i + \frac{1}{2})$, the slope can be approximated by:

$$\frac{\mathrm{d}v}{\mathrm{d}x} = \frac{v_{i+1} - v_i}{\delta x} \tag{10.1}$$

i.e. the differential of v w.r.t. x at any station $(i + \frac{1}{2})$ is approximated by the "difference" of v over δx at $(i + \frac{1}{2})$.

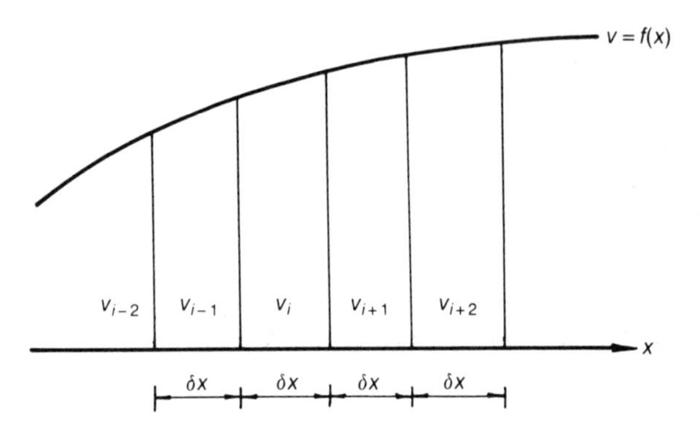

Fig. 10.1. Variation of function with x.

This is known as the first central difference at the station $(i + \frac{1}{2})$, and it is denoted by $\delta v_{i+\frac{1}{2}}$.

It follows, then, that the expression relating the shearing force F and the bending moment M, i.e. $(\mathrm{d}M/\mathrm{d}x = F)$, can be reduced to the expression:

$$F_{i+\frac{1}{2}} = \frac{M_{i+1} - M_i}{\delta x} \tag{10.2}$$

Similarly, from Fig. 10.1, it can be seen that the slope of the slope at any station i can be approximated by:

$$\left(\frac{\mathrm{d}^2 v}{\mathrm{d}x^2}\right)_i = \delta^2 v_i = \frac{\dfrac{v_{i+1} - v_i}{\delta x} - \dfrac{v_i - v_{i-1}}{\delta x}}{\delta x}$$

$$= \frac{v_{i-1} - 2v_i + v_{i+1}}{(\delta x)^2} \tag{10.3}$$

where $\delta^2 v_i$ is known as the second central difference at the station i.

Applying equation (10.3) to the equation relating bending moment M and load intensity w, i.e.

$$\left(\frac{\mathrm{d}^2 M}{\mathrm{d}x^2} = w\right)$$

the following is obtained:

$$\underline{M_{i-1} - 2M_i + M_{i+1} = w_i(\delta x)^2} \tag{10.4}$$

Similarly, if (10.3) is applied to the differential equation relating deflection v and bending moment M, i.e.

$$\left(EI\frac{\mathrm{d}^2 v}{\mathrm{d}x^2} = M\right)$$

the following is obtained:

$$\underline{v_{i-1} - 2v_i + v_{i+1} = M_i(\delta x)^2/(E_i I_i)} \tag{10.5}$$

If (10.4) and (10.5) are applied to a practical problem, several simultaneous equations will be obtained in the form of a tridiagonal band, and solution of these will produce the unknown bending moments M_i and deflections v_i. To obtain

$$\frac{d^4v}{dx^4}$$

consider Fig. 10.1:

$$\delta^3 v_{i-\frac{1}{2}} = \frac{\delta^2 v_i - \delta^2 v_{i-1}}{\delta x}$$

$$= \frac{v_{i-1} - 2v_i + v_{i+1} - v_{i-2} + 2v_{i-1} - v_i}{(\delta x)^3}$$

$$= \frac{-v_{i-2} + 3v_{i-1} - 3v_i + v_{i+1}}{(\delta x)^3}$$

Similarly,

$$\delta^3 v_{i+\frac{1}{2}} = \frac{-v_{i-1} + 3v_i - 3v_{i+1} + v_{i+2}}{(\delta x)^3} \tag{10.6}$$

Now,

$$\delta^4 v_i = \frac{\delta^3 v_{i+\frac{1}{2}} - \delta^3 v_{i-\frac{1}{2}}}{\delta x}$$

$$= \frac{v_{i-2} - 4v_{i-1} + 6v_i - 4v_{i+1} + v_{i+2}}{(\delta x)^4} \tag{10.7}$$

The differential equation relating the displacement of a beam v to the intensity of load w, acting on it, is

$$EI\frac{d^4v}{dx^4} = w \tag{10.8}$$

so that when the fourth central difference equation for a beam, namely equation (10.7), is applied to (10.8), the difference equation at the ith station appears as:

$$E_iI_i\delta^4 v_i = w_i = \frac{v_{i-2} - 4v_{i-1} + 6v_i - 4v_{i+1} + v_{i+2}}{(\delta x)^4} \tag{10.9}$$

10.3.1 FORWARD AND BACKWARD DIFFERENCES

These are useful at boundaries where it is difficult to apply central differences. They are in fact equivalent to applying the boundary conditions in central difference form, to the nearest possible stations to the boundary.

Application of the first *forward difference* to the ith station of Fig. 10.1 gives the following:

$$\left(\frac{dv}{dx}\right)_i = \Delta v_i = \frac{v_{i+1} - v_i}{\delta x}$$

Similarly, for the second forward difference at the ith station,

$$\left(\frac{d^2v}{dx^2}\right)_i = \Delta^2 v_i = \frac{v_i - 2v_{i+1} + v_{i+2}}{(\delta x)^2}$$

Application of the first *backward difference* to the ith station of Fig. 10.1 gives the following:

$$\left(\frac{dv}{dx}\right)_i = \nabla v_i = \frac{v_i - v_{i-1}}{\delta x}$$

Similarly, for the second backward difference at the ith station,

$$\left(\frac{d^2y}{dx^2}\right)_i = \nabla^2 v_i = \frac{v_i - 2v_{i-1} + v_{i-2}}{(\delta x)^2}$$

Table 10.1. Central differences layout

v_{i-3}						
	$\delta v_{i-2\frac{1}{2}}$					
v_{i-2}		$\delta^2 v_{i-2}$				
	$\delta v_{i-1\frac{1}{2}}$		$\delta^3 v_{i-1\frac{1}{2}}$			
v_{i-1}		$\delta^2 v_{i-1}$		$\delta^4 v_{i-1}$		
	$\delta v_{i-\frac{1}{2}}$		$\delta^3 v_{i-\frac{1}{2}}$		$\delta^5 v_{i-\frac{1}{2}}$	
v_i		$\delta^2 v_i$		$\delta^4 v_i$		$\delta^6 v_i$
	$\delta v_{i+\frac{1}{2}}$		$\delta^3 v_{i+\frac{1}{2}}$		$\delta^5 v_{i+\frac{1}{2}}$	
v_{i+1}		$\delta^2 v_{i+1}$		$\delta^4 v_{i+1}$		
	$\delta v_{i+1\frac{1}{2}}$		$\delta^3 v_{i+1\frac{1}{2}}$			
v_{i+2}		$\delta^2 v_{i+2}$				
	$\delta v_{i+2\frac{1}{2}}$					
v_{i+3}						

Table 10.2. Forward differences layout

v_i						
	Δv_i					
v_{i+1}		$\Delta^2 v_i$				
	Δv_{i+1}		$\Delta^3 v_i$			
v_{i+2}		$\Delta^2 v_{i+1}$		$\Delta^4 v_i$		
	Δv_{i+2}		$\Delta^3 v_{i+1}$		$\Delta^5 v_i$	
v_{i+3}		$\Delta^2 v_{i+2}$		$\Delta^4 v_{i+1}$		$\Delta^6 v_i$
	Δv_{i+3}		$\Delta^3 v_{i+2}$		$\Delta^5 v_{i+1}$	
v_{i+4}		$\Delta^2 v_{i+3}$		$\Delta^4 v_{i+2}$		
	Δv_{i+4}		$\Delta^3 v_{i+3}$			
v_{i+5}		$\Delta^2 v_{i+4}$				
	Δv_{i+5}					
v_{i+6}						

For simplicity, it is often convenient to show the layout of finite differences in tabular form, as in Tables 10.1 to 10.3, where the first columns represent the ordinates of Fig. 10.1, and the second, third and other columns represent the first, second, and higher differences, respectively.

Table 10.3. Backward differences layout

v_{i-6}						
	∇v_{i-5}					
v_{i-5}		$\nabla^2 v_{i-4}$				
	∇v_{i-4}		$\nabla^3 v_{i-3}$			
v_{i-4}		$\nabla^2 v_{i-3}$		$\nabla^4 v_{i-2}$		
	∇v_{i-3}		$\nabla^3 v_{i-2}$		$\nabla^5 v_{i-1}$	
v_{i-3}		$\nabla^2 v_{i-2}$		$\nabla^4 v_{i-1}$		$\nabla^6 v_i$
	∇v_{i-2}		$\nabla^3 v_{i-1}$		$\nabla^5 v_i$	
v_{i-2}		$\nabla^2 v_{i-1}$		$\nabla^4 v_i$		
	∇v_{i-1}		$\nabla^2 v_i$			
v_{i-1}		$\nabla^2 v_i$				
	∇v_i					
v_i						

10.3.2

The Operator E is defined by the relation,

$$Ev_i = v_{i+1}$$

Now,

$$\Delta v_i = v_{i+1} - v_i = Ev_i - v_i$$
$$= (E - 1)v_i$$

therefore,

$$\Delta = (E - 1)$$

so that,

$$\underline{E = (1 + \Delta)}$$

Furthermore, if $Ev_{i+2} = E^2 v_i$, then the backward difference,

$$\nabla v_i = v_i - v_{i-1}$$

appears as,

$$\nabla v_i = v_i - E^{-1} v_i = (1 - E^{-1}) v_i$$

i.e.

$$\nabla = 1 - \frac{1}{E}$$

or

$$E = (1 - \nabla)^{-1}$$

Similarly, it can be shown that,

$$\underline{E^n = (1 - \nabla)^{-n}}$$

Although many of these relationships will not be applied in the present text, their use in structural mechanics is invaluable.

10.4.1 EXAMPLE 10.1 SIMPLY-SUPPORTED BEAM WITH A UDL

Calculate the central bending moment and deflection for the simply-supported beam shown in Fig. 10.2, where w is a uniformly distributed load.

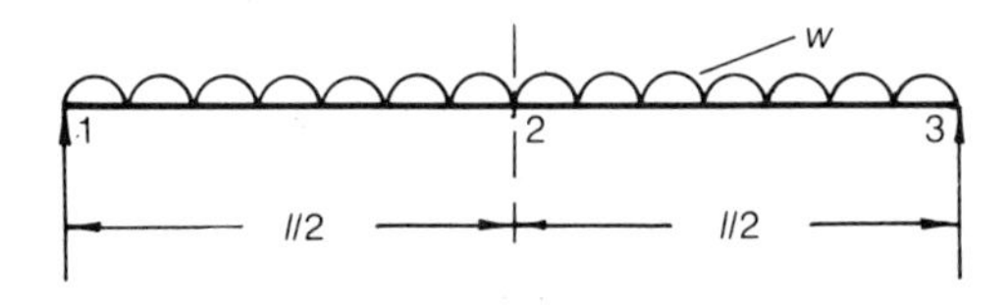

Fig. 10.2.

10.4.2

Application of (10.4) to station 2 gives the following difference equation:

$$M_1 - 2M_2 + M_3 = -w\left(\frac{l}{2}\right)^2$$

but

$$M_1 = M_3 = 0$$

therefore,

$$\underline{M_2 = \frac{wl^2}{8}}$$

The reasons why this value for bending moment was exact was because w was constant over the whole length of the beam.

To find the central deflection, apply (10.5) to station 2:

$$EI(v_i - 2v_2 + v_3) = \frac{wl^2}{8}\left(\frac{l}{2}\right)^2$$

but

$$v_i = v_3 = 0$$

therefore,

$$\underline{v_2 = -\frac{wl^4}{64EI} = -\frac{6wl^4}{384EI}}$$

The exact value is 20% less than this, and greater precision could have been obtained if the beam were divided into more sections.

10.4.3

To investigate the effect of taking a *more refined mesh* for Example 10.1, the beam will now be sub-divided into four equal sections—thus making five stations, as shown in Fig. 10.3.

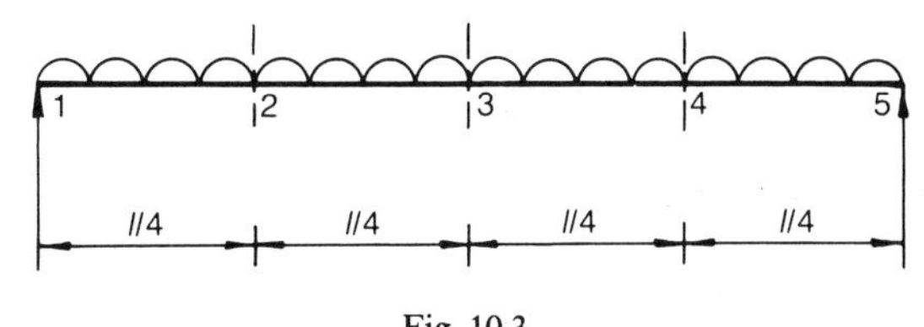

Fig. 10.3.

Application of (10.4) to station 2 gives the following:

$$M_1 - 2M_2 + M_3 = -w\left(\frac{l}{4}\right)^2$$

or

$$-2M_2 + M_3 = -\frac{wl^2}{16} \tag{10.10}$$

Similarly, on applying (10.4) to station 3,

$$M_2 - 2M_3 + M_4 = -\frac{wl^2}{16} \tag{10.11}$$

and to station 4,

$$M_3 - 2M_4 = -\frac{wl^2}{16}$$

or

$$M_4 = \frac{wl^2}{32} + \frac{M_3}{2} \tag{10.12}$$

From (10.10),

$$M_2 = \frac{wl^2}{32} + \frac{M_3}{2} \tag{10.13}$$

Substituting (10.12) and (10.13) into (10.11),

$$\underline{M_3 = \frac{wl^2}{8}}$$

Hence,

$$\underline{M_2 = \frac{3wl^2}{32}}$$

and

$$\underline{M_4 = \frac{3wl^2}{32}}$$

Application of (10.5) to station 2 gives the following:

$$EI(v_1 - 2v_2 + v_3) = \frac{3wl^2}{32}\left(\frac{l}{4}\right)^2$$

or

$$-2v_2 + v_3 = \frac{3wl^4}{512EI} \tag{10.14}$$

Similarly, for station 3,

$$v_2 - 2v_3 + v_4 = \frac{wl^4}{128EI} \tag{10.15}$$

and for station 4,

$$EI(v_3 - 2v_4 + 0) = \frac{3wl^4}{512}$$

or

$$v_3 - 2v_4 = \frac{3wl^4}{512EI} \tag{10.16}$$

From (10.14),

$$v_2 = \frac{-3wl^4}{1024EI} + \frac{v_3}{2} \tag{10.17}$$

and from (10.16),

$$v_4 = \frac{-3wl^4}{1024EI} + \frac{v_3}{2} \tag{10.18}$$

Substituting (10.17) and (10.18) into (10.15),

$$v_3 = \frac{-14wl^4}{1024EI}$$

$$= -\frac{5.25wl^4}{384EI}$$

The error of 5% represents a considerable improvement on the simpler subdivision.

10.5.1 EXAMPLE 10.2 ENCASTRÉ BEAM WITH A UDL

Calculate the central deflection for a beam, encastré at both ends and subjected to a uniformly distributed load, w, as shown in Fig. 10.4.

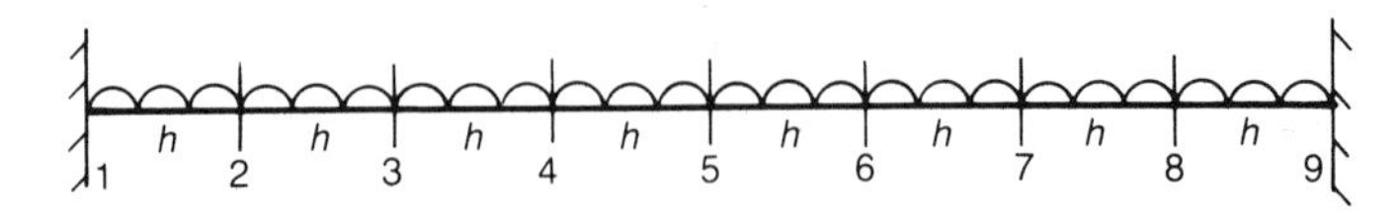

Fig. 10.4

10.5.2

Let,

$$h = l/8$$

The beam is symmetrical about station 5, so that,

$$v_3 = v_7 \quad \text{and} \quad v_4 = v_6$$

Furthermore, as the slope of the beam is zero at station 1, application of the appropriate forward difference equation at this point gives the following:

$$\Delta v_i = \frac{v_2 - v_1}{h} = 0$$

Now, as

$$v_1 = 0, \quad v_2 = 0$$

Hence, as there are only three unknowns, only three equations will be required.

Application of (10.9) to station 3 gives the following:

$$v_1 - 4v_2 + 6v_3 - 4v_4 + v_5 = -wh^4/EI$$

or,

$$6v_3 - 4v_4 + v_5 = -wh^4/EI \tag{10.19}$$

Similarly, at station 4,

$$-4v_3 + 6v_4 - 4v_5 + v_6 = -wh^4/EI$$

or

$$-4v_3 + 7v_4 - 4v_5 = -wh^4/EI \tag{10.20}$$

and at station 5,

$$v_3 - 4v_4 + 6v_5 - 4v_6 + v_7 = -wh^4/EI$$

$$2v_3 - 8v_4 + 6v_5 = -wh^4/EI \tag{10.21}$$

Multiply (10.21) by 2, to give,

$$4v_3 - 16v_4 + 12v_5 = -2wh^4/EI \tag{10.22}$$

and add (10.20) to (10.22)

$$-9v_4 + 8v_5 = -3wh^4/EI \tag{10.23}$$

Multiply (10.21) by -3, to give,

$$-6v_3 + 24v_4 - 18v_5 = 3wh^4/EI \tag{10.24}$$

and add (10.19) to (10.24)

$$20v_4 - 17v_5 = 2wh^4/EI \tag{10.25}$$

Multiply (10.23) by $\frac{20}{9}$, to give,

$$-20v_4 + 17.778v_5 = -20wh^4/3EI \tag{10.26}$$

and add (10.26) to (10.25),

$$0.778v_5 = -14wh^3/(3EI)$$

or

$$v_5 = -6.0wh^4/EI$$

The above value can be seen to underestimate the exact value by about 44%, and the reason for this is due to the assumption that $v_2 = v_8 = 0$. This simple example shows the difficulty of applying the finite difference method to boundary conditions that are more complicated than those associated with simple supports. When conditions such as these are met, it is usually necessary to take a finer mesh, particularly near the encastré ends.

10.6.1 EXAMPLE 10.3 SIMPLY-SUPPORTED BEAM WITH A CONCENTRATED LOAD

Calculate the central bending moment and deflection for the uniform beam shown in Fig. 10.5, subjected to a centrally placed concentrated load.

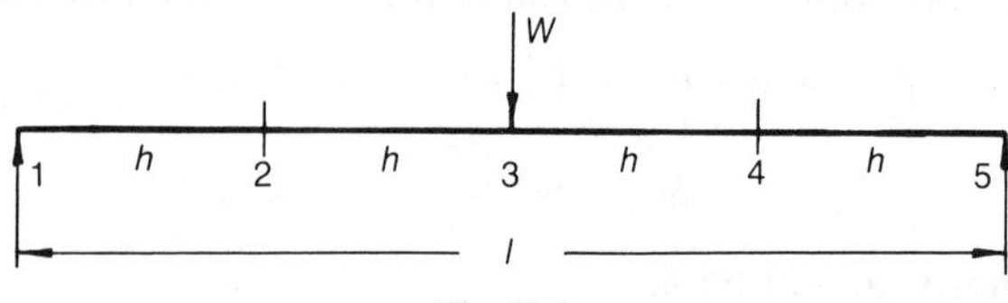

Fig. 10.5.

10.6.2

As it is not possible to apply difference equations to concentrated loads, it will be necessary to replace W with the equivalent triangular load shown in Fig. 10.6.

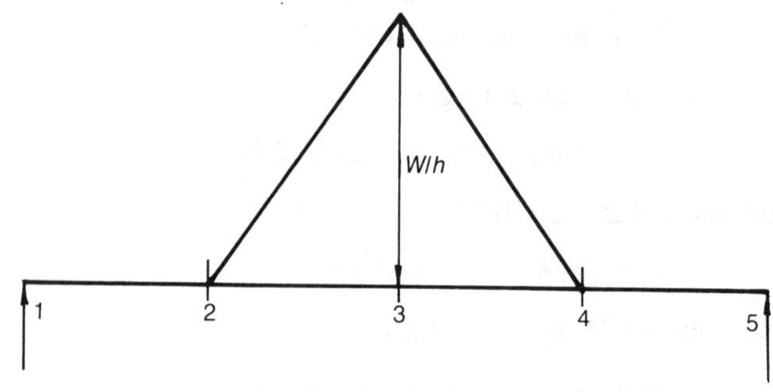

Fig. 10.6.

Furthermore, from symmetry, $M_2 = M_4$ and $v_2 = v_4$.
Now,

$$\frac{d^2M}{dx^2} = -w$$

so that, by applying (10.4) to station 2, the following is obtained.

$$M_1 - 2M_2 + M_3 = 0$$

or

$$M_2 = M_3/2$$

Similarly, for station 3,

$$M_2 - 2M_3 + M_4 = -\frac{W}{h}(h)^2$$

or

$$2M_2 - 2M_3 = -Wh$$

By substitution,

$$M_3 - 2M_3 = -Wh$$

therefore,

$$\underline{M_3 = -Wh = Wl/4}$$

and

$$\underline{M_2 = Wl/8}$$

These are exact values, due probably to the linear distribution of bending moment.

To obtain the central deflection, apply (10.5) to station 3 to give:

$$-2v_2 + v_3 = \frac{Wlh^2}{8EI} \tag{10.27}$$

and

$$+2v_2 - 2v_3 = \frac{Wlh^2}{4EI} \tag{10.28}$$

Adding (10.27) to (10.28),

$$-v_3 = \frac{3Wlh^2}{8EI}$$

or

$$\underline{v_3 = -\frac{Wl^3}{42.7EI}}$$

This is an overestimation of about 12%.

10.7.1 LONGITUDINAL STRENGTH OF SHIPS

In the shipbuilding industry, it is usually necessary to investigate the overall longitudinal strength of a ship by assuming that the ship is a "free–free" beam, where its self-weight is resisted by buoyant forces due to a system of waves. The process, therefore, is to obtain the distribution of the ship's self-weight along the length of the ship and the distribution of the resisting byoyant forces.

The naval architect then finds the difference between these two curves, which can be regarded as a distributed load w of varying intensity, as shown in Fig. 10.7(a).

Solution then follows the following process of integration:

$$M = \iint w \, \mathrm{d}x \, \mathrm{d}x \tag{10.29}$$

and

$$v = \iint \frac{M}{EI} \mathrm{d}x \, \mathrm{d}x \tag{10.30}$$

where,

M = bending moment at any distance x
v = deflection at any distance x
E = elastic modulus at any distance x
I = second moment of area at the ship's section

To solve equations (10.29) and (10.30) by explicit means is a virtually impossible task, because of the complexity of the intensity of the load w and also because the sectional second moment of area varies along the length of the ship.

Thus, the naval architect is faced with solving equations (10.29) and (10.30) by graphical integration or, alternatively by using numerical integration on a computer.

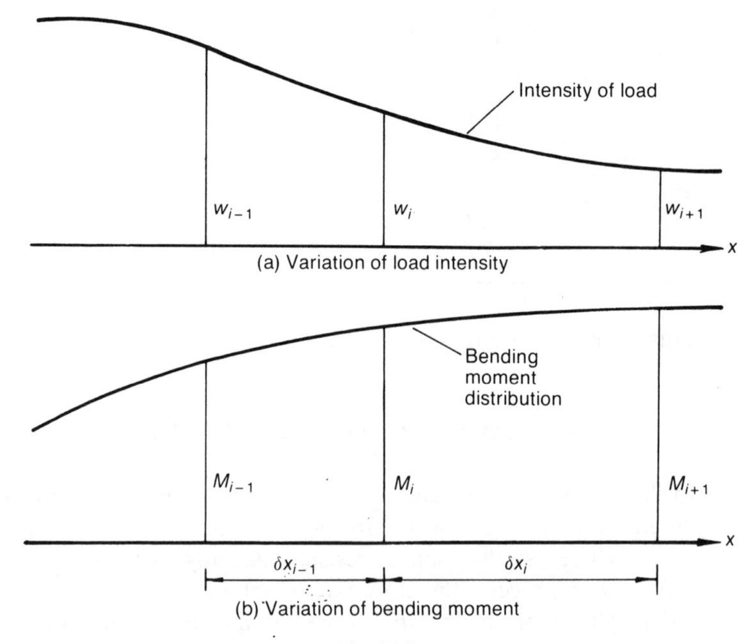

Fig. 10.7.

10.7.2

In both cases, the naval architect assumes that the self-weight of the ship is represented by a series of curves, rectangles and trapeziums, but the present author has shown that if the rectangles are replaced by trapeziums or triangles, the standard longitudinal strength of a ship can be satisfactorily solved by the finite difference method [26, 27]. For this case, the assumption that δx is constant is too restrictive, and it will now be necessary to vary it, as shown in Fig. 10.7.

From Fig. 10.7,

$$\left(\frac{\mathrm{d}^2 M}{\mathrm{d}x^2}\right)_i = \frac{\dfrac{M_{i+1} - M_i}{\delta x_i} - \dfrac{M_i - M_{i-1}}{\delta x_{i-1}}}{0.5(\delta x_{i-1} + \delta x_i)} = -w_i$$

This reduces to the form,

$$\begin{aligned}&\delta x_i M_{i-1} - (\delta x_{i-1} + \delta x_i)M_i + \delta x_{i-1} M_{i+1} \\ &\quad = -0.5\delta x_{i-1}\delta x_i(\delta x_{i-1} + \delta x_i)w_i\end{aligned} \tag{10.31}$$

In a similar manner, the expression for deflection is as follows:

$$\begin{aligned}&\delta x_i v_{i-1} - (\delta x_{i-1} + \delta x_i)v_i + \delta x_{i-1} v_{i+1} \\ &\quad = 0.5\delta x_{i-1}\delta x_i(\delta x_{i-1} + \delta x_i)M_i/(E_i I_i)\end{aligned} \tag{10.32}$$

10.7.3

In matrix form, equations (10.31) and (10.32) appear as follows:

$$\begin{bmatrix} A_{11} & A_{12} & 0 & 0 & 0 & \cdots & 0 \\ A_{21} & A_{22} & A_{23} & 0 & 0 & \cdots & 0 \\ 0 & A_{32} & A_{33} & A_{34} & 0 & \cdots & 0 \\ 0 & & \ddots & \ddots & \ddots & & \\ \vdots & & & & & & A_{n-1,n} \\ 0 & & & & & A_{n,n-1} & A_{nn} \end{bmatrix} \begin{Bmatrix} x_1 \\ x_2 \\ x_3 \\ \vdots \\ x_n \end{Bmatrix} = \begin{Bmatrix} C_1 \\ C_2 \\ C_3 \\ \vdots \\ C_n \end{Bmatrix} \tag{10.33}$$

where, for (10.31),

$$x_1, x_2, x_3, \text{etc.} = M_1, M_2, M_3, \text{etc.}$$

C_i is a function of δx_{i-1}, δx_i and w_i, and for (10.32),

$$x_1, x_2, x_3, \text{etc.} = v_1, v_2, v_3, \text{etc.}$$

C_i is a function of δx_{i-1}, δx_i and M_i.

For both equations, (10.31) and (10.32),

$$[A] \text{ is a function of } \delta x_{i-1} \text{ and } \delta x_i$$

The matrix $[A]$ of equation (10.33) can be seen to be of tridiagonal form, and an efficient method of storing it is in the rectangular form of equation (10.34).

$$\begin{bmatrix} 0 & A_{11} & A_{12} \\ A_{21} & A_{22} & A_{23} \\ A_{32} & A_{33} & A_{34} \\ A_{43} & A_{44} & A_{45} \\ | & | & | \\ A_{n,n-1} & A_{n,n} & 0 \end{bmatrix} \tag{10.34}$$

Solution of (10.34) can then be carried out by elimination, as shown below.

Suppose (10.33) appears as the following set of simultaneous equations:

$$\begin{aligned} A_{11}x_1 + A_{12}x_2 \qquad\qquad &= C_1 \\ A_{21}x_1 + A_{22}x_2 + A_{23}x_3 &= C_2 \\ A_{32}x_2 + A_{33}x_3 + A_{34}x_4 &= C_3 \\ | \qquad | \qquad | \qquad & \\ | \qquad | \qquad | \qquad & \\ A_{n,n-1}x_{n-1} + A_{n,n}x_n \qquad &= C_n \end{aligned} \tag{10.35}$$

Equation (10.35) can be reduced to triangular form, as described in reference [20], by eliminating all the elements of [A] below the leading diagonal, as in (10.36).

$$\begin{aligned} a_{11}x_1 + a_{12}x_2 &= b_1 \\ a_{22}x_2 + a_{23}x_3 &= b_2 \\ | \qquad\qquad & | \\ | \qquad\qquad & | \\ a_{n,n}x_n \qquad &= b_n \end{aligned} \tag{10.36}$$

where,

$$\begin{aligned} a_{11} &= A_{11} & a_{12} &= A_{12} \\ a_{22} &= A_{22} - A_{21}A_{12}/A_{11} & a_{23} &= A_{23} \\ a_{i,i} &= A_{i,i} - A_{i,i-1}A_{i-1,i}/a_{i-1,i-1} & a_{i,i+1} &= A_{i,i+1} \\ | & & & \\ | & & & \\ a_{n,n} &= A_{n.n} - A_{n,n-1}A_{n-1,n}/a_{n-1,n-1} & & \end{aligned} \tag{10.37}$$

$$\begin{aligned} b_1 &= C_1 \\ b_2 &= C_2 - A_{21}b_1/a_{11} \\ | & \\ | & \\ b_n &= C_n - A_{n,n-1}b_{n-1}/a_{n-1,n-1} \end{aligned} \tag{10.38}$$

Hence,

$$\begin{aligned} x_n &= b_n/a_{n,n} \\ x_{n-1} &= b_{n-1}/a_{n-1,n-1} - A_{n-1,n}x_n/a_{n-1,n-1} \\ | & \\ | & \\ x_1 &= b_1/a_{11} - A_{12}x_2/a_{11} \end{aligned} \tag{10.39}$$

This process can be seen to be a simple and efficient way of solving these equations, and the method can be readily adapted to a computer solution.

10.7.4

The method was applied to two ships by Ross and Assheton. These two ships were originally analysed by Turner *et al.*, where they were referred to as ships A and B.

Figs. 10.8 and 10.9 show the weight and buoyancy curves of ships A and B, respectively, and Figs. 10.10 and 10.11 show the approximate weight curves used by Ross and Assheton for ships A and B respectively.

Figs. 10.12 and 10.13 show the resulting bending moment distributions for ships A and B by the standard calculation of Turner *et al.* [28] and also by the finite difference solution of Ross and Assheton [27].

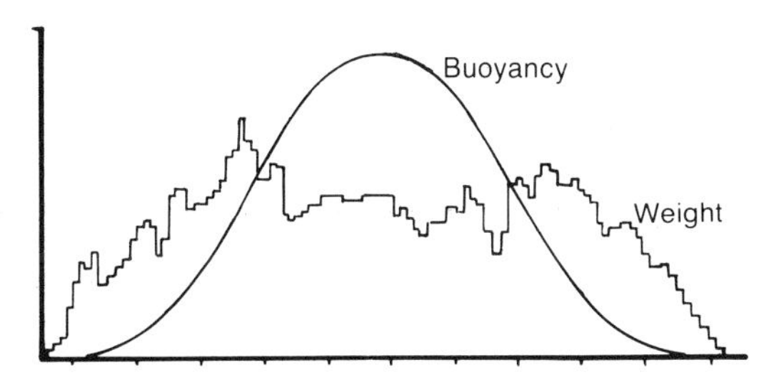

Fig. 10.8. Ship A: weight curve and buoyancy curve

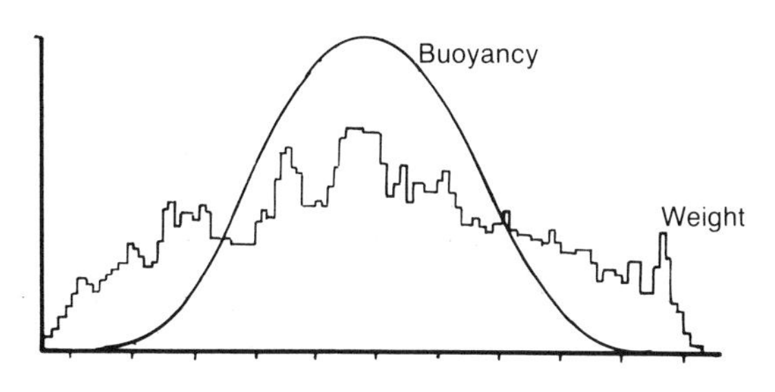

Fig. 10.9. Ship B: weight curve and buoyancy curve

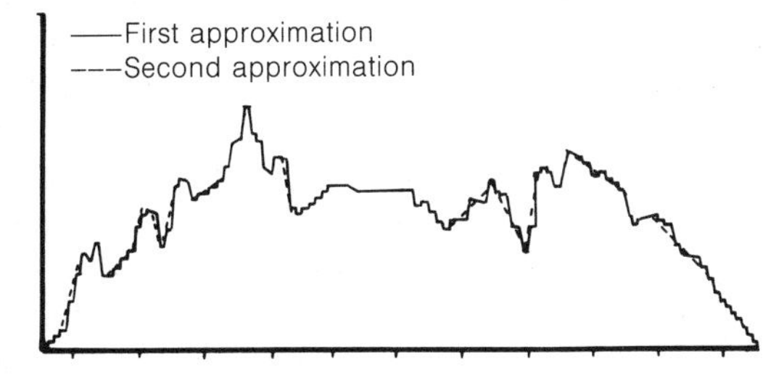

Fig. 10.10. Ship A: weight curve, first and second approximations

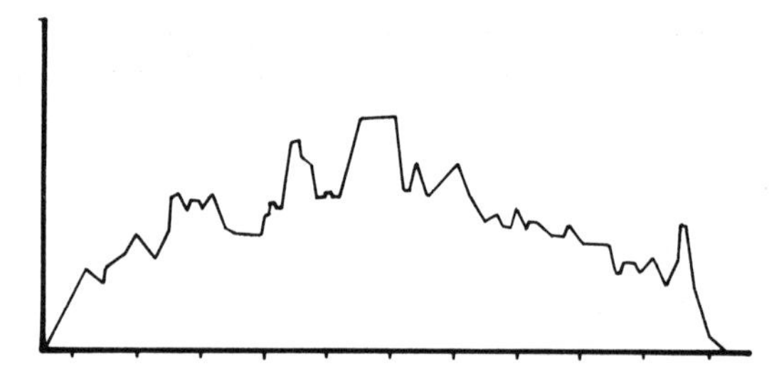

Fig. 10.11. Ship B: weight curve, first approximation

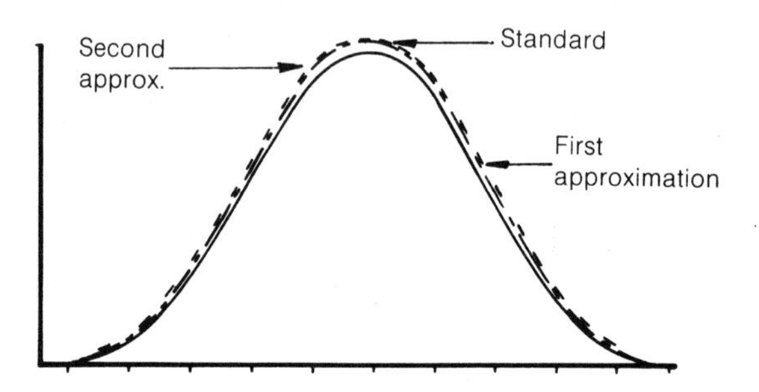

Fig. 10.12. Ship A: bending moment curves

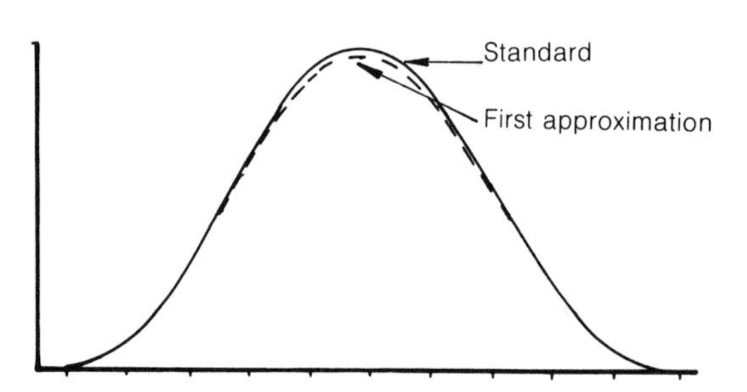

Fig. 10.13. Ship B: bending moment curves

The first approximation for the weight curve of Fig. 10.10 led to 210 simultaneous equations, and the time taken to solve these equations on an Elliot 803 computer was only about five minutes. Similarly, the approximate weight curve of Fig. 10.11 led to 108 simultaneous equations, and solution of these on the same computer took only about three minutes.

Furthermore, from inspection of the bending moment curves of Figs. 10.12 and 10.13, it is evident that the method is sufficiently precise.

These findings clearly indicate that the method is suitable for solution on a microcomputer. The weight curve of Fig. 10.11 also appears to indicate that the standard weight curves of ships can be simplified.

The method of finite differences can also be applied to a number of other problems in structures, and some of these applications are now discussed.

10.8.1 SMALL DEFLECTIONS OF PLATES

The differential equation for a flat rectangular plate in bending [24] is:

$$\frac{\partial^4 w}{\partial x^4}+\frac{2\partial^4 w}{\partial x^2 \partial y^2}+\frac{\partial^4 w}{\partial y^4}=\frac{p}{D} \tag{10.40}$$

where,

w = lateral deflection
p = lateral pressure
D = flexural rigidity
$= Et^3/12(1-\nu^2)$
E = Young's modulus
ν = Poisson's ratio
t = thickness of plate
x, y = rectangular co-ordinates

Solution of (10.40) for plates of complex shape and boundary conditions is virtually impossible. If, however, (10.40) is reduced to a difference form and applied to various points on a plate, the resulting simultaneous equations can be solved by computer.

For example, if the mesh of Fig. 10.14 is taken, then (10.40) can be reduced to the following difference equation:

$$\begin{aligned}&(w_{11}-4w_3+6w_0-4w_1+w_9)+2(w_6-2w_2+w_5-2w_3\\&\quad+4w_0-2w_1+w_7-2w_4+w_8)\\&\quad+(w_{12}-4w_4+6w_0-4w_2+w_{10})=ph^4/D\end{aligned}$$

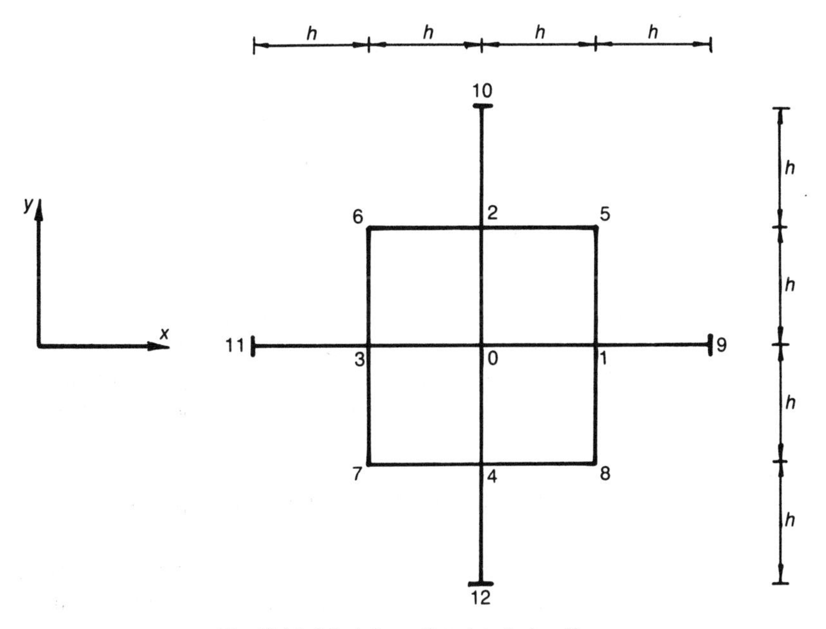

Fig. 10.14. Mesh for a flat plate in bending.

or

$$\frac{ph^4}{D} = 20w_0 - 8(w_1 + w_2 + w_3 + w_4)$$
$$+ 2(w_5 + w_6 + w_7 + w_8)$$
$$+ (w_9 + w_{10} + w_{11} + w_{12}) \tag{10.41}$$

It is evident that application of (10.41) to a practical problem necessitates the use of a computer, and because of this, no attempt will be made to make a detailed calculation in the current text. Solution, of course, follows a similar process to that for the beams, and as the equation cannot be applied to grid points less than two from the boundary, it will be necessary to obtain any further equations by applying boundary conditions using either forward or backward difference equations. Some such boundary conditions are for slope, and others for bending moment.

Once the deflections are known, the bending moment at station 0, in the x direction, can be calculated by transforming the differential equation for bending,

$$M^x = -D\left(\frac{\partial^2 w}{\partial x^2} + v\frac{\partial^2 w}{\partial y^2}\right)$$

to the difference equation,

$$M_0^x = -D[w_3 - 2w_0 + w_1 + v(w_4 - 2w_0 + w_2)]$$

or

$$\underline{M_0^x = -D[-2w_0(1 + v) + w_1 + w_3 + v(w_2 + w_4)]}$$

Similarly, for the relationship,

$$M^y = -D\left(\frac{\partial^2 w}{\partial y^2} + v\frac{\partial^2 w}{\partial x^2}\right)$$

the following difference equation is obtained:

$$M_0^y = -D[w_4 - 2w_0 + w_2 + v(w_3 - 2w_0 + w_1)]$$

or

$$\underline{M_0^y = -D[-2w_0(1 + v) + w_2 + w_4 + v(w_1 + w_3)]}$$

where,

M^x = bending moment in the x direction
M^y = bending moment in the y direction
M_0^x = bending moment at station 0 in the x direction
M_0^y = bending moment at station 0 in the y direction

10.9.1 TORSION OF NON-CIRCULAR SECTIONS

Owing to the unusual shapes of some cross-sections of shafts, the use of finite differences is extremely useful for analysing such structures in torsion.

The differential equation for the torsion of a non-circular section (see Chapter 5) is as follows:

$$\frac{\partial^2 \chi}{\partial x^2}+\frac{\partial^2 \chi}{\partial y^2}=-2 \tag{10.42}$$

where,

$\chi=$ a shear stress function.

If the mesh of Fig. 10.15 is taken, then (10.42) appears as,

$$(\chi_1-2\chi_0+\chi_3)+(\chi_4-2\chi_0+\chi_2)=-2h^2$$

or

$$-4\chi_0+\chi_1+\chi_2+\chi_3+\chi_4=-2h^2 \tag{10.43}$$

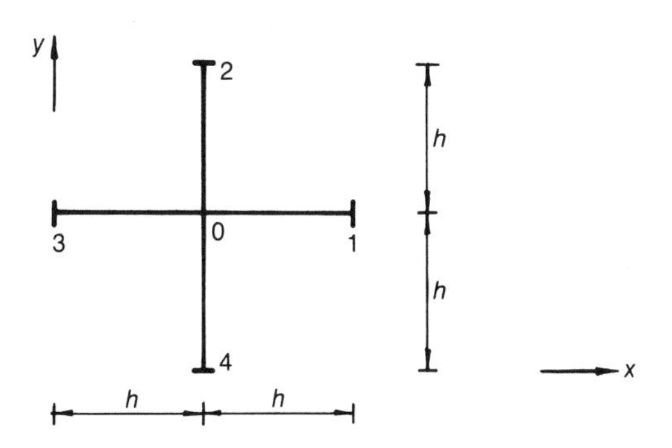

Fig. 10.15. Grid for the torsion of a non-circular section.

It is evident that (10.43) can be applied to grid points up to one from the boundary, and as χ is usually zero at the edge, an adequate number of simultaneous equations can be obtained.

The shear stresses are given by:

$$\tau_{xz}=G\theta\frac{\partial \chi}{\partial y}$$

and

$$\tau_{yz}=-G\theta\frac{\partial \chi}{\partial x}$$

which can usually be obtained by "approximate" central differences. The torsional constant,

$$J=2\iint \chi \,\mathrm{d}x\,\mathrm{d}y$$

can be obtained by numerical integration.

More advanced problems, such as shells {33} and other structures {34}, can also be analysed by this method.

10.10.1 EXMPLE 10.4 TORSION OF A RECTANGULAR SECTION

Calculate the torsional constant for the long thin rectangular section shown in Fig. 10.16, where $b/t > 5$, using the differential equation

$$\frac{\partial^2 \chi}{\partial y^2} = -2$$

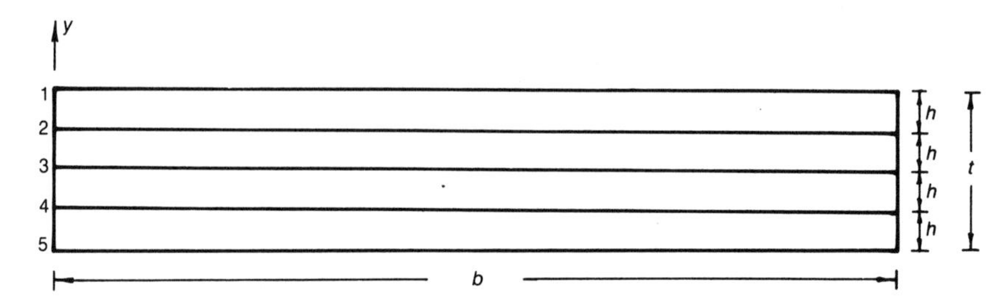

Fig. 10.16. Cross-section of a shaft.

10.10.2

Applying the difference equation,

$$\chi_{i-1} - 2\chi_i + \chi_{i+1} = -2h^2$$

to stations 2, 3 and 4, the following are obtained:

$$2\chi_2 - \chi_3 = \frac{t^2}{8} \quad (10.44)$$

$$\chi_2 - 2\chi_3 + \chi_4 = -t^2/8 \quad (10.45)$$

$$\chi_3 - 2\chi_4 = -t^2/8 \quad (10.46)$$

From (10.44) to (10.46),

$$\chi_2 = \chi_4 = 3t^2/16$$

and

$$\chi_3 = t^2/4$$

Now, $J = 2 \times$ volume under function,

$$= 2 \times b \times \frac{t}{8}\left(0 + 2 \times \tfrac{3}{16}t^2 + 2\frac{t^2}{4} + 2 \times \tfrac{3}{16}t^2 + 0\right)$$

therefore,

$$\underline{J = \frac{bt^3}{3.2}}$$

which is only 6.7% less than the "exact" solution by membrane analogy for long thin rectangular sections. Better precision could have been obtained by sub-dividing the section into a finer mesh.

EXAMPLES FOR PRACTICE 10

1. The differential equation for the axisymmetric deformation of a thin-walled circular cylinder under uniform external pressure is given by,

$$\frac{d^4w}{dx^4} + k_1\frac{d^2w}{dx^2} + k_2w = k_3$$

where

$k_1 = v/a^2 + 6(1 - v^2)ap/(Et^3)$
$k_2 = 12(1 - v^2)/(t^2a^2)$
$k_3 = 12(1 - v^2)(1 - 0.5v)p/(Et^3)$
w = radial deflection (positive inwards)
a = mean radius of cylinder
p = pressure (external positive)
E = elastic modulus
v = Poisson's ratio
t = shell thickness

Determine the difference equation at the ith station using central differences

$$\{w_{i-2} + w_{i-1}(-4 + k_1(\delta x)^2) + w_i(6 - 2k_1(\delta x)^2 + k_2(\delta x)^4) + w_{i+1}(-4 + k_1(\delta x)^2) + w_{i+2} - k_3(\delta x)^4 = 0\}$$

2. Calculate the bending moment at station 2, for the simply-supported beam shown in Fig. Q.10.2, by the method of finite differences, where w equals the maximum intensity of load.

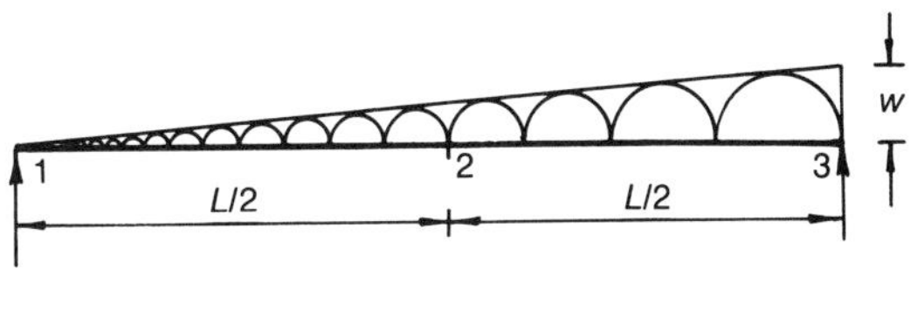

Fig. Q.10.2.

$\{wL^2/16\}$

3. Calculate the bending moments at stations 2 and 3, for the simply-supported beam shown in Fig. Q.10.3, by the method of finite differences, where w is the maximum intensity of load.

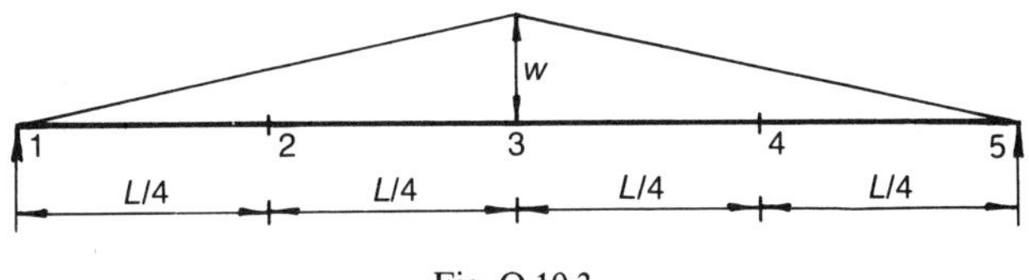

Fig. Q.10.3.

$\{wL^2/16; 3wL^2/32\}$

4. Calculate the bending moments at stations 2, 3 and 4, for the beam shown in Fig. Q.10.4, by the method of finite differences.

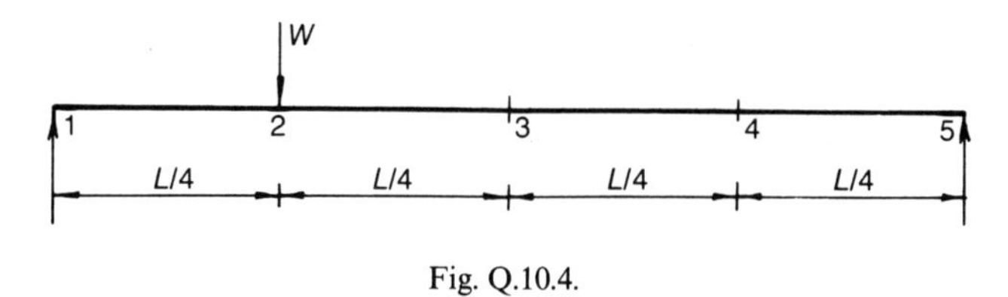

Fig. Q.10.4.

$\{3WL/6; WL/8; WL/16\}$

5. Calculate the deflections at stations 2, 3 and 4, for the beam shown in Fig. Q.10.4, using the results of Example 4.

   $\{-7WL^3/512EI; -WL^3/64EI; -5WL^3/512EI\}$

11

The Matrix Displacement Method

11.1.1 THE FINITE ELEMENT METHOD

The finite element method is one of the most powerful methods of solving partial differential equations, particularly if these equations apply over complex shapes. The method consists of sub-dividing the complex shape into several elements of simpler shape, each of which is more suitable for mathematical analysis. The process then, as far as structural analysis is concerned, is to obtain the elemental stiffnesses of these simpler shapes and then, by considering equilibrium and compatibility at the inter-element boundaries, to assemble all the elements, so that a mathematical model of the entire structure is obtained.

Hence, owing to the application of loads on this mathematical model, the "deflections" at various points of the structure can be obtained through the solution of the resulting simultaneous equations. Once these "deflections" are known, the stresses in the structure can be determined through Hookean elasticity.

Each finite element is described by "nodes" or "nodal points", and the stiffnesses, displacements, loads, etc. are all related to these nodes. Finite elements vary in shape, depending on the systems they have to describe, and some typical finite elements are shown in Fig. 11.1.

Now, the finite element method is a vast topic, covering problems in structural mechanics, fluid flow, heat transfer, acoustics, etc., and because of this, it is beyond the scope of the present book. However, because the finite element method is based on the matrix displacement method, a brief description of the latter will be given in the present chapter.

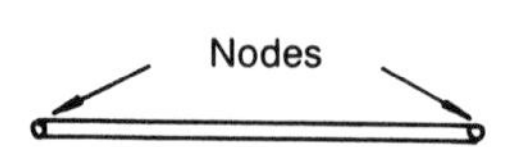

(a) One-dimensional rod element, with end nodes

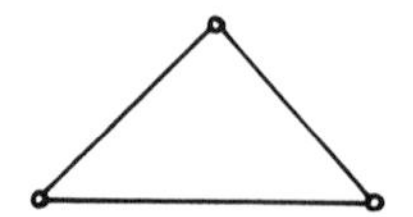

(b) Two-dimensional triangular element with corner nodes

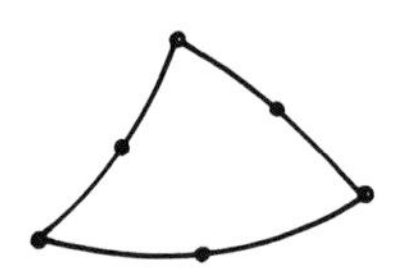

(c) Curved triangular plate, with additional 'mid-side' nodes

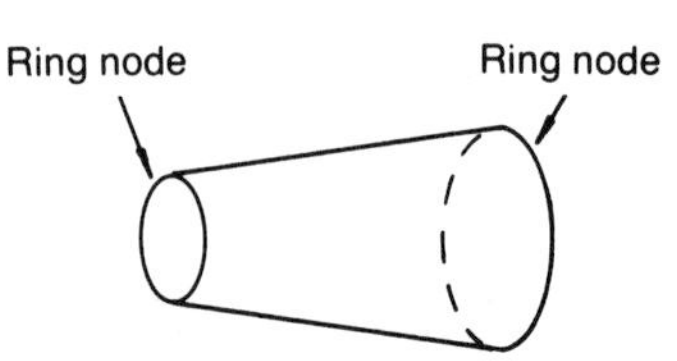

(d) Truncated conical element with ring nodes

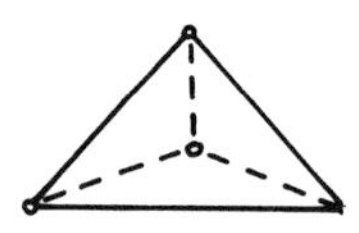

(e) Solid tetrahedral element with corner nodes

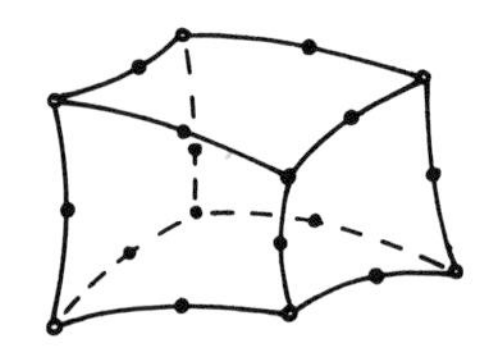

(f) Twenty-node curved brick element

Fig. 11.1. Some typical finite elements.

11.2.1 THE MATRIX DISPLACEMENT METHOD

The matrix displacement method is also known as the stiffness method. It is based on obtaining the stiffness of the entire structure by assembling together all the individual stiffnesses of each member or element of the structure. When this is done, the mathematical model of the structure is subjected to the externally applied loads, and by solving the resulting simultaneous equations, the nodal deflections are determined. Once the nodal deflections are known, the stresses in the structure can be obtained through Hookean elasticity.

Now, in the small deflection theory of elasticity, a structure can be said to behave like a complex spring, where each member or element of the structure can be regarded as an individual spring, with a different type and value of stiffness.

11.2.2

Thus, to introduce the method, let us consider the single elemental spring of Fig. 11.2.

Fig. 11.2. Spring element.

Let,

k = stiffness of spring
= slope of load–deflection relationship for the spring
X_1 = axial force at node 1
X_2 = axial force at node 2
u_1 = nodal displacement at node 1 in the direction of X_1
u_2 = nodal displacement at node 2 in the direction of X_2

1 and 2 in Fig. 11.2 are known as the nodes or nodal points.

From Hooke's law,

$$X_1 = k(u_1 - u_2) \tag{11.1}$$

and from considerations of equilibrium,

$$\begin{aligned} X_2 &= -X_1 \\ &= k(u_2 - u_1) \end{aligned} \tag{11.2}$$

If equations (11.1) and (11.2) are put into matrix form, they appear as shown in equation (11.3):

$$\begin{Bmatrix} X_1 \\ X_2 \end{Bmatrix} = \begin{bmatrix} k & -k \\ -k & k \end{bmatrix} \begin{Bmatrix} u_1 \\ u_2 \end{Bmatrix} \tag{11.3}$$

or,

$$\{P_i\} = [k]\{u_i\}$$

where,

$\{P_i\}$ = a vector of elemental nodal forces
$\{u_i\}$ = a vector of elemental nodal displacements

11.3.1 THE STRUCTURAL STIFFNESS MATRIX [*K*]

Consider a simple structure composed of two elemental springs, as shown in Fig. 11.3, where,

k_a = stiffness of spring 1–2
k_b = stiffness of spring 2–3

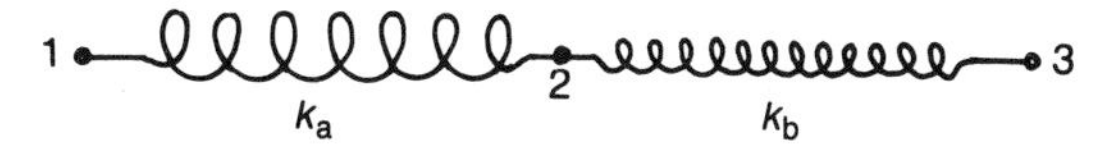

Fig. 11.3. Simple structure.

11.3.2 To determine [*K*]

[*K*] will be of order 3 × 3 because there are three degrees of freedom, namely u_1, u_2 and u_3, and it can be obtained as follows:

Element 1–2

From (11.3), the elemental stiffness matrix for spring 1–2 is given by (11.4), where the components of stiffness are related to the nodal displacements, u_1 and u_2.

$$\begin{array}{c} \\ [k_{1-2}] = \end{array} \begin{array}{c} \begin{array}{cc} u_1 & u_2 \end{array} \\ \left[\begin{array}{cc} k_a & -k_a \\ -k_a & k_a \end{array}\right] \end{array} \begin{array}{c} \\ u_1 \\ u_2 \end{array} \tag{11.4}$$

Similarly for *element 2–3*,

$$\begin{array}{c} \\ [k_{2-3}] = \end{array} \begin{array}{c} \begin{array}{cc} u_2 & u_3 \end{array} \\ \left[\begin{array}{cc} k_b & -k_b \\ -k_b & k_b \end{array}\right] \end{array} \begin{array}{c} \\ u_2 \\ u_3 \end{array} \tag{11.5}$$

Superimposing the components of stiffness corresponding to the displacements, u_1, u_2 and u_3, from (11.4) & (11.5), the stiffness matrix for the entire structure is obtained, as shown by (11.6):

$$\begin{array}{c} \\ [K] = \end{array} \begin{array}{c} \begin{array}{ccc} u_1 & u_2 & u_3 \end{array} \\ \left[\begin{array}{ccc} k_a & -k_a & 0 \\ k_a & (k_a+k_b) & -k_b \\ 0 & -k_b & k_b \end{array}\right] \end{array} \begin{array}{c} \\ u_1 \\ u_2 \\ u_3 \end{array} \tag{11.6}$$

11.3.3 Method of Solution

The above matrix is singular because free body displacements have been allowed, i.e. it is necessary to apply boundary conditions. If the system is fixed at node 3, such that the deflection of 3 is zero, then the equations become:

$$\begin{Bmatrix} Q_1 \\ Q_2 \\ R \end{Bmatrix} = \left[\begin{array}{cc|c} k_a & -k_a & 0 \\ -k_a & (k_a+k_b) & -k_b \\ \hline 0 & -k_b & k_b \end{array}\right] \begin{Bmatrix} u_1 \\ u_2 \\ 0 \end{Bmatrix}$$

or,

$$\begin{Bmatrix} q_F \\ \hline R \end{Bmatrix} = \left[\begin{array}{c|c} K_{11} & K_{12} \\ \hline K_{21} & K_{22} \end{array}\right] \begin{Bmatrix} u_F \\ 0 \end{Bmatrix}$$

i.e. the nodal displacements $\{u_F\}$ are given by:

$$\{u_F\} = [K_{11}]^{-1}\{q_F\} \tag{11.7}$$

and the reactions $\{R\}$ are given by

$$\{R\} = [K_{21}]\{u_F\} \tag{11.8}$$

where,

$\{q_F\}$ = a vector of externally applied loads, corresponding to the free displacements

$[K_{11}]$ = that part of the structural stiffness matrix, corresponding to the free displacements

$$= \begin{bmatrix} k_a & -k_a \\ -k_a & (k_a+k_b) \end{bmatrix}$$

$$\{u_F\} = \text{a vector of free displacements}$$
$$= \begin{Bmatrix} u_1 \\ u_2 \end{Bmatrix}$$

For large stiffness matrices of a banded form, $[K_{11}]$ is, in general, not inverted, and solution is carried out by Gaussian elimination or by Choleski's method.

11.4.1 ELEMENTAL STIFFNESS MATRIX FOR A PLANE ROD

(A rod is defined as a member of a framework which resists its load axially, e.g. a member of a pin-jointed truss.)

Under an axial load X, a rod length l and uniform cross-sectional area A will deflect a distance

$$u = Xl/AE$$

Consider the one-dimensional rod shown in Fig. 11.4.

$X_1, u_1 \rightarrow$ 1 —— 2 $\rightarrow X_2, u_2 \rightarrow x$

Fig. 11.4

$$X_1 = \frac{AE}{l}(u_1 - u_2) = \text{axial force at node 1}$$

$$X_2 = \frac{AE}{l}(u_2 - u_1) = \text{axial force at node 2}$$

or in matrix form:

$$\begin{Bmatrix} X_1 \\ X_2 \end{Bmatrix} = \frac{AE}{l}\begin{bmatrix} 1 & -1 \\ -1 & 1 \end{bmatrix}\begin{Bmatrix} u_1 \\ u_2 \end{Bmatrix}$$

The above can be seen to be of similar form as (11.3). Hence, the elemental stiffness matrix for a plane rod is given by:

$$[k] = \frac{AE}{l}\begin{bmatrix} 1 & -1 \\ -1 & 1 \end{bmatrix} \tag{11.9}$$

Equation (11.9) is the *elemental stiffness matrix for a rod in local* co-ordinates, but in practice it is more useful to obtain the elemental stiffness matrix in global co-ordinates.

Let, Ox^0 and Oy^0 be the global axes and Ox and Oy the local axes, as shown in Fig. 11.5.

In global co-ordinates, both u and v displacements are important; hence, (11.9) must be written as follows.

$$[k] = \frac{AE}{l}\begin{array}{c} \begin{matrix} u_1 & v_1 & u_2 & v_2 \end{matrix} \\ \begin{bmatrix} 1 & 0 & -1 & 0 \\ 0 & 0 & 0 & 0 \\ -1 & 0 & 1 & 0 \\ 0 & 0 & 0 & 0 \end{bmatrix} \end{array} \begin{matrix} u_1 \\ v_1 \\ u_2 \\ v_2 \end{matrix} \tag{11.10}$$

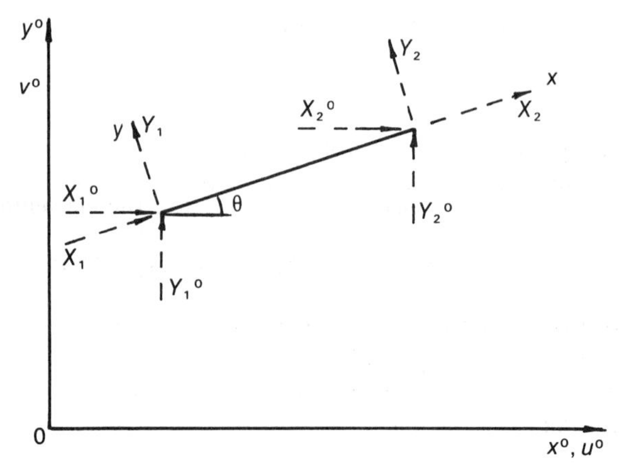

Fig. 11.5. Local and global axes.

From Fig. 11.5, it can be seen that at node 2:

$$X_2 = X_2^0 \cos\theta + Y_2^0 \sin\theta$$
$$Y_2 = -X_2^0 \sin\theta + Y_2^0 \cos\theta \tag{11.11}$$

Similar expressions apply to node 1. Hence, in matrix form:

$$\begin{Bmatrix} X_1 \\ Y_1 \\ X_2 \\ Y_2 \end{Bmatrix} = \left[\begin{array}{cc|cc} \cos\theta & \sin\theta & 0 & 0 \\ -\sin\theta & \cos\theta & 0 & 0 \\ \hline 0 & 0 & \cos\theta & \sin\theta \\ 0 & 0 & -\sin\theta & \cos\theta \end{array}\right] \begin{Bmatrix} X_1^0 \\ Y_1^0 \\ X_2^0 \\ Y_2^0 \end{Bmatrix}$$

i.e.

$$\{P\} = [\Xi]\{P^0\}$$

where,

$[\Xi]$ = a matrix of directional cosines

$\{P^0\}$ = a vector of elemental nodal forces in global co-ordinates

Now, $[\Xi]$ can be seen to be orthogonal, as $\cos^2\theta + \sin^2\theta = 1$, $\cos\theta . -\sin\theta + \sin\theta . \cos\theta = 0$, etc.

Hence,

$$[\Xi]^{-1} = [\Xi]^{\mathrm{T}}$$

therefore

$$\{P^0\} = [\Xi]^{\mathrm{T}}\{P\}$$

Similarly,

$$\{u\} = [\Xi]\{u^0\} \tag{11.12}$$

Now,

$$\{P\} = [k]\{u\}$$

therefore

$$[\Xi]\{P^0\} = [k][\Xi]\{u^0\}$$
$$\{P^0\} = [\Xi]^{\mathrm{T}}[k][\Xi]\{u^0\}$$

hence,

$$[k^0] = [\Xi]^T [k] [\Xi] \tag{11.13}$$

Similarly,

$$[K^0] = [\Xi]^T [K] [\Xi] \tag{11.14}$$

Hence, from equations (11.10) and (11.13), the *elemental stiffness matrix for a rod in global co-ordinates* is given by (11.15):

$$[k^0] = \frac{AE}{l} \begin{array}{c} \begin{array}{cccc} u_1^0 & v_1^0 & u_2^0 & v_2^0 \end{array} \\ \left[\begin{array}{cccc} C^2 & CS & -C^2 & -CS \\ CS & S^2 & -CS & -S^2 \\ -C^2 & -CS & C^2 & CS \\ -CS & -S^2 & CS & S^2 \end{array}\right] \end{array} \begin{array}{c} u_1^0 \\ v_1^0 \\ u_2^0 \\ v_2^0 \end{array} \tag{11.15}$$

where,

$$C = \cos\theta$$
$$S = \sin\theta$$

11.5.1 EXAMPLE 11.1 PLANE PIN-JOINTED TRUSS

Determine the forces in the members of the plane pin-jointed truss of Fig. 11.6. It may be assumed that AE = a constant for all members.

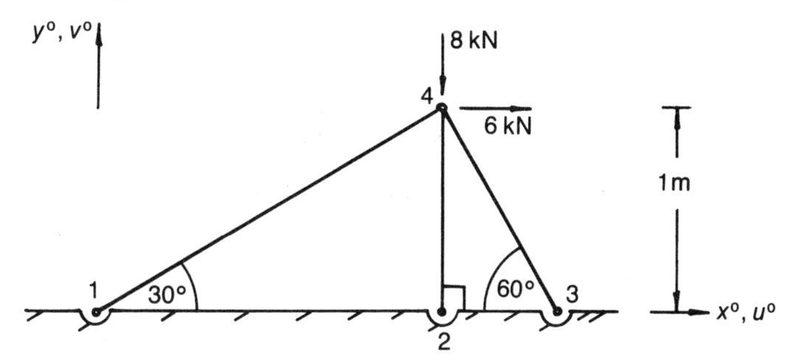

Fig. 11.6. Plane pin-jointed truss.

11.5.2 Member 1–4

$$\theta = 30^\circ \quad C = 0.866 \quad S = 0.5 \quad l = 2\,\text{m}$$

The member points in the direction from node 1 to node 4 are as shown by Fig. 11.7.

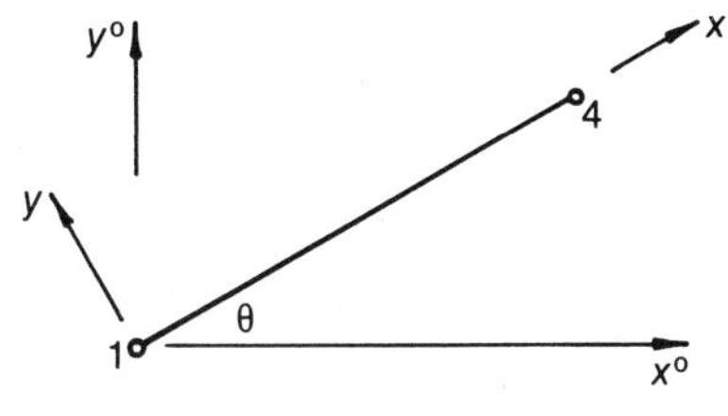

Fig. 11.7. Member 1–4.

Now as the member is firmly pinned at node 1, it is only necessary to consider the components of the stiffness matrix corresponding to the free displacements u_4^0 and v_4^0. Hence, from (11.15):

$$[k_{1\text{-}4}^0] = \frac{AE}{2}\begin{bmatrix} 0.75 & 0.433 \\ 0.433 & 0.25 \end{bmatrix}\begin{matrix} u_4^0 \\ v_4^0 \end{matrix} \quad \text{(columns } u_4^0,\ v_4^0\text{)}$$

$$= AE\begin{bmatrix} 0.375 & 0.216 \\ 0.216 & 0.125 \end{bmatrix} \tag{11.16}$$

11.5.3 Member 2–4

$\theta = 90^\circ \quad C = 0 \quad S = 1 \quad l = 1\,\text{m}$

The member points in the direction from node 2 to node 4 are as shown by Fig. 11.8.

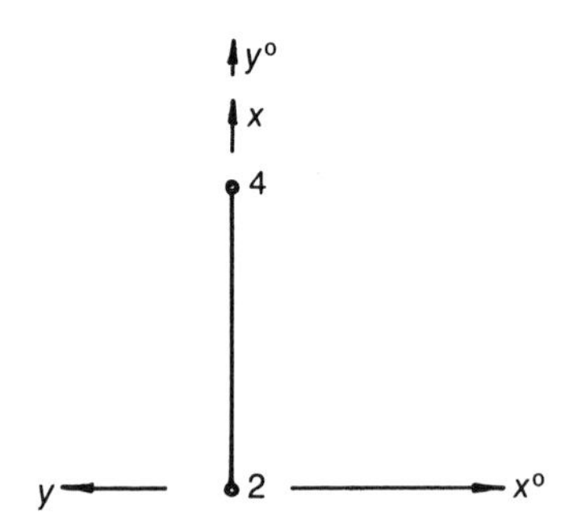

Fig. 11.8. Member 2–4.

$$[k_{2\text{-}4}^0] = \frac{AE}{1}\begin{bmatrix} 0 & 0 \\ 0 & 1 \end{bmatrix}\begin{matrix} u_4^0 \\ v_4^0 \end{matrix} \quad \text{(columns } u_4^0,\ v_4^0\text{)} \tag{11.17}$$

11.5.4 Member 4–3

$\theta = 120^\circ \quad C = -0.5 \quad S = 0.866 \quad l = 1.155\,\text{m}$

The member points in the direction from node 4 to node 3 are as shown by Fig. 11.9.

$$[k_{4\text{-}3}^0] = \frac{AE}{1.155}\begin{bmatrix} 0.25 & -0.433 \\ -0.433 & 0.75 \end{bmatrix}\begin{matrix} u_4^0 \\ v_4^0 \end{matrix} \quad \text{(columns } u_4^0,\ v_4^0\text{)}$$

$$= AE\begin{bmatrix} 0.216 & -0.375 \\ -0.375 & 0.649 \end{bmatrix} \tag{11.18}$$

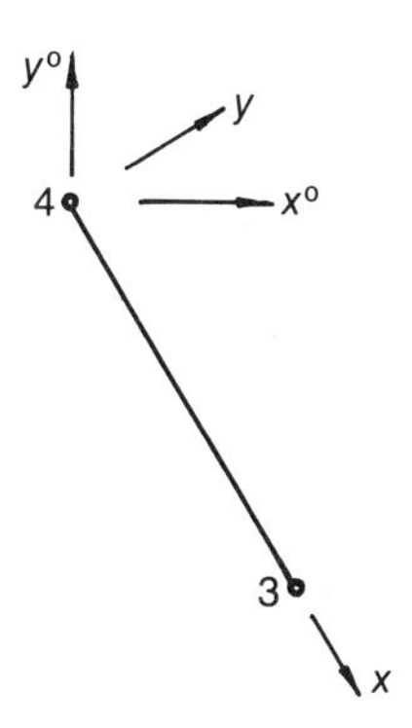

Fig. 11.9. Member 4–3.

11.5.5

From (11.16) to (11.18), the *structural stiffness matrix* $[K_{11}]$ corresponding to the *free displacements* =

$$[K_{11}] = AE\begin{bmatrix} (0.375+0+0.216) & (0.216+0-0.375) \\ (0.216+0-0.375) & (0.125+1+0.649) \end{bmatrix}\begin{matrix} u_4^0 \\ v_4^0 \end{matrix} \quad (\text{columns } u_4^0,\ v_4^0)$$

$$= AE\begin{bmatrix} 0.591 & -0.159 \\ -0.159 & 1.744 \end{bmatrix}\begin{matrix} u_4^0 \\ v_4^0 \end{matrix} \quad (\text{columns } u_4^0,\ v_4^0) \tag{11.19}$$

The *vector of loads* corresponding to the *free displacements* =

$$\{q_F\} = \begin{Bmatrix} 6 \\ -8 \end{Bmatrix}\begin{matrix} u_4^0 \\ v_4^0 \end{matrix} \tag{11.20}$$

Substituting (11.19) and (11.20) into (11.7):

$$\text{or } \{u_F\} = \begin{Bmatrix} u_4^0 \\ v_4^0 \end{Bmatrix} = \frac{1}{AE}\begin{bmatrix} 0.591 & -0.159 \\ -0.159 & 1.774 \end{bmatrix}^{-1}\begin{Bmatrix} 6 \\ -8 \end{Bmatrix}$$

$$= \frac{\dfrac{1}{AE}\begin{bmatrix} 1.774 & 0.159 \\ 0.159 & 0.591 \end{bmatrix}\begin{Bmatrix} 6 \\ -8 \end{Bmatrix}}{(0.591 \times 1.774 - 0.159^2)}$$

$$= \frac{1}{AE}\begin{bmatrix} 1.734 & 0.155 \\ 0.155 & 0.578 \end{bmatrix}\begin{Bmatrix} 6 \\ -8 \end{Bmatrix}$$

or,

$$\underline{u_4^0 = 9.16/AE} \tag{11.21}$$

and,

$$\underline{v_4^0 = -3.69/AE} \tag{11.12}$$

11.5.6

The forces in the members of the framework can be obtained from the theory of Hookean elasticity, if the axial extension or contraction of each element is known. This can be achieved by resolving (11.21) and (11.22) along the local axis of each element, as follows:

Member 1–4

$$u_1 = 0$$

and, u_4 can be obtained from (11.12):

$$u_4 = \lfloor C \quad S \rfloor \begin{Bmatrix} u_4^0 \\ v_4^0 \end{Bmatrix}$$

$$= \lfloor 0.866 \quad 0.5 \rfloor \frac{1}{AE} \begin{Bmatrix} 9.16 \\ -3.69 \end{Bmatrix}$$

$$= 6.088/AE$$

From Hooke's law,

$$F_{1\text{-}4} = \text{axial force in member 1–4}$$

$$= \frac{AE}{l} * (u_4 - u_1)$$

$$= \frac{AE}{2} * \frac{6.088}{AE} = \underline{3.04\,\text{kN}}\ \text{(tensile)}$$

Member 2–4

$$u_2 = 0$$

and,

$$u_4 = \lfloor C \quad S \rfloor \begin{Bmatrix} u_4^0 \\ v_4^0 \end{Bmatrix}$$

$$= \lfloor 0 \quad 1 \rfloor \frac{1}{AE} \begin{Bmatrix} 9.16 \\ -3.69 \end{Bmatrix}$$

$$= -3.69/AE$$

$$F_{2\text{-}4} = \text{axial force in member 2–4}$$

$$= \frac{AE}{1} * (u_4 - u_2)$$

$$= AE * (-3.69/AE) = \underline{-3.69\,\text{kN}\ \text{(compressive)}}$$

Member 4–3

$$u_3 = 0$$

and,

$$u_4 = \lfloor -0.5 \quad 0.866 \rfloor \frac{1}{AE} \begin{Bmatrix} 9.16 \\ -3.69 \end{Bmatrix}$$

$$u_4 = -7.776/AE$$

$$F_{4-3} = \frac{AE}{1.155}(u_4 - u_3)$$

$$= \frac{AE}{1.155} * \left(\frac{-7.776}{AE}\right) = \underline{-6.73\,\text{kN}} \text{ (compressive)}$$

The method can be applied to numerous other problems, which are beyond the scope of this book, but if the reader requires a greater depth of coverage, he/she should consult references [16–20, 29].

EXAMPLES FOR PRACTICE 11

1. Determine the nodal displacements and member forces in the plane pin-jointed strusses of Fig. Q.11.1(a) and Q.11.1(b). For all members, AE = a constant.

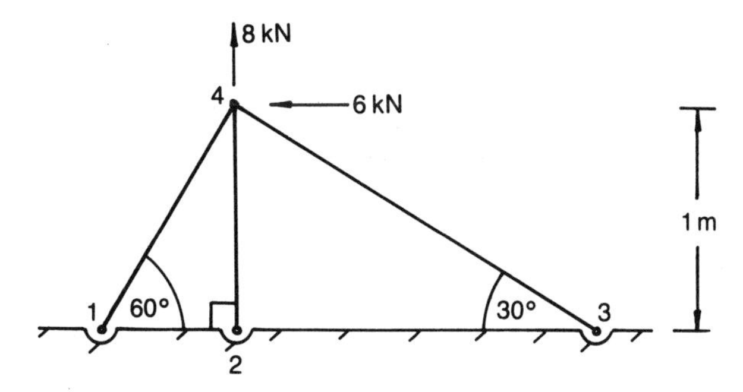

Fig. Q.11.1(a)

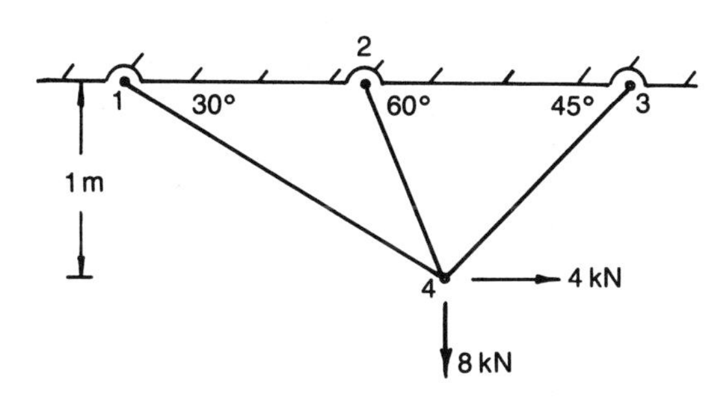

Fig. Q.11.1(b)

{(a) $u_4^0 = -11.692/AE$, $v_4^0 = 5.538/AE$, $F_{1\text{-}4} = -0.858$ kN, $F_{2\text{-}4} = 5.538$ kN $F_{3\text{-}4} = 6.431$ kN; (b) $u_4^0 = 2.585/AE$, $v_4^0 = -6.545/AE$, $F_{1\text{-}4} = 2.756$ kN, $F_{2\text{-}4} = 6.029$ kN, $F_{3\text{-}4} = 1.98$ kN}

References

[1] Ross, C. T. F., *Applied Stress Analysis*, Ellis Horwood, 1986.
[2] Pipes, L. A. and Harvill, L. R., *Applied Mathematics for Engineers and Physicists*, 3rd edn, McGraw-Hill, 1970.
[3] Alexander, J. M., *Strength of Materials—Vol. 1: Fundamentals*, Ellis Horwood, 1981.
[4] Fenner, R. T., *Engineering Elasticity*, Ellis Horwood, 1986.
[5] Saada, A. S., *Elasticity*, Pergamon, 1974.
[6] Ford, H. and Alexander, J. M., *Advanced Mechanics of Materials*, Ellis Horwood, 1977.
[7] Budynas, R. G., *Advanced Strength and Applied Stress Analysis*, McGraw-Hill/Kogakusha, 1977.
[8] Megson, T. H. G., *Aircraft Structures*, Edward Arnold, 1972.
[9] Kazinsky, G., Kiserletek Befalazott Tartokal, *Betonszemele*, **2**, 68, 1914.
[10] Baker, J. F., A review of Recent Investigations into the Behaviour of Steel Frames in the Plastic Range, *J. Inst. Civil. Engrs*, **31**, 188, 1949.
[11] Baker, J. F., Horne, M. R. and Heyman, J., *The Steel Skeleton*, Volume 2, Cambridge University Press, 1956.
[12] Neal, B. G., *The Plastic Methods of Structural Analysis*, Chapman and Hall, 1965.
[13] Coates, R. C., Coutie, M. G. and Kong, F. H., *Structural Analysis*, Van Nostrand Reinhold, 1980.
[14] Johnson, W. and Mellor, P. B., *Engineering Plasticity*, Ellis Horwood, 1983.
[15] Calladine, C. R., *Plasticity for Engineers*, Ellis Horwood, 1985.
[16] Zienkiewicz, O. C., *The Finite Element Method*, 3rd edn, McGraw-Hill, 1977.
[17] Irons, B. and Ahmad, *Techniques of Finite Elements*, Ellis Horwood, 1980.

[18] Fenner, R. T., *Finite Element Methods of Engineers*, Macmillan, 1975.
[19] Cope, R. J., Sawko, F. and Tickell, R. G., *Computer Methods for Civil Engineers*, McGraw-Hill, 1982.
[20] Ross, C. T. F., *Finite Element Methods in Structural Mechanics*, Ellis Horwood, 1985.
[21] Ross, C. T. F., The Second Polar Moment of Area by Membrane Analogy, *The Shipbuilder and Marine Engine-Builder*, 151–153, Mar. 1962.
[22] Benham, P. P. and Warnock, F. V., *Mechanics of Solids and Structures*, Pitman, 1976.
[23] Ryder, G. H., *Strength of Materials*, Macmillan, 1969.
[24] Timoshenko, S. and Woinowsky-Kreiger, S., *Theory of Plates and Shells*, McGraw-Hill/Kogakusha, 1959.
[25] Ross, C. T. F., *Finite Element Programs for Axisymmetric Problems in Engineering*, Ellis Horwood, 1984.
[26] Ross, C. T. F., Longitudinal Strength by Method of Finite Differences, *Shipping World and Shipbuilder*, **159**, 496–498, Oct., 1966.
[27] Ross, C. T. F. and Assheton, P., Finite Difference Solution for Longitudinal Strength, *Shipping World and Shipbuilder*, **162**, 453–454, Mar. 1969.
[28] Turner, R. V., Harper, M. and Moor, D. I., Some Aspects of Passenger Liner Design, *Trans. R.I.N.A.*, 1963.
[29] Rockey, K. C., Evans, H. R., Griffiths, D. W. and Nethercott, D. A., *The Finite Element Method*, 2nd Edn, Collins, 1983.
[30] Way, S. Bending of Circular Plates with Large Deflection, *A.S.M.E.*, *APM-56-12*, 56, 1934.
[31] Hewitt, D. A. and Tannent, J. O., Large Deflections of Circular Plates, Portsmouth Polytechnic Report M195, 1973/74.
[32] Bolotin, V. V., The Dynamic Stability of Elastic Systems, Holden Day, 1964.
[33] Zienkiewicz, O. C. and Cheung, Y. K., Finite Element Method of Analysis for Arch Dam Shells and Comparison with Finite Difference Procedures, Proc. Symp. Theory of Arch Dams, Southampton Univ., 1964.
[34] Brotton, D. M., The Application of Digital Computers to Structural Engineering Problems, Spon, 1962.

Appendix I

Computer program for determining the roots of a cubic equation

```
100 REM COPYRIGHT OF DR C.T.F.ROSS
110 REM DEPT OF MECH ENG.,
120 REM PORTSMOUTH POLYTECHNIC,
130 REM PORTSMOUTH PO1 3DJ
140 PRINT"       ***************************"
150 PRINT"       *PROGRAM TO FIND THE ROOTS*"
160 PRINT"       *   OF A CUBIC EQUATION   *"
170 PRINT"       *       OF THE FORM       *"
180 PRINT"       *    AX↑3+BX↑2+CX+D=0     *"
190 PRINT"       ***************************"
200 PRINT"
210 PRINT"TYPE COEFF OF X↑3 (A)";
220 INPUT A
230 PRINT"TYPE COEFF OF X↑2 (B)";
240 INPUT B
250 PRINT"TYPE COEFF OF X (C)";
260 INPUT C
270 PRINT"TYPE THE VALUE OF D";
280 INPUT D
290 A2=B/A
300 A1=C/A
310 AZ=D/A
320 Q=-A1+A2↑2/3
330 R=-AZ+A1*A2/3-2*A2↑3/27
340 RS=R
350 R=ABS(R)
360 IFQ<0 THEN 650
370 QH=SQR(Q)
380 QH=2*QH/1.732051
390 CN=(3/Q)↑1.5
400 CN=CN*R/2
410 KN=27*R↑2
420 K1=4*Q↑3
430 IF KN>K1 THEN 650
440 C1=CN/(SQR(1-CN↑2))
450 PH=-ATN(C1)+π/2
460 X1=QH*COS(PH/3)
470 X2=-QH*COS((π-PH)/3)
480 X3=-QH*COS((π+PH)/3)
490 IF RS>0 THEN 530
500 X1=-X1
510 X2=-X2
520 X3=-X3
530 X1=X1-A2/3
540 X2=X2-A2/3
550 X3=X3-A2/3
```

```
560 PRINT"■■■
570 PRINT"THERE ARE 3 REAL ROOTS"
580 PRINT"■■
590 PRINT"1ST ROOT=";X1
600 PRINT"■
610 PRINT"2ND ROOT=";X2
620 PRINT"■
630 PRINT"3RD ROOT=";X3
640 GOTO 940
650 IF Q<=0 THEN 740
660 C1=2*CN
670 XL=(C1+SQR(C1↑2-4))/2
680 PH=LOG(XL)
690 KO=(EXP(PH/3)+EXP(-PH/3))/2
700 X1=QH*KO
710 IF RS<0 THEN X1=-X1
720 X1=X1-A2/3
730 GOTO 850
740 C1=(-3/Q)↑1.5
750 C1=C1*R
760 XL=(C1+SQR(C1↑2+4))/2
770 PH=LOG(XL)
780 SY=(EXP(PH/3)-EXP(-PH/3))/2
790 QH=-Q
800 QH=SQR(QH)
810 QH=2*QH/1.732051
820 X1=QH*SY
830 IF RS<0 THEN X1=-X1
840 X1=X1-A2/3
850 PRINT"■■■
860 PRINT"1ST ROOT=";X1
870 X2=-(B+A*X1)/(2*A)
880 X3=(SQR(-B↑2+3*A↑2*X1↑2+2*A*B*X1+4*A*C))/(2*A)
890 PRINT"■
900 PRINT"REAL PARTS OF 2ND & 3RD"
910 PRINT"ROOTS=";X2:PRINT"■
920 PRINT"IMAGINARY PARTS OF 2ND & 3RD"
930 PRINT"ROOTS=+OR-";X3
940 PRINT:PRINT"DO YOU WISH TO SOLVE ANOTHER EQUATION?.IF YES, TYPE (Y)";
950 PRINT", ELSE TYPE ANY OTHER KEY":PRINT
960 GETA$:IFA$=""THEN 960
970 IFA$="Y"THEN 200
```

Appendix II

Computer program for determining the roots of a quartic equation

```
100 REM COPYRIGHT OF DR C.T.F.ROSS
110 REM DEPT OF MECH ENG..
120 REM PORTSMOUTH POLYTECHNIC,
130 REM ANGLESEA ROAD,
140 REM PORTSMOUTH PO1 3DJ
150 PRINT"        ****************************"
160 PRINT"        *PROGRAM TO FIND THE ROOTS*"
170 PRINT"        *  OF A QUARTIC EQUATION  *"
180 PRINT"        *       OF THE FORM        *"
190 PRINT"        * AX↑4+BX↑3+CX↑2+DX+E=0   *"
200 PRINT"        ****************************"
210 PRINT"[illegible]
220 MX=-1000000
230 PRINT"TYPE COEFF OF X↑4 (A)";
240 INPUT C(1)
250 PRINT"TYPE COEFF OF X↑3 (B)";
260 INPUT C(2)
270 PRINT"TYPE GOEFF OF X↑2 (C)";
280 INPUT C(3)
290 PRINT"TYPE COEFF OF X (D)";
300 INPUT C(4)
310 PRINT"TYPE THE VALUE OF E";
320 INPUT C(5)
330 Z3=C(2)/C(1)
340 Z2=C(3)/C(1)
350 Z1=C(4)/C(1)
360 Z0=C(5)/C(1)
370 Z=Z3/2
380 A=1
390 B=-Z2
400 C=Z1*Z3-4*Z0
410 D=Z0*(4*Z2-Z3*Z3)-Z1*Z1
420 A2=B/A
430 A1=C/A
440 AZ=D/A
450 Q=-A1+A2↑2/3
460 R=-AZ+A1*A2/3-2*A2↑3/27
470 RS=R
480 R=ABS(R)
490 IFQ<0 THEN 760
500 QH=SQR(Q)
510 QH=2*QH/1.732051
520 CN=(3/Q)↑1.5
530 CN=CN*R/2
540 KN=27*R↑2
550 K1=4*Q↑3
```

```
560 IF KN>K1 THEN 760
570 O=ABS(CN)
580 IFABS(O-1)<1E-5THENC1=1E3*CN
590 IFABS(O-1)<1E-5THEN610
600 C1=CN/(SQR(1-CN↑2))
610 PH=-ATN(C1)+π/2
620 X1=QH*COS(PH/3)
630 X2=-QH*COS((π-PH)/3)
640 X3=-QH*COS((π+PH)/3)
650 IF RS>0 THEN 690
660 X1=-X1
670 X2=-X2
680 X3=-X3
690 X1=X1-A2/3
700 X2=X2-A2/3
710 X3=X3-A2/3
720 IFX1>MXTHENMX=X1
730 IFX2>MXTHENMX=X2
740 IFX3>MXTHENMX=X3
750 GOTO 980
760 IF Q<=0 THEN 860
770 C1=2*CN
780 XL=(C1+SQR(C1↑2-4))/2
790 PH=LOG(XL)
800 KO=(EXP(PH/3)+EXP(-PH/3))/2
810 X1=QH*KO
820 IF RS<0 THEN X1=-X1
830 X1=X1-A2/3
840 MX=X1
850 GOTO 980
860 C1=(-3/Q)↑1.5
870 C1=C1*R
880 XL=(C1+SQR(C1↑2+4))/2
890 PH=LOG(XL)
900 SY=(EXP(PH/3)-EXP(-PH/3))/2
910 QH=-Q
920 QH=SQR(QH)
930 QH=2*QH/1.732051
940 X1=QH*SY
950 IF RS<0 THEN X1=-X1
960 X1=X1-A2/3
970 MX=X1
980 PRINT"[illegible]
990 A=Z3/2
1000 B=MX/2
1010 IF(B↑2-Z0)<=0THEND=0
1020 IF(B↑2-Z0)<=0THEN1040
1030 D=SQR(B↑2-Z0)
1040 IFD=0THENC=SQR(A↑2-Z2+MX)
1050 IFD<>0THENC=-(Z1/2-A*B)/D
1060 B2=(A-C)↑2
1070 AC=4*(B-D)
1080 IFAC>B2THEN1130
1090 Y1=(C-A+SQR(B2-AC))/2
1100 Y2=(C-A-SQR(B2-AC))/2
1110 PRINT:PRINT"TWO REAL ROOTS ARE:";Y1;"AND";Y2
1120 GOTO1170
1130 Y1=(C-A)/2
1140 Y2=(SQR(AC-B2))/2
1150 PRINT:PRINT"REAL PART OF IMAGINARY ROOTS IS:";Y1
1160 PRINT:PRINT"IMAGINARY PARTS OF IMAGINARY ROOTS ARE:+OR-";Y2
1170 B2=(A+C)↑2
1180 AC=4*(B+D)
1190 IFAC>B2THEN1240
1200 Y1=(-A-C+SQR(B2-AC))/2
1210 Y2=(-A-C-SQR(B2-AC))/2
1220 PRINT:PRINT"TWO REAL ROOTS ARE:";Y1;"AND";Y2
1230 GOTO1280
1240 Y1=(-A-C)/2
1250 Y2=(SQR(AC-B2))/2
1260 PRINT:PRINT"REAL PART OF IMAGINARY ROOTS IS:";Y1
1270 PRINT:PRINT"IMAGINARY PARTS OF IMAGINARY ROOTS ARE:+OR-";Y2
1280 END
```

Index